D1698342

Seizure Prediction in Epilepsy

Edited by
Björn Schelter, Jens Timmer,
and Andreas Schulze-Bonhage

Related Titles

Bevelacqua, J. J.

Health Physics in the 21 Century

2008
ISBN: 978-3-527-40822-1

Andrä, W., Nowak, H. (eds.)

Magnetism in Medicine

A Handbook

2007
ISBN: 978-3-527-40558-9

Hendee, W. R., Ritenour, E. R.

Medical Imaging Physics

2002
ISBN: 978-0-471-38226-3

Seizure Prediction in Epilepsy

From Basic Mechanisms to Clinical Applications

Edited by
Björn Schelter, Jens Timmer,
and Andreas Schulze-Bonhage

WILEY-VCH

WILEY-VCH Verlag GmbH & Co. KGaA

The Editors

Dr. Björn Schelter
University of Freiburg
Center for Data Analysis (FDM)
Freiburg, Germany
schelter@fdm.uni-freiburg.de

Dr. Jens Timmer
University of Freiburg
Center for Data Analysis (FDM)
Freiburg, Germany
jeti@fdm.uni-freiburg.de

Dr. Andreas Schulze-Bonhage
University Hospital Freiburg
Epilepsy Center
Freiburg, Germany
andreas.schulze-bonhage@
uniklinik-freiburg.de

Cover
based on an illustration of a human head taken
from ''Gray's Anatomy'', first published in
1918

■ All books published by Wiley-VCH are
carefully produced. Nevertheless, authors,
editors, and publisher do not warrant the
information contained in these books,
including this book, to be free of errors.
Readers are advised to keep in mind that
statements, data, illustrations, procedural
details or other items may inadvertently be
inaccurate.

Library of Congress Card No.: applied for
**British Library Cataloguing-in-Publication
Data**
A catalogue record for this book is available
from the British Library.

**Bibliographic information published by
the Deutsche Nationalbibliothek**
Die Deutsche Nationalbibliothek lists this
publication in the Deutsche National-
bibliografie; detailed bibliographic data are
available in the Internet at
<http://dnb.d-nb.de>.

© 2008 WILEY-VCH Verlag GmbH & Co.
KGaA, Weinheim

Typesetting Laserwords Private Ltd,
Chennai, India
Printing Strauss GmbH, Mörlenbach
Binding Litges & Dopf GmbH,
Heppenheim

Printed in the Federal Republic of Germany
Printed on acid-free paper

ISBN: 978-3-527-40756-9

Contents

Seizure Prediction in Epilepsy. Edited by Björn Schelter, Jens Timmer and Andreas Schulze-Bonhage
Copyright © 2008 WILEY-VCH Verlag GmbH & Co. KGaA, Weinheim
ISBN: 978-3-527-40756-9

Preface

The field of seizure prediction in epilepsy has joined theoretical aspects of time series analysis and clinical applications but has had its ups and downs over recent years. Public perception of this field has grown and awareness has increased of the usefulness of a clinical system based on seizure prediction for warning and for new intervention strategies. Rigorous statistical evaluations demonstrated that the performance of present-day available prediction methods has to be improved to warrant a clinical applicability.

There is therefore reason to reflect on our understanding of the mechanisms underlying interictal–ictal transitions and to analyze the factors limiting the present-day performance of algorithms. The hope is that a better understanding of mechanisms contributing to the variability of cerebral dynamics will offer new chances to improve prediction methods.

In parallel, new intervention devices are being developed which could greatly profit from the effectiveness of closed-loop systems based on seizure prediction. The analysis of intervention techniques resetting the dynamics of the preictal and early ictal period may also offer opportunities to better understand preictal and ictal dynamics, thereby giving impetus to the development of new prediction approaches.

This book comprises a wide range of current topics in the field of seizure prediction ranging from basic mechanisms to clinical applications. This covers the whole spectrum of studies from modeling neuronal networks in silicon to closed-loop intervention strategies. We greatly appreciate the fact that leading scientists from various fields including electrophysiology, computational neuroscience, mathematics, statistics, time series analysis, engineering, physics and clinical experts who have participated in the 3rd International Workshop on Epileptic Seizure Prediction in Freiburg, Germany, have contributed to this book. It presents an up-to-date survey of the state of the art in seizure prediction including related research areas. This book provides guidance for all those working on seizure prediction, from students to experienced investigators.

For help with the layout of the book we would like to thank Dr. Michael Jachan, Hinnerk Feldwisch, David Feess, Wolfgang Mader, Jakob Nawrath, and Raimar Sandner. Moreover, we would like to thank Kathrin Henschel, Ariane Schad and Raimar Sandner for the design of the cover.

Freiburg
January 2008

Björn Schelter, Jens Timmer,
and Andreas Schulze-Bonhage

Thanks to the Sponsors of the Workshop

On behalf of the Organizing Board of the 3rd International Workshop on Epileptic Seizure Prediction, we would like to thank the sponsors of this workshop:

- Deutsche Forschungsgemeinschaft
- National Institutes of Health
- Deutsche Gesellschaft für Epileptologie
- Bernstein Center for Computational Neuroscience Freiburg (Bundesministerium für Bildung und Forschung)
- epilepScio

Furthermore, we would like to thank NeuroVista Corporation, Cyberonics, Inc., Pfizer Inc. and UCB Pharma GmbH for their support of the workshop.

Please note:
Funding for this conference was made possible (in part) by Grant Number R13NS060623 from the National Institute of Neurological Disorders And Stroke (NINDS) and National Institute of Biomedical Imaging and Bioengineering (NIBIB). The views expressed in written conference materials or publications and by speakers and moderators do not necessarily represent the official views of the NINDS, NIBIB or NIH and do not necessarily reflect the official policies of the Department of Health and Human Services; nor does mention by trade names, commercial practices, or organizations imply endorsement by the U.S. Government.

List of Contributors

William S. Anderson
Harvard Medical School
Department of Neurosurgery
Brigham and Women's Hospital
15 Francis Street PBB 3
Boston, MS 02115
USA

Ralph G. Andrzejak
Universitat Pompeu Fabra
Departament de Tecnologia
Passeig de Circumvallació 8
08003 Barcelona
Spain

Gerold Baier
University of Manchester
Manchester Interdisciplinary
Biocentre
131 Princess Street
Manchester M1 7DN
United Kingdom

Gregory K. Bergey
Johns Hopkins University School of
Medicine
Johns Hopkins Epilepsy Center
Department of Neurology
600 North Wolfe Street
Meyer 2-147
Baltimore, MD 21287
USA

Stephan Bialonski
University of Bonn
Department of Epileptology
Sigmund-Freud-Str. 25
53105 Bonn
Germany

Paul Boon
Gent University Hospital
Department of Neurology
De Pintelaan 185
9000 Gent
Belgium

Anatol Bragin
Seizure Disorder Center
Brain Research Institute
UCLA
Box 951761
Los Angeles, CA 90095-1761
USA

Alex J. Cadotte
University of Florida
McKnight Brain Institute
Department of Pediatrics and
Biomedical Engineering
P.O. Box 100296
Gainesville, FL 32610-0296
USA

Seizure Prediction in Epilepsy. Edited by Björn Schelter, Jens Timmer and Andreas Schulze-Bonhage
Copyright © 2008 WILEY-VCH Verlag GmbH & Co. KGaA, Weinheim
ISBN: 978-3-527-40756-9

Paul R. Carney
University of Florida
McKnight Brain Institute
Departments of Pediatrics,
Neurology, Neuroscience and
Biomedical Engineering
P.O. Box 100296
Gainesville, FL 32610-0296
USA

Anton Chernihovskyi
University of Bonn
Department of Epileptology
Sigmund-Freud-Str. 25
53105 Bonn
Germany

Thomas B. DeMarse
University of Florida
Department of Biomedical
Engineering
147 Biomedical Engineering Bldg
Gainesville, FL 32611-6131
USA

William Ditto
University of Florida
Department of Biomedical
Engineering
147 Biomedical Engineering Bldg
Gainesville, FL 32611-6131
USA

Christian E. Elger
University of Bonn
Department of Epileptology
Sigmund-Freud-Str. 25
53105 Bonn
Germany

Jerome Engel Jr.
Seizure Disorder Center
Brain Research Institute
UCLA
Box 951761
Los Angeles, CA 90095-1761
USA

Hinnerk Feldwisch genannt Drentrup
University of Freiburg
Bernstein Center for Computational
Neuroscience
Hansastr. 9a
79104 Freiburg
Germany

Philipp Fischer
Johann-Wolfgang-Goethe University
Institute of Applied Physics
Max-von-Laue-Str. 1
60438 Frankfurt

Piotr J. Franaszczuk
Johns Hopkins University
School of Medicine
Johns Hopkins Epilepsy Center
Department of Neurology
600 North Wolfe Street
Meyer 2-147
Baltimore, MD 21287
USA

Mark G. Frei
Flint Hills Scientific L.L.C.
5040 Bob Billings PKWY
Suite A
Lawrence, KS 66049
USA

Carolin Gierschner
Universitätsklinikum Freiburg
Sektion Epileptologie
Ltd. MTA
Neurozentrum
Breisacher Str. 64
79106 Freiburg
Germany

Frank Gollas
Johann-Wolfgang-Goethe University
Institute of Applied Physics
Max-von-Laue-Str. 1
60438 Frankfurt

Veerle De Herdt
University of Ghent
Department of Internal Medicine
De Pintelaan 185
3000 Ghent
Belgium

Marie-Therese Horstmann
University of Bonn
Department of Epileptology
Sigmund-Freud-Str. 25
53105 Bonn
Germany

Michael Jachan
University of Freiburg
Freiburg Center for Data
Analysis and Modeling
Eckerstr. 1
79104 Freiburg
and
University Hospital of Freiburg
Epilepsy Center
Breisacher Str. 64
79106 Freiburg

John G. R. Jefferys
University of Birmingham
Division of Neuroscience
School of Medicine
Vincent Drive
Birmingham B15 2TT
United Kingdom

Premysl Jiruska
University of Birmingham
Division of Neuroscience
School of Medicine
Vincent Drive
Birmingham B15 2TT
United Kingdom

Kevin M. Kelly
Drexel University
Allegheny-Singer-Resarch Institute
Department of Neurology
940 South Tower of Allegheny
General Hospital
320 E. North Avenue
Pittsburgh, PA 15212
USA

Dieter Krug
University of Bonn
Department of Epileptology
Sigmund-Freud-Str. 25
53105 Bonn
Germany

Pawel Kudela
Johns Hopkins University
School of Medicine
Johns Hopkins Epilepsy Center
Department of Neurology
600 North Wolfe Street
Meyer 2-147
Baltimore, MD 21287
USA

Roland Kunz
Wilhelm-Fay-Str. 30-34
65936 Frankfurt

Anne Kühn
University Hospital of Freiburg
Epilepsy Center
Breisacher Str. 64
79106 Freiburg

Ying-Cheng Lai
Department of Electrical
Engineering
Arizona State University
Tempe, AZ 85287-5706
USA

Klaus Lehnertz
University of Bonn
Department of Epileptology
Sigmund-Freud-Str. 25
53105 Bonn
Germany

Fernando H. Lopes da Silva
University of Amsterdam
Center of Neuroscience
Swammerdam Institute for
Life Sciences
Kruislaan 320
1098 SM, Amsterdam
The Netherlands

Thomas H. Mareci
University of Florida
McKnight Brain Institute
Department of Biochemistry
and Molecular Biology
P.O.Box 100245
Gainesville, FL 32610-0245
USA

Florian Mormann
University of Bonn
Department of Epileptology
Sigmund-Freud-Str. 25
53105 Bonn
Germany

Andy Müller
University of Bonn
Department of Epileptology
Sigmund-Freud-Str. 25
53105 Bonn
Germany

Markus Müller
Universidad Autonoma del Estado
de Morelos
Facultad de Ciencias
62209 Cuernavaca
Morelos
Mexico

Sandeep P. Nair
Allegheny-Singer-Resarch Institute
Department of Neurology
940 South Tower of Allegheny
General Hospital
320 E. North Avenue
Pittsburgh, PA 15212
USA

Christian Niederhöfer
Johann-Wolfgang-Goethe University
Institute of Applied Physics
Max-von-Laue-Str. 1
60438 Frankfurt

Ivan Osorio
University of Kansas
Medical Center
Department of Neurology
Landon Center of Aging
3599 Rainbow Blvd
Mailstop 2012
Kansas City, KS 66160
USA

Hannes Osterhage
University of Bonn
Department of Epileptology
Sigmund-Freud-Str. 25
53105 Bonn
Germany

Jens Prusseit
University of Bonn
Department of Epileptology
Sigmund-Freud-Str. 25
53105 Bonn
Germany

Robrecht Raedt
University of Ghent
Department of Surgery
De Pintelaan 185
3000 Ghent
Belgium

Hermine Reichau
Johann-Wolfgang-Goethe University
Institute of Applied Physics
Max-von-Laue-Str. 1
60438 Frankfurt

Alexander Rothkegel
University of Bonn
Department of Epileptology
Sigmund-Freud-Str. 25
53105 Bonn
Germany

Steven M. Rothman
Department of Pediatrics
(Clinical Neuroscience)
University of Minnesota Medical
School – MMC 486
420 Delaware Street S.E.
Minneapolis, MN 55455-0374
USA

Christian Rummel
University of Bern
Department of Neurology
Inselspital
3010 Bern
Switzerland

J. Chris Sackellares
Optima Neuroscience, Inc.
101 SE 2nd Place, Suite 201-A
Gainesville, Florida 32601
USA

Björn Schelter
University of Freiburg
Bernstein Center for Computational
Neuroscience
Hansastr. 9a
79104 Freiburg
and
Freiburg Center for Data Analysis
and Modeling
Eckerstr. 1
79104 Freiburg
Germany

Kaspar Schindler
University of Bern
Department of Neurology
Inselspital
3010 Bern
Switzerland

Andreas Schulze-Bonhage
University Hospital of Freiburg
Epilepsy Center
Breisacher Str. 64
79106 Freiburg
and
University of Freiburg
Bernstein Center for Computational
Neuroscience
Hansastr. 9a
79104 Freiburg
Germany

Justus Schwabedal
University of Potsdam
Institut für Physik und Astronomie
(Haus 28)
Karl-Liebknecht-Str. 24/25
14476 Potsdam-Golm

Deng-Shan Shiau
Optima Neuroscience, Inc.
101 SE 2nd Place, Suite 201-A
Gainesville, Florida 32601
USA

Yousheng Shu
Institute of Neuroscience
State Key Laboratory of
Neuroscience
Shanghai Institutes for
Biological Sciences
Chinese Academy of Sciences
320 Yueyang Road
Shanghai 200031
P.R. of China

Matthäus Staniek
University of Bonn
Department of Epileptology
Sigmund-Freud-Str. 25
53105 Bonn
Germany

Ulrich Stephani
Christian-Albrechts University
Department of Neuropediatrics
Schwanenweg 20
24105 Kiel
Germany

Riem El Tahry
University of Ghent
Department of Internal Medicine
De Pintelaan 185
3000 Ghent
Belgium

Ronald Tetzlaff
Technical University of Dresden
Faculty of Electrical Engineering and
Information Technology
Mommsenstr. 12
01069 Dresden
Germany

Jens Timmer
University of Freiburg
Bernstein Center for Computational
Neuroscience
Hansastr. 9a
79104 Freiburg
and
Freiburg Center for Data Analysis
and Modeling
Eckerstr. 1
79104 Freiburg
Germany

Andrew J. Trevelyan
University of Newcastle
Institute of Neuroscience
Medical School
Framlington Place
Newcastle upon Tyne NE2 4HH
United Kingdom

Annelies Van Dycke
University of Ghent
Department of Internal Medicine
De Pintelaan 185
3000 Ghent
Belgium

Dirk Van Roost
University of Ghent
Department of Surgery
De Pintelaan 185
3000 Ghent
Belgium

Demetrios N. Velis
Epilepsy Institutes of the
Netherlands (SEIN)
Department of Clinical Neurophysi-
ology and Epilepsy Monitoring Unit
'Meer en Bosch' Campus
Achterweg 5
2103 SW Heemstede
The Netherlands

Baba C. Vemuri
University of Florida
Department of Computer and Infor-
mation Science and Engineering
P.O. Box 116120
Gainesville, FL 32611-6131
USA

Kristl Vonck
University of Ghent
Department of Internal Medicine
De Pintelaan 185
3000 Ghent
Belgium

Wytse J. Wadman
University of Amsterdam
Faculty of Science
Swammerdam Institute for
Life Sciences
Kruislaan 3201098 SM Amsterdam

Tobias Wagner
University of Bonn
Department of Epileptology
Sigmund-Freud-Str. 25
53105 Bonn
Germany

Liesbeth Waterschoot
Gent University Hospital
Department of Neurology
De Pintelaan 185
9000 Gent
Belgium

Richard Wennberg
University of Toronto
Toronto Western Hospital
Krembil Neuroscience Center
399 Bathurst Street
Toronto M5T 2S8
Canada

Tine Wyckhuys
Gent University Hospital
Department of Neurology
De Pintelaan 185
9000 Gent
Belgium

Hitten P. Zaveri
Yale University
Department of Neurology
333 Cedar Street
New Haven, CT 06520-8018
USA

Color Plates

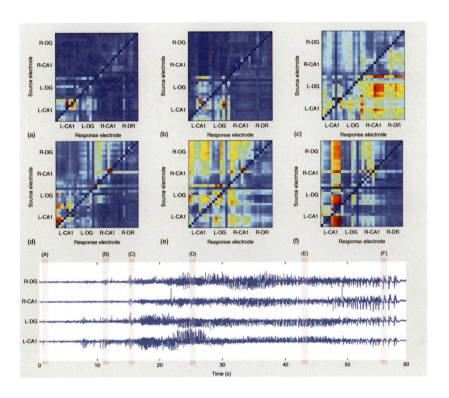

Seizure Prediction in Epilepsy. Edited by Björn Schelter, Jens Timmer and Andreas Schulze-Bonhage
Copyright © 2008 WILEY-VCH Verlag GmbH & Co. KGaA, Weinheim
ISBN: 978-3-527-40756-9

Fig. 4.3 Progression of Granger causality analyses through a single seizure event. The bottom electrode voltage traces are from four of the 32 electrodes from each of the four major hippocampal regions including the L-CA1, L-DG, R-CA1, and R-DG. The Granger analyses in panels A through F correspond to the red shaded windows of data highlighted over the electrode traces (A through F, respectively). In each Granger causality plot the driving brain area (source) is on the y-axis and the target area (response) is on the x-axis. For example, the driving influence from the R-DG to the L-CA1 in panel D is located in the upper left corner of the plot. The magnitude of the Granger causality interaction is represented by color ranging through blue (near zero), light blue, green, yellow, to red (highly causal). The interictal activity about 15 seconds prior to seizure in panel A shows a strong interaction within the L-CA1. Stage 1 in panel B shows this relationship increasing and spreading into the L-DG as well. Stage 1 is the beginning of the tonic behavioral state for the rat that persists until the Stage 4 transition. This driving influence from the CA1 to the DG is abnormal and persists until the Stage 4 transition. Panel C shows a directional transfer in Stage 2 from the left hemisphere to the right hemisphere. These transfers reverberate directionally several times across hemispheres. Panel D shows intrahemisphere activity common to Stage 3. Panel E shows across the board synchronization for the entire array for the 'transition stage' or Stage 4. Stage 4 is called the transition stage because behaviorally the rat moves from tonic to clonic activity and the primary driving relationship transitions from CA1 driving DG to DG driving CA1. Panel F shows activity from Stage 5 after the transition. In this stage the rat exhibits clonic behavior marked by rhythmic shaking of the limbs. (This figure also appears on page 51.)

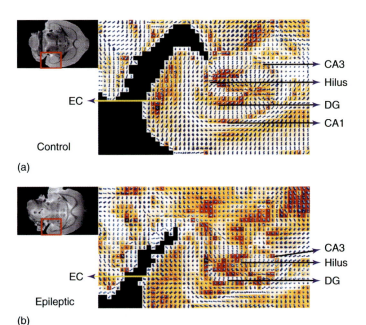

Fig. 4.5 Probability maps of coronal images of a control and an epileptic hippocampus. The upper left corner shows the corresponding reference images where the rectangle regions enclose the hippocampus. In the control hippocampus, the molecular layer and stratum radiatum fiber orientations paralleled the apical dendrites of granule cells and pyramidal neurons, respectively; whereas in the stratum lacunosum, moleculare orientations paralleled Schaffer collaterals from CA1 neurons. In the epileptic hippocampus, the overall architecture is notably altered; the CA1 subfield is lost, while an increase in crossing fibers can be seen in the hilus and dentate gyrus (dg). Increased crossing fibers can also be seen in the entorhinal cortex (ec). Fiber density within the statum lacunosum moleculare and statum radiale is also notably reduced, although fiber orientation remains unaltered (from NeuroImage 37 (2007) 164–176, used with permission). (This figure also appears on page 57.)

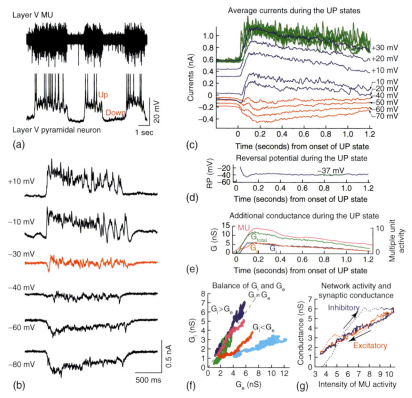

Fig. 6.1 Recurrent cortical activity is generated by a balance of excitation and inhibition. (a) The prefrontal cortical slice maintained *in vitro* spontaneously generates slow oscillations. Simultaneous extracellular multiple-unit (MU) recording and intracellular recording from a layer V pyramidal neuron reveal the two intermittent states; the Up and Down state. The action potentials are truncated. (b) Voltage-clamp recording demonstrates that the Up state currents have a reversal potential around −30 mV. (c) Averaged currents during the Up state at holding potentials from −70 to +30 mV. Several raw current traces at +30 mV are shown with the average for comparison. The neuron was recorded with sharp electrode filled with CsAc and QX314 to minimize the contribution from K$^+$ and Na$^+$ currents. (d) Reversal potential is relatively stable during the Up state. (e) Calculation of the total conductance (Gtotal), excitatory (Ge) and inhibitory conductance (Gi). Note the relationship between the intensity of multiple unit activity and the changes in these conductances. (f) Linear relationship between Ge and Gi indicates the proportionality between excitation and inhibition. (g) Relationship between the average intensity of MU activity and the amplitude of Ge and Gi. In this neuron, inhibitory conductance lagged the excitatory conductance as the network transitioned into the Up state. Following the onset of the Up state the Ge and Gi were proportional and correlated strongly with the intensity of MU activity recorded locally. (Figure reproduced from 22.) (This figure also appears on page 87.)

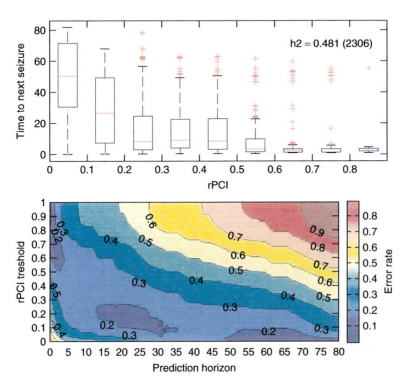

Fig. 7.7 rPCI as function of time preceding a seizure. Illustration (averages of six data from six patients shown in box plots) of the time-course of the rPCI (horizontal axis) in relation to the time to the next seizure (vertical axis, hours). The lower panel is a pseudo-color and contour plot of the prediction error rates, with respect to predicting a seizure within a certain time, as a function of the rPCI threshold (vertical axis) and the prediction horizon (in hours, horizontal axis). Note, for example, that for values of rPCI > 0.6 it may be predicted with 80 % accuracy that a seizure will occur in less than 2 h. (Adapted from 25.) (This figure also appears on page 105.)

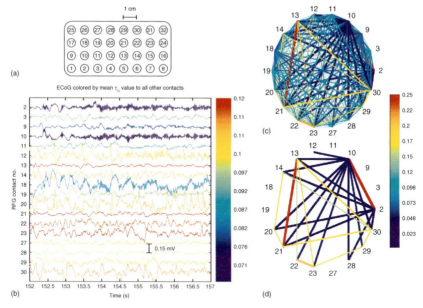

Fig. 8.3 (a) Grid used in recording ECoG signals. (b) Referential ECoG during a 5 s epoch colored according to mean phase synchronization level between the respective channel and all other intact contacts. (c) Graphical depiction of the bivariate phase synchronization measure between all pairs of channels during this epoch, with color and linewidths corresponding to phase synchronization level. (d) Illustration of the most significant pair-wise synchronization within the epoch. (This figure also appears on page 113.)

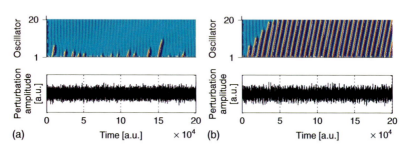

Fig. 10.1 Examples for the induction of spatial-temporal patterns in an excitable medium (20 *FitzHugh–Nagumo* oscillators) due to a perturbation of the first oscillator. The range of amplitude values of this medium is encoded with colors ranging from blue (minimum amplitudes) to red (maximum amplitudes). (b) A noisy signal containing a profound rhythmic component leads to coherent periodic patterns. (a) Only few excitation waves are induced if the medium is perturbed with a non-periodic or, in general, a non-correlated signal (here white noise). (This figure also appears on page 136.)

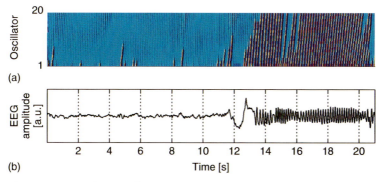

(a)

(b)

Fig. 10.2 (a) Spatial-temporal pattern induced in an excitable medium (20 *FitzHugh–Nagumo* oscillators due a perturbation of the first oscillator with an EEG signal (b). The range of amplitude values of this medium is encoded with colors ranging from blue (minimum amplitudes) to red (maximum amplitudes). The EEG signal was recorded intracranially from a patient suffering from medial temporal lobe epilepsy. Data were sampled at 200 Hz within the frequency band of 0.5–85 Hz using a 16 bit analog-to-digital converter. (This figure also appears on page 137.)

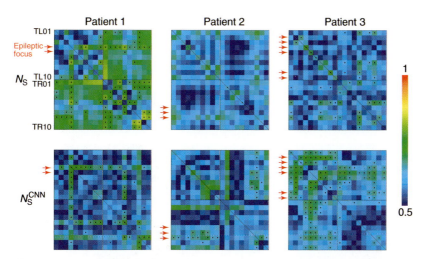

Fig. 10.5 Across-subject generalization properties of the adaptive CNN-based concept of measuring the strength of generalized synchronization. Color-coded ROC-areas obtained from numerically derived (upper row) and approximated (lower row) profiles of the symmetric nonlinear interdependence N_s for all electrode combinations. Black dots indicate a significant ROC-area value (using 19 seizure time surrogates) for a given electrode combination. Data from a single channel combination from patient #1 was used for optimizing the CNN. This optimized CNN was used to estimate N_s^{CNN} values for the remaining channel combinations in this patient, and for the data from patient #2 and #3. Arrows indicate seizure onset area. (This figure also appears on page 144.)

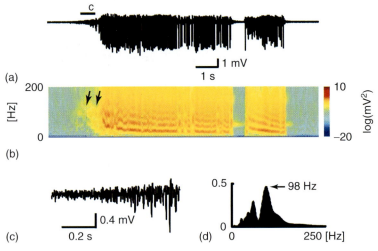

Fig. 13.1 Hippocampal slice perfused in low-calcium artificial cerebrospinal fluid leads to development of spontaneous recurrent electrographic seizures. (a) Recording from CA1 region (DC removed). (b) Seizure onset is characterized by occurrence of low-amplitude high-frequency activity, which is well demonstrated in corresponding wavelet spectrogram (arrows). (c) Detail of seizure onset. (d) Corresponding wavelet power spectrum demonstrating peak frequency at 98 Hz. (This figure also appears on page 171.)

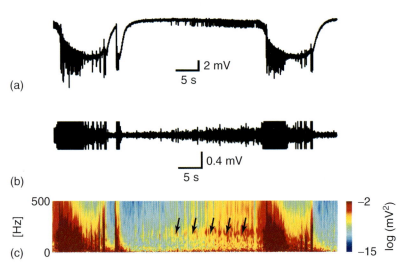

Fig. 13.2 Hippocampal slice perfused with high-potassium ACSF. (a) Recording from CA1 region shows presence of repeated electrographic seizures. Seizures are superimposed on large DC shifts. (b) Band-pass filtered data (80–250 Hz) show that seizures are preceded by gradual build-up of high-frequency activity. (c) Corresponding wavelet spectrogram demonstrates that the build-up of high-frequency activity has a peak frequency of 200 Hz (arrows). (This figure also appears on page 171.)

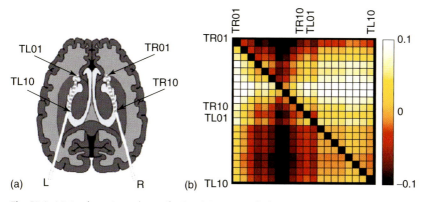

Fig. 15.3 (a) Implantation scheme for intrahippocampal electrodes. (b) Directionality matrix $M(\hat{T})$ obtained from the transfer entropy. Matrix entries are color coded and represent the temporal average of $\hat{T}$ for each combination of electrode contacts. (This figure also appears on page 198.)

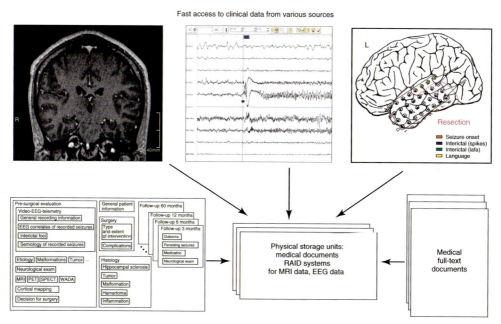

Fig. 20.3 Scheme of data types and relational connectivity in a multi-dimensional database containing raw data and metadata for epileptic seizure prediction. (This figure also appears on page 266.)

1

Unpredictability of Seizures and the Burden of Epilepsy

Andreas Schulze-Bonhage, Anne Kühn

1.1
Introduction

About 0.8 % of the world's population suffer from epilepsy. The classical definitions [1] of epilepsy deem the 'spontaneous occurrence' of recurrent seizures to be essential for a seizure disorder to be diagnosed as 'epilepsy'. Spontaneous seizures stand in contrast to situation-related seizures which are thought to be triggered by precipitants of the ictal event. The term 'spontaneous' reflects our present lack of understanding about the mechanisms underlying interictal–ictal transitions. Indeed, the relative contribution of genetic make-up in determining the individual 'seizure threshold', of intrinsic fluctuations in EEG dynamics, and of external factors, toward the development of an ictal state are unknown. It follows from this that it is currently not possible to predict the critical time points at which the interictal–ictal transition and the manifest 'seizure' take place.

Overt symptoms in epilepsy are virtually absent during the interictal period (as long as no elaborated diagnostic procedures are performed). This means that patients might be thought to suffer from the disease only during the brief paroxysmal episodes. A mean seizure frequency of three such events per month in the average adult pharmacoresistant patient [2] and an average seizure duration of 1–2 minutes [3–5] corresponds to a symptomatic period that effectively lasts less than one hour per year. Even if impairments during the postictal period are taken into consideration, most epilepsy patients are in a functional interictal state without obvious impairments for 95–99 % of the time.

Despite the relatively low percentage of absolute time in the ictal state, the unpredictability of seizures still overshadows the life of most epilepsy patients. The simple fact of not knowing when an interictal–ictal transition may occur can greatly accentuate the subjective impact of potentially imminent seizures on the patient's everyday life [6–10]. Seizure unpredictability has major implications for patients, including its impact on medical diagnosis, its current role in determining the therapeutic approach, and its practical clinical consequences for the patients,

Seizure Prediction in Epilepsy. Edited by Björn Schelter, Jens Timmer and Andreas Schulze-Bonhage
Copyright © 2008 WILEY-VCH Verlag GmbH & Co. KGaA, Weinheim
ISBN: 978-3-527-40756-9

ranging from the risk of complications secondary to epileptic seizures to socio-legal constraints and psychiatric comorbidity of the disease.

1.2
Medical Implications of Unpredictability

1.2.1
Diagnostic Uncertainty

Based on the intermittent character and brief duration of epileptic seizures, the patient will meet a physician in the vast majority of cases in the interictal state only, even if treated for years. This may delay or considerably reduce the chances of correct diagnosis. This is particularly true when additional features of epileptic attacks, such as partial or complete loss of consciousness during a seizure and retrograde amnesia for ictal events, render it difficult to obtain sufficient circumstantial information on the paroxysmal events, from the patient's history [11]. Documentation and analysis of paroxysmal events using video-EEG-monitoring is the gold standard for determining the diagnosis of epilepsy [12–14], but the unpredictability of the events may make this diagnostic method unfeasible in cases with low frequency.

1.2.2
Treatment Options

Whereas in other paroxysmal diseases, such as migraine, an effective acute intervention is possible in the early phase of an attack, the brief duration of most epileptic seizures places a severe limitation on the potential effect of acute treatment. Medical treatment will not take effect if systemically applied before a seizure has spontaneously ended because of the pharmacokinetic delays between application and efficacy at their targets associated with drug absorption and distribution. Brain stimulation offers advantages in this respect. But at present, stimulation is only used, albeit widely, in the form of vagus nerve stimulation, which can be interactively triggered by the patient him or herself or another person by using a magnet, once the clinical symptoms become overt [15]. So far, the efficacy of ictal vagus nerve stimulation has yet to be studied in detail in the human; limited efficacy may be related to rapid spread of ictal activity, particularly if clinical seizure onset precedes its activation [16]. Rapid detection methods may provide new opportunities in a closed-loop setting [17].

The unpredictability of seizure occurrence generally results in the treatment being performed continuously over time in order to prevent interictal–ictal transitions. In treatments using brain stimulation via the vagus nerve or directly using intracranially implanted depth electrodes, more or less continuous stimulation patterns with variable duty cycles are applied [18, 19]. The vast majority of patients have to take medication every day over a period of years, independent of actual seizure frequency. As the aim is to achieve a steady drug concentration at the target site, many drugs require an intake of at least twice a day, in some cases up to four times a day, to avoid breakthrough seizures [20], which may pose problems for patient compliance [21]. Drug level fluctuations in the case of irregular intake or of intercurrent alterations in resorption, metabolism or pharmacodynamic interactions may themselves trigger 'withdrawal' seizures [22–25].

The unpredictability of seizure occurrence is not the only burden for patients because of the regular intake requirement of antiepileptic medication. A continuously high level of antiepileptic medication is necessary for an optimal control of seizure frequency but is often accompanied by side effects. These typically encompass unspecific CNS-related effects like tiredness, dizziness, blurred vision or headaches, but may also include more specific impairments in cognition including difficulties with attention, concentration or language functions [26]. In turn, such dose-dependent side effects during continuous intake often place constraints on the maximal efficacy a drug can achieve; the need for continuous long-term administration thus affects both efficacy and tolerability of present-day pharmacotherapy. Another problem related with the continuous long-term treatment is the loss of efficacy in certain drugs (e.g., benzodiazepines) which are effective only in an acute setting but not for a protracted period, this being due to the development of tolerance [27]. At maximally tolerated dosages with continuous systemic drug administration, about one third of current epilepsy patients continue to have epileptic seizures and are regarded as 'pharmacoresistant' [28]. Seizure prediction would therefore open new avenues for drug treatment; for example, using short-acting drugs like lorazepam, or the transient application of high drug dosages for acute seizure abortion which cannot be used in the long term.

1.2.3
Physical Risks

Patients with epilepsy frequently suffer from seizure-related injuries [29–31]. These result in part from the loss of control over the motor system during a seizure, as in the case of falls during atonic or tonic seizures. Often, a limited reactivity to external stimuli plays a major role in epilepsy-related injures. The loss of consciousness or delayed reactivity may therefore lead to an increased accident risk of patients exposed to demanding road traffic, particularly if there is no preceding warning symptom [29]. Similarly, the risk of accidental physical harm is increased in a spectrum of sport activities [32]. Even at home, the risk of sustaining seizure-related injuries is greatly increased during everyday activities; examples of these are increased frequencies of burns during cooking or showering [33, 34], or an increased risk of drowning in the bathtub [35], which is a major cause of death in epilepsy patients. The absence of any warning signal preceding a seizure may in fact increase the risk of accidental physical harm considerably.

1.2.4
Risks Associated with Continuous Long-term Antiepileptic Treatment

Although most antiepileptic drugs are remarkably well tolerated over many years, there is a spectrum of substance-dependent risks associated with continuous long-term application. The relative importance of such side effects may be different depending on age. Cognitive side effects of phenobarbitone and pro-apoptotic actions of several drugs may be particularly important in the developing brain [36], the hormonal effects of valproate [37] may have their greatest impact in fertile females, and the induction of ostepenia by enzyme-inducing drugs [38, 39] may play a particular role in increasing the risk in the elderly for the development of pathological fractures. Other side effects of chronic intake like the development of cerebellar atrophy and polyneuropathy with phenytoin [40, 41], disturbances of hormonal metabolism [42], the induction of mood disorders by several drugs [43] and the development of visual field constrictions with vigabatrin [44] are consequences of long-term intake and pose problems at any age. The task of controlling the development of side effects related to long-term therapy imposes considerable costs on the healthcare system, even if these effects are as rare as certain idiosyncratic reactions, like liver failure or bone marrow aplasia.

1.3
Psychosocial Consequences of Unpredictability

The unpredictability of seizure occurrence has psychological and social consequences that are often closely interrelated. Seizures that reoccur frequently and unpredictably are accompanied by a patient's objective loss of control associated with the reduced ability to steer motor behavior and by the subjective loss of control associated with feeling overwhelmed by the effects of the disease. This experience is particular to epilepsy, the consequences of which are reflected in the psychological concept of 'locus of control' and in the development of psychiatric symptoms of anxiety and depression that have, in turn, social implications.

1.3.1
Loss of Control

The concept of 'locus of control' reflects the cognitive style of attributing events and actions to either internal or external factors. Internal factors encompass the person's own behavior, abilities or characteristics, whereas external factors include chance or misfortune on the one hand and actions of other persons on the other. This psychological concept thus deals with an individual's tendency to perceive events as being controlled either by themselves or by external forces [45]. This idea was originally formulated by Julian B. Rotter and recognises the importance of the individual's perception of causality as attributable to both intrapersonal determinants and the social context of behavior. Individuals tend to categorise situations according to their perception of success or failure and, in particular, according to the reason for their outcome. Depending on the individual's perception of causality, an internal locus of control reflects the belief in a positive reinforcement by his or her own action; whereas an external locus of control is associated with the expectancy of reinforcement by chance or by uncontrollable factors – as would apply in the case of unpredictably occurring seizures.

The concept of locus of control is crucial for achievement motivation and actions and for emotional reactions to social events. It may therefore influence compliance, in terms of outpatient attendance at epilepsy clinics and regular intake of antiepileptic medication, and be a contributory factor particularly in the occurrence of depressed mood and anxiety. Coping strategies and the well-being of patients with chronic diseases are generally associated with perceived locus of control [46], though not only in patients with epilepsy [47, 48]. Children with epilepsy are, already at this early age, more likely to attribute control of events to external factors than are chronically ill children with diabetes or healthy controls. Children with epilepsy show also a lower self-esteem and a higher level of anxiety compared with their peers [49, 50].

Clinical research uses condition-specific locus of control scales like the Multidimensional Health Locus of Control scales by Wallston and colleagues [51]. Health locus of control mirrors the patients beliefs regarding perceived control over the disease and determines health-related behavior. The Multidimensional Health Locus of Control (MHLC) scale comprises three subscales: internality (I-HLC), chance (C-HLC), doctors and powerful others (P-HLC) [52]. Studies examining the attitudes of patients with epilepsy revealed weak perception of internal and strong perception of external health locus of control [50, 53, 54]. This pattern of internality and externality may result in a less effective adaptation of these patients with epilepsy to their illness and a lower engagement in beneficial health behavior and active coping strategies [54].

Recent studies additionally addressed associations between locus of control and self-efficacy or self-confidence, showing that a patient's internal locus of control correlates with his or her self-efficacy. Both are described as mastery variables influencing the patient's quality of life [55] and as predictors for psychological distress [56]. Accordingly, happiness with life and self-confidence are particularly low in patients with high seizure frequency [57].

Even though most patients with epilepsy do not feel that they have control over their seizures, more than fifty percent of patients believe that they can identify seizure precipitants like stress and fatigue correctly, and many have developed strategies by which they try to prevent occurrence of seizures or to abort them [58, 59]. Self-control of seizures would 'elevate' the individual to the position of being able to regulate events and this shift in control expectancy may likely have a positive psychosocial impact. 'High controllers' and 'low controllers' can thus be distinguished according to their belief in their ability to exert control over their seizures, and this, again, correlates with scores on health locus of control scales [53, 60]. Importantly, the degree of self-perceived seizure control practically manifests itself in patients who seek low-risk-for-seizure situations, avoid high-risk-for-seizure situations, and make attempts at seizure prevention.

1.3.2
Problems with Coping Strategies

Strategies such as seeking out low-risk-for-seizure situations and avoiding high-risk situations are considered behavioral coping strategies. These correspond at a cognitive level with propositions like 'Try to maintain some control over the situation' or 'Hope things will get better' [61], and may lead to a search for information, contact with other patients and support groups, and to keeping a seizure diary [62]. Whereas coping strategies may activate patients' resources and contribute to psychosocial adjustment and health [46], avoidance behavior may have negative social implications. Feeling stigmatised, patients with epilepsy are often ashamed of having publicly 'displayed' unpredictable seizures, and they may therefore tend to avoid leaving home. This can result in social withdrawal, isolation, lack of positive social interaction and experience, and, finally, in a loss of self-efficacy and a decline in quality of life.

1.3.3
Depression and Anxiety

Depression and anxiety are the most prevalent psychiatric disorders in adults with epilepsy [63] and already appear in one-third of children suffering from epilepsy. Both depression and anxiety may result in suicidal ideation and behavior, even in childhood [64]. About thirty percent of adult patients with epilepsy report suicide attempts [65], and the suicide rate of epilepsy patients is at least three times higher than that of the general population [66, 67], particularly in women [68]. Loss of control is a major psychological factor leading to depression and anxiety [54, 59]. Epilepsy patients have low internal control beliefs and medium beliefs in the role of chance. The patient's tendency to attribute power to others correlates with the degree of anxiety. Depressivity is related with a more external attributional style and with loss of internal control beliefs [46, 69, 70]. This corresponds to the learned helplessness model of depression by Seligman [70]. Herein, the symptoms of depression such as passiveness, cognitive deficits, and problems with self-esteem, are due to a lack of contingency between a person's behavior and its consequences [71]. Unpredictably occurring seizures, particularly if accompanied by loss of motor control or consciousness, are paradigmatic for such a helpless situation. Correspondingly, patients with pharmacoresistant temporal lobe epilepsy undergoing surgical therapy show, preoperatively, a significant relationship between self-reported depression and external locus of control [72]. Postoperatively, the early anticipation of seizure freedom may already improve mood.

1.3.4
Immobility and Vocational Restrictions

The unpredictable occurrence of seizures results in the imposition of driving restrictions on epilepsy patients, necessarily so because the patient represents a danger to him or herself and to all other traffic participants. Most patients with epilepsy are therefore wholly reliant on public transport or on their personal social environment. Immobility may lead to vocational problems due either to problems accessing the place of employment or to driving a car being a job requirement [73].

Vocational restrictions may also be encountered by patients as a result of many indirect consequences of seizure unpredictability. The occurrence of seizures may be stigmatising in itself, but cognitive impairments related with the disease and side effects of antiepileptic medication often also place limitations on a patient's level of work performance and achievement. More than one-third of patients with active epilepsy are unemployed, while this applies to about ten percent of patients in remission. This serves only to heighten the dependency on the social security system [57] and to reinforce the feeling of subjective handicap. The range of vocational possibilities depends on seizure severity: the unpredictable loss of consciousness, falls, inadequate behavior and loss of motor control are particularly unfavorable [74].

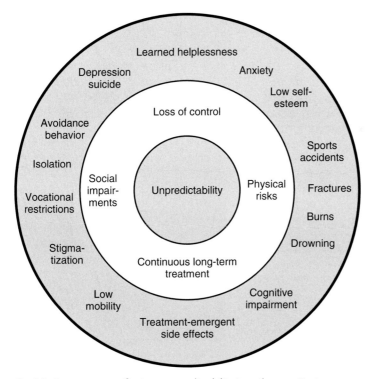

Fig. 1.1 Consequences of seizure unpredictability in epilepsy patients.

1.4
Conclusion

The development of techniques designed to predict epileptic seizure occurrence could make a considerable contribution to improving the well-being of patients in all areas discussed here and summarised in Figure 1.1. Progress in this field could facilitate medical diagnosis, and new timely drug delivery or stimulation techniques could be specifically targeted to intervene in the preictal brain dynamics that lead to a seizure. Intermittent therapy could offer major advantages both in efficacy and in long-term tolerability. Finally, reliable warning systems would reduce patients' risk of physical harm, might offer new windows of occupational opportunity and leisure activities, and could contribute to a change in the patient's perception of external locus of control, the feeling of helplessness and to secondary psychiatric problems.

References

1 *Epilepsia* **22**, 489 (**1981**).
2 J. Bauer and W. Burr, *Seizure* **10**, 239 (**2001**).
3 S. Jenssen, E. J. Gracely and M. R. Sperling, *Epilepsia* **47**, 1499 (**2006**).
4 W. H. Theodore, R. J. Porter, P. Albert, K. Kelley, E. Bromfield, O. Devinsky and S. Sato, *Neurology* **44**, 1403 (**1994**).
5 O. Devinsky, K. Kelley, R. J. Porter and W. H. Theodore, *Neurology* **38**, 1347 (**1988**).
6 F. Gilliam, R. Kuzniecky, E. Faught, L. Black, G. Carpenter and R. Schrodt, *Epilepsia* **38**, 233 (**1997**).
7 M. Hayden, C. Penna and N. Buchanan, *Seizure* **1**, 191 (**1992**).
8 R. Ortiz and J. Liporace, *Epilepsy Behavior* **6**, 620 (**2005**).
9 I. M. Elliott, L. Lach and M. L. Smith, *Epilepsy Behavior* **7**, 664 (**2005**).
10 A. Jacoby, *Soc. Sci. Med.* **34**, 657 (**1992**).
11 *Rev. Neurol. (Paris)* **156**, 481 (**2000**).
12 S. Kipervasser and M. Y. Neufeld, *Acta Neurol. Scand.* **116**, 221 (**2007**).
13 B. J. Steinhoff and A. Schulze-Bonhage, *EEG* (Thieme, Stuttgart, **2006**), chap. Langzeit-EEG, pp. 79–83.

14 M. Manford, *J. Neurol. Neurosurg. Psychiatry* **70** Suppl 2, I13 (**2001**).
15 G. L. Morris, *Epilepsy Behavior* **4**, 740 (**2003**).
16 K. Götz-Trabert, C. Hauck, K. Wagner, S. Fauser and A. Schulze-Bonhage, *Epilepsia* in press (**2008**).
17 I. Osorio, M. G. Frei, S. Sunderam, J. Giftakis, N. C. Bhavaraju, S. F. Schaffner and S. B. Wilkinson, *Ann. Neurol.* **57**, 258 (**2005**).
18 C. Heck, S. L. Helmers and C. M. DeGiorgio, *Neurology* **59**, S31 (**2002**).
19 W. H. Theodore and R. S. Fisher, *Lancet Neurol.* **3**, 111 (**2004**).
20 M. Bialer, *CNS Drugs* **21**, 765 (**2007**).
21 J. A. Cramer, R. H. Mattson, M. L. Prevey, R. D. Scheyer and V. L. Ouellette, *JAMA* **261**, 3273 (**1989**).
22 J. C. DeToledo, R. E. Ramsay, M. R. Lowe, M. Greiner and E. A. Garofalo, *Ther. Drug Monit.* **22**, 753 (**2000**).
23 H. Martínez-Cano, A. Vela-Bueno, M. de Iceta, R. Pomalima and I. Martínez-Gras, *Pharmacopsychiatry* **28**, 257 (**1995**).
24 J. S. Duncan, S. D. Shorvon and M. R. Trimble, *Epilepsia* **31**, 324 (**1990**).

25 F. Buchthal, O. Svensmark and
    H. Simonsen, *Arch. Neurol.* **19**, 567
    (**1968**).
26 B. Gomer, K. Wagner, L. Frings,
    J. Saar, A. Carius, M. Härle,
    B. J. Steinhoff and A. Schulze-
    Bonhage, *Epilepsy Behavior* **10**, 486
    (**2007**).
27 W. Löscher and D. Schmidt, *Epilepsia*
    47, 1253 (**2006**).
28 P. Kwan and M. J. Brodie, *N Engl.
    J. Med.* **342**, 314 (**2000**).
29 E. C. Wirrell, *Epilepsia* **47** Suppl. 1,
    79 (**2006**).
30 P. Vestergaard, S. Tigaran,
    L. Rejnmark, C. Tigaran, M.
    Dam and L. Mosekilde, *Acta.
    Neurol. Scand.* **99**, 269 (**1999**).
31 H. B. I. Persson, K. A. Alberts, B. Y.
    Farahmand and T. Tomson, *Epilepsia*
    43, 768 (**2002**).
32 G. M. Howard, M. Radloff and T. L.
    Sevier, *Curr. Sports Med. Rep.* **3**, 15
    (**2004**).
33 I. C. Josty, V. Narayanan and W. A.
    Dickson, *Epilepsia* **41**, 453 (**2000**).
34 M. C. Spitz, J. A. Towbin, D. Shantz
    and L. E. Adler, *Epilepsia* **35**, 764
    (**1994**).
35 C. A. Ryan and G. Dowling, *CMAJ*
    148, 781 (**1993**).
36 P. Bittigau, M. Sifringer, K. Genz,
    E. Reith, D. Pospischil,
    S. Govindarajalu, M. Dzietko, S.
    Pesditschek, I. Mai, K. Dikranian
    et al., *Proc. Natl. Acad. Sci. USA* **99**,
    15089 (**2002**).
37 A. Verrotti, R. Greco, G. Latini
    and F. Chiarelli, *J Pediatr. Endo-
    crinol. Metab.* **18**, 423 (**2005**).
38 A. M. Pack and M. J. Morrell,
    Epilepsy Behavior **5** Suppl. 2, S24
    (**2004**).
39 R. D. Sheth, B. E. Gidal and B. P.
    Hermann, *Epilepsy Behavior* **9**, 601
    (**2006**).
40 F. A. D. Marcos, E. Ghizoni, E.
    Kobayashi, L. M. Li and F. Cendes,
    Seizure **12**, 312 (**2003**).
41 A. Bono, E. Beghi, G. Bogliun,
    G. Cavaletti, N. Curtó, L. Marzorati
    and L. Frattola, *Epilepsia* **34**, 323
    (**1993**).
42 J. I. T. Isojärvi, E. Taubøll and A. G.
    Herzog, *CNS Drugs* **19**, 207 (**2005**).
43 B. Schmitz, *Epilepsia* **47** Suppl. 2, 28
    (**2006**).
44 L. V. Wilton, M. D. Stephens and
    R. D. Mann, *BMJ* **319**, 1165 (**1999**).
45 J. B. Rotter, *Psychol. Monogr.* **80**, 1
    (**1966**).
46 K. Krakow, K. Bühler and
    H. Haltenhof, *Seizure* **8**, 111 (**1999**).
47 M. C. Chung, E. Preveza,
    K. Papandreou and N. Prevezas,
    J. Affect. Disord. **93**, 229 (**2006**).
48 M. C. Chung, E. Preveza, K.
    Papandreou and N. Prevezas,
    Psychiatry Res. **152**, 253 (**2007**).
49 S. Correa, *Psychol. Rep.* **60**, 9 (**1987**).
50 W. S. Matthews, G. Barabas and
    M. Ferrari, *Epilepsia* **23**, 671 (**1982**).
51 K. A. Wallston, B. S. Wallston and
    R. DeVellis, *Health Educ. Monogr.* **6**,
    160 (**1978**).
52 K. A. Wallston, M. J. Stein and C. A.
    Smith, *J. Pers. Assess* **63**, 534 (**1994**).
53 S. Spector, C. Cull and L. H.
    Goldstein, *Epilepsia* **42**, 556 (**2001**).
54 A. A. Asadi-Pooya, C. A. Schilling,
    D. Glosser, J. I. Tracy and M. R.
    Sperling, *Epilepsy Behavior* **11**, 347
    (**2007**).
55 M. Amir, I. Roziner, A. Knoll and
    M. Y. Neufeld, *Epilepsia* **40**, 216
    (**1999**).
56 A. M. S. Wu, C. S. K. Tang and T. C.
    Y. Kwok, *Aging. Ment. Health* **8**, 21
    (**2004**).
57 M. F. O'Donoghue, D. M. Goodridge,
    K. Redhead, J. W. Sander and J. S.
    Duncan, *Br. J. Gen. Pract.* **49**, 211
    (**1999**).
58 C. A. Cull, M. Fowler and S. W.
    Brown, *Seizure* **5**, 131 (**1996**).
59 M. R. Sperling, C. A. Schilling,
    D. Glosser, J. I. Tracy and A. A.
    Asadi-Pooya, *Seizure* (**2007**).
60 M. Rosenbaum and N. Palmon,
    J. Consult. Clin. Psychol. **52**, 244
    (**1984**).
61 M. Snyder, *Int. Disabil. Stud.* **12**, 100
    (**1990**).
62 J. Murray, *Seizure* **2**, 167 (**1993**).
63 M. F. Mendez, R. C. Doss, J. L.
    Taylor and P. Salguero, *J. Nerv.
    Ment. Dis.* **181**, 444 (**1993**).

64 R. Caplan, P. Siddarth, S. Gurbani, R. Hanson, R. Sankar and W. D. Shields, *Epilepsia* **46**, 720 (**2005**).

65 M. F. Mendez, J. L. Cummings and D. F. Benson, *Arch. Neurol.* **43**, 766 (**1986**).

66 L. Nilsson, T. Tomson, B. Y. Farahmand, V. Diwan and P. G. Persson, *Epilepsia* **38**, 1062 (**1997**).

67 J. E. Jones, B. P. Hermann, J. J. Barry, F. G. Gilliam, A. M. Kanner and K. J. Meador, *Epilepsy Behavior* 4 Suppl. **3**, S31 (**2003**).

68 V. V. Kalinin and D. A. Polyanskiy, *Epilepsy Behavior* **6**, 424 (**2005**).

69 B. P. Hermann, M. R. Trenerry and R. C. Colligan, *Epilepsia* **37**, 680 (**1996**).

70 D. W. Dunn, J. K. Austin and G. A. Huster, *J. Am. Acad. Child Adolesc. Psychiatry* **38**, 1132 (**1999**).

71 M. E. P. Seligman, *Learned Helplessness* (Freeman, San Francisco, **1975**).

72 B. P. Hermann and A. R. Wyler, *Epilepsia* **30**, 332 (**1989**).

73 H. R. Stöckli, *Ther. Umsch.* **64**, 429 (**2007**).

74 P. Bülau and A. zur Verbesserung der Eingliederungschancen von Personen mit Epilepsie, *Rehabilitation (Stuttg.)* **40**, 97 (**2001**).

2

The History of Seizure Prediction

M. Jachan, H. Feldwisch genannt Drentrup, B. Schelter, J. Timmer

2.1
Introduction

Seizure-prediction methods, which are based on EEG recordings only, could be an integral part of future therapeutical devices in epilepsy. The current focus of research is on a seizure-control system which can suppress an upcoming seizure (closed-loop intervention), or at least warn the patient prior to an expected event. However, this goal seems to be very far off but important progress has been made in neurology, biology, network modeling, microsystems engineering, and time series analysis. The state of the art in seizure prediction is that, to date, a clinically applicable seizure-prediction algorithm does not exist, but several studies have shown that one can in fact predict seizures significantly better than a random predictor. Much too optimistic results obtained in the last two decades have been shown not to be specific for seizure precursors, resulting in frequent false alarms of the methods. A patient will certainly not accept a seizure-control system which produces too many false alarms, because it will hamper the patient instead of helping him. This review chapter presents the history, the current state of the art, and also an outlook on future developments in seizure prediction; a research field which has gained much attention in the last thirty years. We discuss the possible existence of a preictal phase, review several-prediction methods, elaborate on statistical evaluation, address open problems and caveats, and discuss the design of seizure-prediction studies, referring to the chapters of this book contributed by specialists in the field.

2.2
Motivation

Epilepsy is one of the most common neurological disorders with a prevalence of up to 1 % of the world's population. Quality of life is severely affected due to the sudden and unforeseeable occurrence of epileptic seizures. Not only are many patients not allowed to drive a car and have severe problems in finding appropriate employment, but also other parts of their daily lives, such as sports, social aspects, etc., are influenced. Approximately two-thirds of epilepsy patients can become seizure-free by continuous anti-convulsive medication, whereas one-third continue to suffer from seizures due to non-effectiveness of the therapy at tolerable doses. For a fraction of these medical refractory patients, resective brain surgery is an adequate cure. For this aim the seizure-generating brain area can be localized first by a simple diagnostic means (on-scalp EEG, semiology of the seizure). An exact localization of the seizure focus can be strongly supported by functional medical imaging techniques and intracranial long-term video-EEG monitoring. Minimal-invasive surgery techniques allow for a precise placement of intracranial EEG electrodes. More than one hundred channels of EEG can be acquired simultaneously and are used for focus localization by visual inspection. After the focus has been identified by correlating imaging, video, EEG, and possibly other information, it can be removed by computer-aided brain surgery. Also, care is taken to predict possible shortcomings of functionality of the patient. Most resected patients become seizure-free (about 70 %) or experience a certain alleviation.

For the residual amount of epilepsy patients (25–30 %) no sufficient treatment is currently known. This group contains patients suffering from a generalized epilepsy, where surgery is impossible, as well as patients with a focal epilepsy who cannot be operated for a spectrum of reasons. Thus, for the latter group of patients who do not achieve complete seizure control, new therapeutic methods have to be set up [1, 2].

2.2.1
The Need for a Seizure-prediction Device

From the earliest time of EEG analysis there has been interest in finding EEG patterns characteristic for and preceding epileptic seizures. In the seventies of the 20th century, physicists, physicians and engineers have addressed the issue of predicting an upcoming seizure by use of the patients' EEG recordings (see the review articles [3–6]). Either from surface or intracranial EEG, certain measures which are supposed to be able to detect an assumed preictal state can be derived. An alarm generated by such a seizure prediction device can be utilized to simply trigger a warning for the patient, or, for more advanced methods, to trigger a closed-loop intervention system (see Figure 2.1). Such a closed-loop intervention could be based on the injection of fast-acting medication, on electrical stimulation of the brain [7], or on other intervention techniques, e.g., cooling of the epileptogenic area [8] (cf. also Chapter 21 of this book). It is expected that temporally and spatially targeted drug application can also be effective for patients who are refractory up to now, because with a much lower dosage a much higher local drug level can be achieved. Also present-day available stimulation techniques, like vagus nerve stimulation, could be specifically triggered by a seizure-prediction device. Electrical stimulation of brain structures has proved to be highly effective against other neurological diseases such as tremor; thus the question arises of whether seizure control can be achieved by deep-brain stimulation for epilepsy patients.

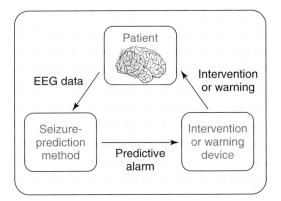

Fig. 2.1 Scheme of a future warning device or closed-loop seizure control system. From the patients' EEG, features are extracted online, which can detect the transition of the brain into the preictal phase. Thus, an alarm can be raised before the seizure manifests. The patient can be warned of the seizure, or in more advanced methods, automatic countermeasures can be applied which prevent the evolution of the seizure.

We next discuss the terms: (i) seizure detection; (ii) seizure anticipation; and (iii) seizure prediction in more detail [5]. If an alarm is generated at the beginning of or during the seizure, we speak about seizure detection [9–11]. Detection methods can be of use for offline analysis of EEG data as well as for seizure-abortion devices. However, here we will focus on the prediction of seizures. If a precise horizon of seizure occurrence cannot be stated, one speaks about seizure anticipation. Seizure prediction means to determine its occurrence in advance with a certain temporal precision called the occurrence period. In addition, the expected event may not occur immediately after the alarm, because this would not be a true prediction, but rather a detection. The time window starting at the alarm and ranging to the earliest allowed occurrence of the seizure may be called the warning or *intervention time* [12, 13]. For epileptic seizure prediction, meaningful intervention times range from several seconds up to several hours. Shorter intervention times might be sufficient for warning devices, where it is necessary for the patient to have some minutes before the actual seizure onset to stop a dangerous activity. Also, for triggering brain stimulation, relatively short intervention times may suffice. Longer intervention times are appropriate for pharmacologically acting seizure-prevention devices, to ensure that injected drugs arrive at their targets and become effective. Usually, the terms prediction and anticipation are used interchangeable, but here we are going to use the term prediction only.

Having the outline for an automatic seizure prediction or prevention device as described above, the question arises of how it can act for the patient in a practical and safe way. Clearly, the patient needs to carry sensors for long-term monitoring of brain activity and an effector in closed-loop treatment systems. As a first idea, people came up with surface-EEG-based analysis systems. Neurosurgical implantation of deep-brain stimulation devices, as well as vagus nerve stimulator implantation have evolved to relatively safe methods and both open-loop and closed-loop implantable devices have been available for years. An EEG sensor, perhaps including chemical or electrical actors, could be implanted permanently in the identified brain region, and a control and supply device, perhaps including a drug reservoir, could be connected from the trunk by subcutane wiring. Such an implanted solution may not only be the ultimate cure for untreatable patients, but may also offer advantages over existing treatments for other patients. However, a necessary condition for the realization of such a device is the ability to accurately predict the patients' seizures. The heterogeneity of the disease of epilepsy, the complexity of the human brain, and the unavailability of intracranial data from healthy control subjects render this goal ambitious.

2.2.2
The Assumed Preictal Phase

Ictogenesis could take place in two different scenarios, either abrupt or gradual [14–16]. A completely unforeseeable and sudden initiation of a seizure, which is a suitable scenario for seizures from a generalized epilepsy, would not allow for its prediction at all. The brain would change immediately from the stable interictal state to the unstable ictal state that is characterized by paroxysmal occurrence of synchronous discharges [5]. If a seizure was generated gradually, e.g., by evolution of the stable brain state into a seizure-prone state, a prediction would in principle be possible. A sudden or gradual change of internal dynamics could modify the threshold for seizure generation and could make the brain vulnerable to seizures. This evolutionary phase is usually termed the preictal phase. The second gradual scenario is more appropriate as clinical prodromi are felt by about 6 % of the patients, on average, 90 minutes prior to the seizure [17]; thus they occur at similar time horizons prior to seizures as predictive EEG changes have been reported using methods from time series analysis [4]. Another type of clinical precursor often felt or experienced immediately before the seizure from patients suffering from a focal epilepsy, are called auras. An aura represents ongoing epileptic activity only in parts of the brain and causes subjective symptoms, such as psychic or sensory feelings [17]. Preictal phenomena can be of different nature, depending on the affected brain area and they might or might not be reproducible.

We next review some results from EEG analysis which suggest the existence of a preictal phase. Due to the fact that an epileptic seizure is generated by pathological hyper-synchronous firing of neurons, changes of synchronization between EEG channels in the preictal phase have been investigated [18]. Another approach to detect the preictal phase is to compare the actual EEG to a preselected neutral period showing no epileptiform activity [19]. Other findings support the fact that seizure precursors, i.e., localized prolonged bursts of complex epileptiform discharges, may be present up to seven hours in advance of its onset [20]. As a final example, we mention the occurrence of high-frequency oscillations with a frequency of up to hundreds of Hertz and of short duration [21–24] in the epileptogenic area or possibly prior to a seizure (see also Chapter 13).

Given the wide range of types of epileptic seizures, the question arises of whether all of them can be predicted in principle. Reflex epilepsies, where seizures are triggered by external stimuli such as light, sound, touch, etc., are not in the focus of seizure prediction. The seizure-prediction community focuses on patients suffering from focal epilepsies for two reasons: because there is strong evidence that these seizures can be predicted; and intracranial data are available almost exclusively from such patients. Attention has to be paid in classifying seizure types according to localization, cause of the disease; e.g., tumors, lesions, anomalies of the brain tissue, toxic brain effects, and whether the seizure becomes generalized or not. Another particular problem is posed by electrographic ictal events which remain subclinical. At present these events are generally not targeted by seizure prediction algorithms and their evaluation, although identical mechanisms may underlie their generation.

The hypothesized preictal state has thus been characterized by clinical as well as by many algorithmic methods which are based on physiological and pathological findings. Another view of preictal phenomena assumes a phase of increased probability of seizure occurrence; thus not every such preictal phase needs to be accompanied by a seizure manifestation [25]. The occurrence of seizure clusters, i.e., several seizures within a short period of time, can also be explained by the existence of a state of increased seizure probability. Following this idea, the problem arises up that the detection of the preictal phase might be perfectly right, but the specificity, in the sense of the occurrence of a subsequent seizure, is limited.

These findings on the preictal state naturally lead towards seizure-prediction methods, which rely on the automatic detection of preictal changes in the EEG. However, the interictal occurrence of preictal phenomena turn the design of sensitive, as well as specific, seizure-prediction methods into a challenge.

2.3
A Historical Overview

In this section we review the conceptions of studies for seizure prediction from EEG recordings which have been presented in the last thirty years. Research on the predictability of seizures started in the 1970s, where mainly linear methods, such as spectral analysis and autoregressive modeling, have been applied to detect the preictal phase from surface recordings. It has been reported that seizures can be predicted several seconds, minutes or hours before their occurrence [3, 5]. Also, pattern-recognition techniques such as the analysis of spiking rates have been applied, but they have not been shown to possess predictive value [26]. In [5], five types of studies have been classified. The first three study types are based on evaluation on short data segments: preictal-only studies, studies with interictal control periods, and controlled studies. Later, continuous long-term data has been used to run studies, which can be of retrospective or of prospective nature. The first three types of study are from today's point of view with the availability of modern computational power no longer sufficient, while the latter two study types can produce results on predictability which are of clinical interest (cf. Chapters 18, 19).

2.3.1
Older Types of Studies

Preictal-only studies were limited to rather short data segments taken prior to seizures due to the lack of storage and processing capabilities available some decades ago. In the 1990s, as the theory of nonlinear dynamics evolved, several measures characterizing nonlinear properties have been proposed. The largest Lyapunov exponent [27] was utilized to measure chaoticity of an EEG signal, which was found to drop before seizures. The bivariate measure mean phase coherence has been introduced, which quantifies the relation between two EEG channels in terms of their phase difference [18, 28, 29]. It has been reported that changes of synchronization occur during the preictal phase. By analyzing the correlation density in a larger group of patients it was concluded that the spatio-temporal complexity drops in the preictal phase [30]. Other approaches were to compare the dynamics of the actual EEG window to a fixed reference window by use of the dynamical similarity index [19, 31] or to analyze the correlation dimension [32], the accumulated energy [20], i.e., the running sum of the EEG variance. Later, interictal data segments have been used to provide a control on whether the hypothesized preictal changes also occur in interictal periods [18, 33, 34]. Studies which took interictal control periods into account have been run on selected data and are thus of exemplary nature but of practically limited value. Some measures show some ability to distinguish the preictal phase from the interictal phase suggesting the actual existence of a preictal phase. For the assessment of statistical significance of prediction performance several approaches have been introduced. Applied methods range from the receiver operating characteristic, the seizure prediction characteristic [35], to surrogate-based methods such as seizure-time surrogates [15] or the measure profile surrogates [36]. Statistical evaluation is of central importance and has been addressed very often recently.

2.3.2
Modern Types of Studies

While the first three study types on predictability are insufficient from today's point of view, only long-term retrospective or prospective studies can now be accepted. However, smaller data sets can still be valuable to introduce novel prediction features. As both long-term recordings and sufficient hardware capabilities are available, these more advanced study types should be the underlying conception for current work. In 2002 several groups together created a joint multi-center database of long-term recordings of five patients [25]. This database, created for the First International Collaborative Workshop on Seizure Prediction (Bonn, 2002) can be regarded as the starting point of modern research on seizure prediction. It comprises intracranial EEG data of mesial temporal lobe epilepsy patients with at least 48 hours duration, 30 to 80 channels from subdural strip and amygdalo-hippocampal depth electrodes, and a minimum of three recorded seizures. The EEG was digitized with at least 200 Hz at a resolution of 10 bit or more. All patients were seizure-free after surgery (Engel Class I outcome), which indicates that recordings as close as possible to the epileptic focus were included. To make long-term data also available for researchers without access to an epilepsy monitoring unit, we specially emphasize the need for an international database (cf. [20]).

Studies based on larger data collectives have shown that the promising results of predictability by EEG analysis are not specific [4, 34, 37]. Doubts have been raised on the performance of the similarity index [35, 38], the correlation dimension [39, 40], the largest Lyapunov exponent [41, 42], the accumulated energy [34, 43], and the correlation density [44]. Almost all introduced measures were not able to distinguish the interictal phase from the preictal phase if evaluation was performed on long-term data. Reasons for this sobering result are found in the fact that the optimization of algorithm parameters was performed on selected and too small data sets. However, some bi- and multivariate measures have proven to be indeed better than random level as shown in [4, 45]. Bi- and multivariate measures are discussed in Chapters 15 and 16. It was found that preictal phenomena are not a global phenomenon, but they are rather limited in the spatial extension, although not restricted to the seizure-generation area [20]. Truly prospective studies were carried out for the first time starting in 2003 [3, 46–49], but no sufficient analysis of whether the performance is above chance level was included. The results which have been obtained by long-term studies have nevertheless led to valuable insights on ictogenesis, such as preictal changes of synchronization and high-frequency content.

2.3.3
Survey of Prediction Methods

In this section we will give an overview of algorithmic seizure-prediction methods, which have been applied up to now [4]. The dynamics of EEG signals can be modeled in various different ways, resulting in a broad range of candidate analysis methods. The EEG is a nonstationary nonlinear stochastic process. From a clinical point of view, the EEG is divided into several rhythms, i.e., delta (0.5–4 Hz), theta (4–8 Hz), alpha (8–13 Hz), beta (13–30 Hz), and gamma (30–70 Hz) [50]. Beyond the gamma range, the ripple band (100–200 Hz), and fast ripple band (250–500 Hz) [24, 51] have been defined. Given the high complexity of the EEG signal, a single prediction feature can only quantify some of its properties. A seizure-prediction method usually maps the measured property onto a one-dimensional feature time series, which is subsequently compared to a threshold. Either rising, declining, or changes in the instantaneous value of the feature can be exploited to detect the preictal phase. Thus, the EEG can be searched for many, possibly independent, causes of epileptic seizures. Before EEG analysis can be started, one has to consider artifact removal [52], especially in on-scalp recordings. Typical artifacts are mains noise (50 or 60 Hz), plateaus resulting from saturation or zero-lines during periods of disconnection and muscle or eye-blink artifacts. In intracranial recordings artifacts are rare compared to surface recordings.

Seizure prediction methods can be classified either into linear or nonlinear methods or into uni-, bi-, or multivariate approaches [4]. Univariate measures quantify the state of a single cortical region while multivariate measures quantify interaction of different brain regions. All these measures are inspired by linear and nonlinear dynamical systems theory, and they try to quantify effects which are hypothesized based on clinical and biological results. Univariate linear methods are, e.g., the statistical moments and the relative power of the different EEG spectral bands (delta, theta, alpha, beta, gamma). Other measures are based on the autocorrelation function, such as the decorrelation time. Univariate nonlinear measures are: the effective correlation dimension; the largest Lyapunov exponent; local flow; algorithmic complexity; and loss of recurrence [4, 37]. These measures are able to assess various dynamic properties inherent to signals originating from nonlinear systems. Bivariate linear measures are based on the cross-correlation, or its spectral analogue, while the coherence function and bivariate nonlinear methods are based on the phase or on information theoretic aspects. They rely on a representation of the signals' phase expressed by the Hilbert transform [53] or on the wavelet transform [54]. In a practical implementation, the measures are computed in a block-processing or sliding window fashion, where the block length is in the range of several seconds. Smoothing of the feature profile, mostly implemented as a median filter, seems necessary to allow for a narrower amplitude distribution and to remove outliers, which cause false alarms.

The time profiles of the features can be analyzed according to several evaluation schemes [4]. To find possible global effects all features stemming from the given channels or channel combinations can be pooled and their distribution in the inter- and preictal phases can be compared. If, however, the seizure precursors are present in only some channels or channel combinations, a pooling would not be appropriate. In this case the features have to be analyzed separately in order to find those exhibiting the best prediction power. In particular, if more than one seizure type per patient is present, an analysis of each single preictal phase in comparison to the entire interictal time, may give insight. Circadian rhythms, drifts due to medication changes, and possibly other external or internal influences may affect the baseline of the respective feature. True seizure precursors may be small components which are buried in larger and slower fluctuations, thus an adaptive baseline can correct for these slow changes in the feature dynamics.

Finally, meta-parameters such as window length and occurrence period have to be optimized to achieve maximum performance. These are the following constraints on such meta-parameters. The window length may not be too small because most of the features express a certain average property of the actual EEG segment; it may not be too large because an online application can only allow for a certain maximum delay and computational requirements usually increase with polynomial order with the window length. Constraints on the occurrence period are given by practical aspects of the prediction device for the patient. Very small horizons can be appropriate for warning as well as for some intervention techniques while too-large horizons are impractical because the time under (false) warning should not exceed a certain limit.

2.4
The State of the Art in Seizure Prediction

The related problem of seizure detection can, in principle, be regarded as solved, but a complete solution is not available. Based on the inspection of spectral power in higher frequency bands or on the extraction of characteristic waveforms, epileptic seizures can be detected accurately as they come up [9–11]. Seizure prediction with a time horizon of minutes to hours remains an open issue. For a clinically applicable seizure-prediction method, the prediction performances have to be improved considerably. However, scientific effort in the field has produced numerous important intermediate goals, which will be reviewed in the sequel.

2.4.1
Partially Solved Issues

A central and partially solved problem in the field of seizure prediction is the statistical evaluation of prediction schemes (cf. Chapter 18 of this book). Analysis methods based on the receiver-operating-characteristic used to compare the feature distributions of the preictal and the interictal phases, have been proposed [4, 12, 35]. Performance measures are the sensitivity, i.e., the percentage of correct predicted seizures and the false prediction rate, i.e., the complementary specificity. Whether the prediction performance is above chance level can be quantified either based on an analytical random predictor [12, 35], or on numerical bootstrap techniques, such as seizure time surrogates [15]. A random prediction scheme generates alarms either periodically or based on a Poisson process, thus not using any information contained in the EEG. It exhibits increased sensitivity for larger occurrence periods and larger false alarm rates. If a best feature or appropriate channels are going to be selected from a pool of features or channels, a correction for multiple testing has to be incorporated, which again increases the power of the random predictor. Seizure time surrogates are a prominent method of Monte Carlo evaluation methods. The original seizure-onset times are replaced by randomly generated seizure-onset times, which are obtained by shuffling the original onset times while leaving the EEG data unchanged. If the performance of the measure under consideration is higher on the original onset times than on the surrogate times, it can be considered significantly better than a random prediction [48, 49].

Because a certain prediction method usually exploits only one special dynamical property of the EEG, a combination of independent methods can produce increased performance (cf. Chapter 17 of this book). Given that two independent prediction methods are available, their combination in the sense of a logical 'OR' can increase the sensitivity, but the specificity may be decreased. If the combination is based on a logical 'AND' operation, false alarms can be reduced, but an alarm is only raised if both methods issue an alarm within a given time frame. Thus, the sensitivity of the 'AND' combination will be lower than that of the individual methods.

The incorporation of confounding side effects such as circadian rhythms, especially sleep-wake cycles [55], drug level [56], heart rate, cerebral blood flow and other effects has been studied [5]. Taking physiological as well as environmental side effects into account has the potential to lower the amount of false alarms and to raise sensitivity. Much research has been performed in other areas next to seizure prediction. Recent developments in the field of synchronization clusters [57, 58] allows for the mapping of epileptogenic activity onto the brain. Also, source-localization techniques have been applied to on-scalp EEG and their usefulness is critically reviewed in [59]. On the experimental side, animal models of epilepsy have been set up [60], which are required to test and evaluate intervention techniques. Animal models are addressed in Chapters 4, 8 and 11. Results on the interictal/ictal transition mechanism based on statistical modeling using computational network models (see Chapters 3, 5, 7, 10, 14) have been reported [16], where data from animal, human *in vivo*, and *in vitro* models have been utilized. Also cellular mechanisms such as synaptic transmissions (cf. Chapters 6 and 9) have been studied.

Neuro-stimulation in epilepsy has made huge steps forward [61], (cf. Chapters 22–24). By stimulation of different target areas in some patients a decrease in seizure frequency has been achieved. Based on alarms raised by seizure detection, a stimulation-based seizure abortion device has already been conceived [62]. Last but not least, the actual seizure onset has been examined in detail and possible trigger mechanisms have been described [63, 64].

2.4.2
Unsolved Issues

Let us now pay attention to the central unsolved problems. Until now it has not been possible to report a successful prospective prediction of epileptic seizures including a sound statistical analysis [5]. The proposed algorithm needs to show prediction performance when evaluated on unselected long-term EEG data from several patients and *a posteriori* information may not be taken into account. For the establishment of high-performance prediction algorithms it seems to be necessary that the nature of the preictal phase has to be explored in more detail. The diverse results obtained by time-series analysis should be combined in an optimal manner. On the more product-oriented side severe technical shortcomings also have to be improved. An implantable device has very high requirements on size, power consumption, bio-compatibility, maintenance, and reliability. Existing algorithms are run offline, usually on computer clusters, but for an on-chip solution, computational power is an extremely restricted resource. Attention has to be paid for channel selection to keep the amount of raw data to a minimum. The lifetime of the device must be in the range of years and service access should be enabled from an external controller, via a wireless connection. The issue of power supply, especially for stimulation-based systems, remains to be solved. In the case of a chemically acting intervention system, further issues such as drug refill and overdose prevention come into play. Future aspects of seizure prediction in epilepsy are discussed in Chapters 25 and 26 of this book.

2.5
Seizure Prediction in the Future

Future studies on epileptic seizure prediction will show significant improvements in sensitivity and specificity, which need to approach the level of clinical applicability. In particular, measures which are based on the incorporation of information of several cortical network structures have the potential to yield the desired improvements. Further, research on ictogenesis is assumed to deliver more insight into the mechanisms which generate seizures, and in this way the predictability of seizure could be improved. Concentrating on seizure prediction from the EEG, we restate the guidelines for designing seizure-prediction studies as defined in [5]:

- **Data collective.** The testing of prediction algorithms should be done on unselected long-term recordings over several days, to incorporate all kinds of pathological and pathophysiological states of a patient. Today, data of duration from one to two weeks is available almost exclusively from epilepsy monitoring units, which are recorded for focus localization. This data can show severe dependencies on, e.g., changing drug levels to provoke seizures. In the future, data could be gathered from patients carrying a permanent implant. This latter situation has the potential to deliver data stemming from a real daily-life situation as, e.g., stable medication can be guaranteed and much longer acquisition periods are available. Further, algorithm performance should be evaluated on several unselected patients to avoid over-fitting of the meta-parameters.

- **Assessment of sensitivity and specificity.** Long-term data, comprising many seizures, allow for the estimation of both performance measures together. Results should be stated including the temporal parameters' intervention time and occurrence period. False-prediction rates (e.g., false alarms per hour) have always to be stated together with the duration of the subsequent occurrence period, because the false prediction rate times the occurrence period results in the time under false warning, which is a parameter of central importance for the patient.

- **Statistical validation.** The performance parameters' sensitivity and specificity need to be estimated from long-time recordings. Whether or not the algorithm under consideration performs better than random has to be analyzed using a statistical evaluation method. For this purpose, two families of validation techniques are available: Monte Carlo simulations and naïve prediction schemes (cf. Section 2.3).

- **Algorithm training.** Patient-dependent algorithm parameters have to be optimized on a training set (in-sample) while prediction performance has to be estimated on an independent test data set (out-of-sample), which has not been used to optimize the algorithm. Algorithm performance has to be reported for the test data set.

By following the above study design rules, results on the actual prediction performance of the candidate method can be stated in a statistically sound manner and a valid comparison between methods is possible. Effects related to external influences as well as inter- and intra-subject variabilities can be studied.

References

1 A. G. Stein, H. G. Eder, D. E. Blum, A. Drachev and R. S. Fisher, *Epilepsy Res.* **39**, 103 (**2000**).

2 Y. Li and D. J. Mogul, *J. Clin. Neurophysiol.* **24**, 197 (**2007**).

3 J. S. Ebersole, *Clin. Neurophysiol.* **116**, 489 (**2005**).

4 F. Mormann, T. Kreuz, C. Rieke, R. G. Andrzejak, A. Kraskov, P. David, C. E. Elger and K. Lehnertz, *Clin. Neurophysiol.* **116**, 569 (**2005**).

5 F. Mormann, R. G. Andrzejak, C. E. Elger and K. Lehnertz, *Brain* **130**, 314 (**2007**).

6 K. Lehnertz, F. Mormann, H. Osterhage, A. Muller, J. Prusseit, A. Chernihovskyi, M. Staniek, D. Krug, S. Bialonski and C. E. Elger, *Clin. Neurophysiol.* **24**, 147 (**2007**).

7 C. Raftopoulos, K. van Rijckevorsel, B. Abu Serieh, M. de Tourtchaninoff, A. Ivanoiu, G. Mary, C. Grandin and T. Duprez, *Neuromodulation* **8**, 236 (**2005**).

8 S. Rothman and X.-F. Yang, *Epilepsy Curr.* **3**, 153 (**2003**).

9 H. Qu and J. Gotman, *IEEE Trans. Biomed. Eng.* **44**, 115 (**1997**).

10 K. K. Jerger, T. I. Netoff, J. T. Francis, T. Sauer, L. Pecora, S. L. Weinstein and S. J. Schiff, *Clin. Neurophysiol.* **18**, 259 (**2001**).

11 I. Osorio, M. G. Frei, J. Giftakis, T. Peters, J. Ingram, M. Turnbull, M. Herzog, M. T. Rise, S. Schaffner, R. A. Wennberg *et al.*, *Epilepsia* **43**, 1522 (**2002**).

12 B. Schelter, M. Winterhalder, T. Maiwald, A. Brandt, A. Schad, A. Schulze-Bonhage and J. Timmer, *Chaos* **16**, 013108 (**2006**).

13 B. Schelter, M. Winterhalder, H. F. genannt Drentrup, J. Wohlmuth, J. Nawrath, A. Brandt, A. Schulze-Bonhage and J. Timmer, *Epilepsy Res.* **73**, 213 (**2007**).

14 F. H. L. da Silva, W. Blanes, S. N. Kalitzin, J. Parra, P. Suffczynski and D. N. Velis, *IEEE Trans. Biomed. Eng.* **50**, 540 (**2003**).

15 R. G. Andrzejak, F. Mormann, T. Kreuz, C. Rieke, A. Kraskov, C. E. Elger and K. Lehnertz, *Phys. Rev. E* **67** (**2003**).

16 P. Suffczynski, F. H. Lopes da Silva, J. Parra, D. N. Velis, B. M. Bouwman, C. M. van Rijn, P. van Hese, P. Boon, H. Khosravani, M. Derchansky *et al.*, *IEEE Trans. Biomed. Eng.* **53**, 524 (**2006**).

17 A. Schulze-Bonhage, C. Kurth, A. Carius, B. J. Steinhoff and T. Mayer, *Epilepsy Res.* **70**, 83 (**2006**).

18 F. Mormann, K. Lehnertz, P. David and C. E. Elger, *Physica D* **144**, 358 (**2000**).

19 M. Le Van Quyen, J. Martinerie, M. Baulac and F. Varela, *Neuroreport* **10**, 2149 (**1999**).

20 B. Litt, R. Esteller, J. Echauz, M. D'Alessandro, R. Shor and T. H. *et al.*, *Neuron* **30**, 51 (**2001**).

21 R. S. Fisher, W. R. Webber, R. P. Lesser, S. Arroyo and S. Uematsu, *Clin. Neurophysiol.* **9**, 441 (**1992**).

22 A. Bragin, C. L. Wilson, R. J. Staba, M. Reddick, I. Fried and J. Engel Jr., *Ann. Neurol.* **52**, 407 (**2002**).

23 G. A. Worrell, L. Parish, S. D. Cranstoun, R. Jonas, G. Baltuch and B. Litt, *Brain* **127**, 1496 (**2004**).

24 J. D. Jirsch, E. Urrestarazu, P. LeVan, A. Olivier, F. Dubeau and J. Gotman, *Brain* **129**, 1593 (**2006**).

25 K. Lehnertz and B. Litt, *Clin. Neurophysiol.* **116**, 493 (**2005**).

26 J. Gotman and M. G. Marciani, *Ann. Neurol.* **17**, 597 (**1985**).

27 L. D. Iasemides and J. C. Sackellares, *Brain Topography* **2**, 187 (**1990**).

28 F. Mormann, T. Kreuz, R. G. Andrzejak, P. David, K. Lehnertz and C. E. Elger, *Epilepsy Res.* **53**, 173 (**2003**).

29 F. Mormann, R. G. Andrzejak, T. Kreuz, C. Rieke, P. David, C. E. Elger and K. Lehnertz, *Phys. Rev. E* **67** (**2003**).

30 J. Martinerie, C. Adam, M. Le Van Quyen, M. Baulac, S. Clemenceau, B. Renault and F. J. Varela, *Nature Medicine* **4**, 1173 (**1998**).

31 M. Le Van Quyen, J. Martinerie, V. Navarro, M. Baulac and F. J. Varela, *Clin. Neurophysiol.* **18**, 191 (**2001**).

32 K. Lehnertz and C. E. Elger, *Phys. Rev. Lett.* **80**, 5019 (**1998**).

33 V. Navarro, J. Martinerie, M. Le Van Quyen, S. Clemenceau, C. Adam, M. Baulac and F. Varela, *Brain* **125**, 640 (**2002**).

34 T. Maiwald, M. Winterhalder, R. Aschenbrenner-Scheibe, H. U. Voss, A. Schulze-Bonhage and J. Timmer, *Physica D* **194**, 357 (**2004**).

35 M. Winterhalder, T. Maiwald, H. U. Voss, R. Aschenbrenner-Scheibe, J. Timmer and A. Schulze-Bonhage, *Epilepsy Behavior* **4**, 318 (**2003**).

36 T. Kreuz, R. G. Andrzejak, F. Mormann, A. Kraskov, H. Stögbauer, C. E. Elger, K. Lehnertz and P. Grassberger, *Phys. Rev. E* **69**, 061915 (**2004**).

37 R. Andrzejak, F. Mormann, G. Widman, T. Kreuz, C. Elger and K. Lehnertz, *Epilepsy Res.* **69**, 30 (**2006**).

38 W. De Clercq, P. Lemmerling, S. Van Huffel and W. Van Paesschen, *The Lancet* **361**, 970 (**2003**).

39 R. Aschenbrenner-Scheibe, T. Maiwald, M. Winterhalder, H. U. Voss, J. Timmer and A. Schulze-Bonhage, *Brain* **216**, 2616 (**2003**).

40 M. A. Harrison, I. Osorio, M. G. Frei and Y.-C. Lai, *Chaos* **15**, 33106 (**2005**).

41 Y.-C. Lai, M. A. Harrison, M. G. Frei and I. Osorio, *Phys. Rev. Lett.* **91**, 068102 (**2003**).

42 Y.-C. Lai, M. A. Harrison, M. G. Frei and I. Osorio, *Chaos* **14**, 603 (**2004**).

43 M. A. Harrison, M. G. Frei and I. Osorio, *Clin. Neurophysiol.* **116**, 527 (**2005**).

44 P. E. McSharry, L. A. Smith and L. Tarassenko, *Nature Medicine* **9**, 241 (**2003**).

45 M. Le Van Quyen, J. Soss, V. Navarro, R. Robertson, M. Chavez, M. Baulac and J. Martinerie, *Clin. Neurophysiol.* **116**, 559 (**2005**).

46 L. D. Iasemidis, D.-S. Shiau, W. Chaovalitwongse, J. C. Sackellares, P. M. Pardalos, J. Principe, P. R. Carney, A. Prasad, B. Veeramani and K. Tsakalis, *IEEE Trans. Biomed. Eng.* **50**, 616 (**2003**).

47 M. D'Alessandro, G. Vachtsevanos, R. Esteller, J. Echauz, S. Cranstoun, G. Worrell, L. Parish and B. Litt, *Clin. Neurophysiol.* **116**, 506 (**2005**).

48 W. Chaovalitwongse, L. D. Iasemidis, P. M. Pardalos, P. R. Carney, D. S. Shiau and J. C. Sackellares, *Epilepsy Res.* **64**, 93 (**2005**).

49 M. Winterhalder, B. Schelter, A. Schulze-Bonhage and J. Timmer, *Epilepsy Res.* **72**, 80 (**2006**).

50 S. Zschocke, *Klinische Elektroenzephalographie* (Springer, **2002**).

51 R. J. Staba, C. L. Wilson, A. Bragin, I. Fried and J. Engel Jr., *J. Neurophysiol.* **88**, 1743 (**2002**).

52 T.-P. Jung, S. Makeing, C. Humphries, T.-W. Lee, M. J. Mckeown, V. Iragui and T. J.

Sejnowski, *Psychophysiology* **37**, 163 (**2000**).

53 A. V. Oppenheim and R. W. Schafer, *Digital Signal Processing* (Prentice Hall, Englewood Cliffs, NJ, **1975**).

54 S. G. Mallat, *A Wavelet Tour of Signal Processing* (Academic Press, San Diego, **1998**).

55 B. Schelter, M. Winterhalder, T. Maiwald, A. Brandt, A. Schad, J. Timmer and A. Schulze-Bonhage, *Epilepsia* **47**, 2058 (**2006**).

56 K. Lehnertz and C. E. Elger, *Clin. Neurophysiol.* **103**, 376 (**1997**).

57 S. Bialonski and K. Lehnertz, *Phys. Rev. E* **74**, 051909 (**2006**).

58 G. J. Ortega, L. Menendez de la Prida, R. G. Sola and J. Pastor, *Epilepsia* **49**, 269 (**2007**).

59 C. Plummer, A. S. Harvey and M. Cook, *Epilepsia* (**2007**).

60 A. Pitkänen, P. A. Schwartzkroin and S. L. Moshe (editors) *Models of Seizures and Epilepsy* (Elsevier, **2006**).

61 V. Salanova and R. Worth, *Current Neurology and Neuroscience Reports* 7 (**2007**).

62 K. N. Fountas and J. R. Smith, *Acta Neurochir. Suppl.* **97**, 357 (**2007**).

63 A. Bragin, C. L. Wilson, T. Fields, I. Fried and J. Engel Jr., *Epilepsia* **46**, 59 (**2005**).

64 A. Bragin, P. Claeys, K. Vonck, D. Van Roost, C. Wilson, P. Boon and J. Engel Jr., *Epilepsia* 48 (**2007**).

3
Impact of Computational Models for an Improved Understanding of Ictogenesis: From Single Neurons to Networks of Neurons

Marie-Therese Horstmann, Andy Müller, Alexander Rothkegel, Justus Schwabedal, Christian E. Elger, Klaus Lehnertz[1])

3.1
Introduction

The nervous system is a complex network composed of a huge number of neurons [1]. The human brain contains approximately 100 billion neurons and 10 million kilometers of wiring. Neurons couple to networks. They are organized in different morphological structures and perform different functions. Like other cells neurons consist of a cell membrane which encloses the cytoplasm and the cell nucleus. Neurons can transmit electrical signals over long distances. The size and shape of neurons varies over a broad range depending on their location and special role in the nervous system. Regardless of this variability the basic functionality is always the same: a neuron receives input signals, processes this input, and transfers an output signal to other neurons. Accordingly, the basic structure is the same for every neuron. The cell body (soma) has appendages that are responsible for the input and output of signals. A neuron usually has many input appendages (the dendrites), and one output appendage (the axon). A neuron is typically connected with approximately 10 000 other neurons via synapses which are located at the end of the neuron's axon. Generally, a negative electric potential difference exists between the extracellular and the intracellular space (extracellular potential set to zero) [1, 2], which is caused by differences in ion concentrations between the inner and the outer side of the cell membrane. Inputs received via synaptic connections cause transmembrane currents that change the membrane potential, and eventually can cause the generation of an action potential or spike '[. . .] an abrupt and transient change of membrane voltage' [3]. Action potentials can propagate along the axon to other neurons. Time sequences of these action potentials are considered the basis for encoding information and for communication between neurons [4].

Computational neuroscience is an interdisciplinary field of research connected with neuroscience, applied mathematics, physics, and computer science. This

---

1) This work was supported by the Deutsche Forschungsgemeinschaft, Grant No. LE660/4-1

discipline discusses neurophysiogically or neurobiologically relevant mathematical models and simulation methods that contribute to our understanding of neural mechanisms. It is a rapidly growing field, mainly because of the necessity of integrating structural, functional, and dynamic approaches in the study of the brain [5]. A number of specialized journals and textbooks (Amazon lists 48 books when searching for the term *computational neuroscience*) are now available, and there are even research institutes whose main efforts lies in setting up models of neuronal systems. But what can we learn from computational models?

Models can form a bridge between experiments and theoretical understanding. They can help to combine multiple observations derived from different experiments to a compound view. Working with a model is much easier than working with living cells *in vitro* or *in vivo*. Computational models thus provide a convenient way to explore the impact of some mechanisms underlying normal or disturbed neuronal functioning. This becomes especially important when the mechanism under consideration is hard to isolate experimentally. Using models, parameter ranges can be explored, which are hard or even impossible to access experimentally. By evaluating a model, one can even figure out some prerequisites for certain dynamics, such as finding boundary conditions for the synchronizability of neural networks.

The crucial question in modeling is how to find an *appropriate* model. But what does *appropriate* mean? The answer is: 'it depends on the problem you want to address'. On the one hand, a model should be detailed enough to cover all important aspects under consideration. On the other hand, it should be simple enough to improve understanding of the underlying mechanisms. Modeling can be done with arbitrary complexity and richness of detail. Modeling the whole brain on the level of single neurons and synapses requires huge computational effort (see, e.g., [6]). But even if we had a computational model that could model the signal processing of the whole brain, which would represent a computational representation of a real brain, what would we learn? Of course, one could conduct experiments more easily than in a real brain, but in principal our understanding of fundamental principles would not improve. Modeling is more than an insufficient representation of reality. The art of modeling is to find a model that is as detailed as necessary and as simple as possible.

Consider the probably most famous model in computational neuroscience: the Hodgkin–Huxley model. In 1952 the ionic mechanisms underlying excitability and propagation of action potentials were investigated mainly by Hodgkin and Huxley [7–10] using the squid giant axon. They postulated voltage-dependent ion conductivities of the membrane. Nowadays it is known that these conductivities originate from voltage-gated macromolecular pores in cell membranes, called ion channels. The electric properties of a membrane patch can be represented as an equivalent electric circuit. The membrane separates the intracellular from the extracellular space. Because of different ionic concentrations of the intra- and extra-cellular space there is a voltage gradient over the membrane and the membrane can thus be regarded as a capacitor. The membrane incorporates ion channels through which ions can diffuse. These can be represented as resistances or conductances. Ion channels can be in an open or a closed state, and the total conductivity of a membrane patch depends on the fraction of open ion channels. This fraction in turn depends on the membrane voltage and can, in general, also be dependent on ion concentrations, temperature, and other factors. There exists a variety of ion channels, and most of them are permeable for a specific ion only, e.g., sodium or potassium.

The total ionic current flowing through a patch of membrane is given by

$$i = c\frac{dE}{dt} + i_{Na} + i_K + i_{leak}, \tag{3.1}$$

where i_{Na}, i_K and i_{leak} represent the sodium current, the potassium current, and the leak current respectively. E denotes the membrane potential, t the time, and c the membrane capacitance. The single currents are given by

$$i_{Na} = m^3 h \bar{g}_{Na}(E - E_{Na}), \tag{3.2}$$

$$i_K = n^4 \bar{g}_K(E - E_K), \tag{3.3}$$

$$i_{leak} = g_{leak}(E - E_{leak}), \tag{3.4}$$

where m and h represent the fraction of gating particles of the sodium and n of the potassium channel being in a state that allows the channel to be open. E_{Na}, E_K and E_{leak} are the equilibrium potentials specific for different ions. The dynamics of the gating particles is modeled as

$$\frac{dx}{dt} = \frac{x_\infty - x}{\tau_x}, \tag{3.5}$$

where x denotes the fraction of gating particles being in a state that allows the channel to be open, x_∞ denotes the equilibrium state and τ_x the time constant of the dynamics. x_∞ and τ_x are functions that depend on the membrane voltage.

The Hodgkin–Huxley model had, and still has, an important influence on neuroscience in general and on the modeling of neurons in particular. It clearly indicates that it is not necessary to model every single ion crossing the membrane and the opening and closing of ion channels, but that it is sufficient to model the statistical properties. This again shows that the choice of an appropriate scale is important.

3.2
Single Neuron Models

According to [11] neuron models can be roughly divided into three groups. Since the assignment of some model to a specific group may depend on the state-of-the-art of research, the division is somewhat arbitrary. Nevertheless, it provides a classification, which can be helpful for an overview.

- **Canonical Models.** These are abstract, mathematical models that only capture some prototype features of neural dynamics in critical parameter regimes [11]. They are mainly useful to investigate particular properties of the dynamical system and transitions between different dynamical regimes, e.g., bifurcations. They have the major advantage that they can describe systems, which are not known in detail, but from which some general features are available. The disadvantage is that they are only applicable near the critical regimes. A typical example is the Landau–Stuart oscillator [12], which is the canonical model for systems near a Hopf-bifurcation.

- **Empirical Models.** These are models constructed from a minimal number of phenomenological observations. For example, the McCulloch–Pitts model [13] is the easiest model used in computational neuroscience. With this model one assumes that neurons are bistable systems with a certain threshold. Other popular empirical models are the *integrate and fire* model and the *resonate and fire* model [3], which are based on the fact that spiking neurons can act as integrators or resonators, respectively. Last but not least, we mention the *spike response models* [14, 15]. These models are typically used to model the activity of large populations of neurons (see also Section 3.3) for which the detailed dynamics of single neurons is of minor interest as, e.g., in the Wilson–Cowan model [16].

- **Comprehensive Models.** These models take into account (nearly) all known physiological facts. The most famous model of this class is the Hodgkin–Huxley model, already mentioned above. In general, all models that are based on the conductance of ion channels, belong to this class of models.

In the following we focus on conductance based models, since they are usually the stating point for the investigation of neuron dynamics. In succession to such investigations, the canonical models emerge out of a bifurcation analysis of these models. Section 3.3 incorporates empirical models.

3.2.1
Conductance-based Models

There are a number of excellent text books dealing with this type of model. We here mention the book of Hille [2], which provides a comprehensive account on the biophysical mechanism of ion channels in excitable membranes. Cronin's book [17] presents a mathematical treatment of the Hodgkin–Huxley model, while the textbook of Koch [1] addresses, besides the Hodgkin–Huxley formalism, various aspects of neuronal modeling and provides an excellent and comprehensive list of references. During the last decades several modeling tools were invented, which allow one to conveniently build and compile Hodgkin–Huxley type models without detailed programming knowledge. Two well known examples are NEURON [18] and GENESIS [19].

The Hodgkin–Huxley model can serve as a general formalism for conductance-based models. For ionic mechanisms, which have to deal with low ionic concentrations like, e.g., calcium, the Hodgkin–Huxley formalism can be extended to the Goldman–Hodgkin–Katz equation, which – instead of Ohm's law – describes the current through the membrane. The morphology can be represented by connecting cylindrical membrane patches via an axial resistivity. This can be made arbitrarily accurate but, in most cases, a simplification of the neuronal morphology to the soma, the basal and apical dendrites, and possibly to an axon is sufficient. Representing the real morphology with connected equivalent cylinders is called 'compartmental modeling'. A number of such models are already available and can be found in the model database of the NEURON Project, currently located at *http://senselab.med.yale.edu/modeldb/*.

A complex hippocampal CA1 neuron model was presented by Warman and colleagues [20]. The authors combined results from multiple observations to a synthesized view of neuron dynamics. The resulting model served as a basis for other studies. Among these studies we here mention the work of Stacey and Durand [21–23] who addressed the question of whether noise can play a constructive role in signal processing of hippocampal neurons. Neurons receive tens of thousands of input signals. Most of them are subthreshold signals, i.e., signals that are too weak to elicit an action potential. A possible mechanism to cope with this bunch of subthreshold signals is *stochastic resonance* [24, 25]. The term stochastic resonance describes a counterintuitive phenomenon that can be observed in a variety of artificial and natural, but mostly nonlinear, systems; namely an *improved* detection or transmission of weak (subthreshold) signals in the presence of noise. In this context the hippocampal region is of special interest, not only because it is responsible for memory and learning processes [26] and therefore receives signals from different regions of the brain, but also because the hippocampus is crucially involved in diseases of the central nervous system, like Alzheimer's disease and epilepsy. Using both a computational model and experimental *in vitro* studies in rats, Stacey and Durand showed that CA1 neurons are indeed able to exhibit stochastic resonance. The model enabled them to exploit parameter ranges that are difficult or even impossible to access experimentally, e.g., rather high levels of noise. Moreover, they investigated stochastic resonance in a small network of neurons under controlled conditions. Such a study is hard to perform in *in vitro* slices or *in vivo* studies. The authors showed that model simulations and experimental measurements can complement each other. In addition, these studies clearly indicate that modeling can help to improve understanding of basic principles, which may not be fully uncovered in experimental studies due to a bunch of uncontrollable factors.

3.2.2
Single Neuron Models and Epilepsy

Studies in animal models of epilepsy and in resected tissue from epilepsy patients provide us with a detailed knowledge of altered physiological properties of neurons from epileptic tissue, such as ion channel dynamics, ion kinetics, gating, morphology, synapses, or gap junctions (for an overview see [27, 28] and references therein). Despite these advances it is still an unsolved issue which of these alterations actually underly ictogenesis in humans. In addition, it has not yet been clarified how the effects of these alterations can be distinguished from those that occur due to alterations in network properties. Bursting behavior is a striking difference between neurons from 'epileptic' and from healthy tissue [29, 30]. Bursting can be defined as the firing of two or more action potentials followed by a period of quiescence [3, 31]. The actual burst pattern, such as attenuation of action potential amplitude in the course of a burst, the number of action potentials riding on top of the paroxysmal depolarizing shift (PDS) – a long-lasting membrane depolarization–depends on many factors (brain region, the individual cell, stimulus properties, etc.). There is a variety of bursting neurons, both in healthy and pathological tissue. Some neurons exhibit bursting behavior in response to a strong stimulation, others generate bursts when being stimulated with a weak input. In some neurons bursts appear to have a special role in synaptic plasticity and information processing [32, 33]. Bursting of single neurons has to be differentiated from bursting as a network phenomenon.

Traub proposed whole-cell (multi-compartment) models for neocortical and hippocampal pyramidal cells [34, 35], in which ionic channels in each compartment are represented by differential equations describing their properties. Using these models, the author identified processes at the membrane level that may be responsible for the generation of intrinsic epileptiform bursts in single cells.

Heilman and Quattrochi recently studied calcium-dependent bursting, in a computational model with increasing physiological and morphological complexity, of a hippocampal pyramidal neuron [36]. They aimed at determining the minimal complexity needed to enable the model to show epileptiform behavior and explored the influence of ionic mechanisms and of the morphological structure. From their findings they concluded that epileptiform behavior resides mainly in the membrane channels and not in a specific morphology or network interactions. However, a more complex than point-like morphology is required to generate epileptiform behavior.

Golomb and colleagues investigated bursting in a CA1 hippocampal neuron model with physiologically reasonable properties [37]. They aimed at identifying mechanisms underlying the variety of different firing and bursting patterns that can be observed experimentally. Although three ionic channels – sodium current, delayed rectifier potassium current, and M-type potassium current – are sufficient for a minimal bursting model [3], the authors also investigated the impact of the persistent sodium current. By adding calcium dynamics, calcium and calcium-gated channels, they studied the influence of the extracellular calcium concentration on the dynamics of their neuron model. Combining their results obtained from experimental and from modeling studies, they concluded that bursting in CA1 pyramidal cells can be explained by a single compartment 'square bursting' mechanism with one slow variable, namely the activation of M-type potassium current.

The role of potassium accumulation in generating epileptiform activity has been discussed in the literature for many years (see, e.g., [38] and references therein). In a recent study, Park and Durand hypothesized that spontaneous neuronal activity can be generated by a coupling of neurons via lateral diffusion of potassium ions [38]. Using two single-compartment Hodgkin–Huxley type models the authors showed that this coupling mechanism can lead to self-sustained activity if one neuron is excited by a long-lasting suprathreshold stimulus. After termination of the stimulus both neurons fired regularly and phase-locked. Similar results could also be obtained in a network of four cells. The authors concluded that potassium lateral diffusion can play an important role in the synchronization and generation of nonsynaptic epileptiform activity.

In addition to coupling mechanisms between neurons, there is now strong evidence for the coupling between neurons and surrounding glia cells (astrocytes) to play an important role in physiological and pathophysiological functions (for an overview see, e.g., [39]). Using a Hodgkin–Huxley model, Nadkarni and Jung investigated neuron–glia coupling as a possible mechanism for the generation of epileptiform activity [40]. Upon stimulation the model neuron releases quantal amounts of neurotransmitters, which bind to receptors of the astrocytes and trigger the release of inositol trisphosphate (IP3). This in turn triggers the release of calcium from internal buffers of the astrocytes. The calcium dynamics was modeled following [41]. In addition, the authors connected the calcium concentration in the astrocytic environment to an inward current into the neuron model. This simplification can be regarded as a good example of abstraction, which is necessary in modeling. The authors retained the basic effect, namely an inward current and thus the depolarization of the membrane, but left out details, namely calcium and calcium-gated channels. Depending on the coupling strength between neuron and astrocyte (defined as the IP3 production rate in response to an action potential generated by the neuron) the authors observed spontaneous oscillations in their neuron model. Enhanced neuron–glia coupling can thus be regarded as a possible mechanism to generate epileptiform activity, which is in line with experimental results [42, 43]. Using a far more complex and detailed neuron model (in terms of morphological geometry and a more realistic simulation of ion channel dynamics) Kager and colleagues [44] recently confirmed the findings presented in [40] and explored the electrophysiological properties which lead to self-regenerating afterdischarges in much more detail.

These few examples show that it is important to understand the intrinsic properties of neurons. Working with detailed models of single neurons can help us to explore pathologic alterations in cellular and/or synaptic properties of neurons or a small number of neurons. Such models, however, may not suffice to account for large-scale network phenomena such as those implied in epileptic discharges.

3.3
Neural Networks

Brain dynamics, as observed on the EEG, is much more complicated than the dynamics of single neurons. Major aspects of this complexity can be attributed to the arrangement of neurons and of synapses between them, i.e., the underlying network topology. As regards epilepsy, there is now growing evidence for seizures to be network phenomena [27, 45–47], and as such they might emerge – at least in part – from the underlying network topology. Epileptic seizures are usually characterized by hypersynchrony (see, however, [48–50]). From the reductionist point of view, as early as 1975 Peskin gave a mathematical proof that two integrate-and-fire neurons synchronize completely (i.e., the phase difference vanishes for large times) if they are coupled excitatory [51]. Netoff and colleagues observed this synchronization phenomenon for two neurons from a slice of rat hippocampus [52]. However, the situation gets much more complicated for more than two neurons that are arranged in some network with a complicated topology. Moreover, in order to derive an adequate description of large-scale epileptic phenomena it is generally not clear how the notion of synchronization can be generalized to brain networks. So how can brain networks be characterized? Which network topology should be chosen when modeling epileptic phenomena? Does the network topology of an epileptic brain differ from a healthy one? How can we differentiate physiological synchronization, which is believed to be crucial for normal brain functionality, from pathophysiological synchronization?

Modeling neural networks can provide some answers to these questions if it is combined with experimental investigations on the network topology of the human brain. Understanding seizures means to understand the circumstances which lead to synchronization of neurons in neural networks. The investigation of neural-network models can provide some clues as to which properties of network topology favor synchronization. Combined with findings from an evaluation of brain networks this can then help us to accept or reject the hypothesis that these properties do indeed play an important role in the human brain.

When exploring network dynamics, it may not be necessary to use the most physiological accurate neuron model available but to use a model which comprises only the main dynamical features. The task at hand is surely not to model the whole brain but to reproduce the observed dynamics. Both the neuron model and the network topology have to be chosen to be as simple as possible for these features to occur in order to improve understanding (for a methodological overview of simulation of neural networks we refer the reader to [53] and [54]).

3.3.1
Network Characteristics

In the following we regard a neural network as a graph where the vertices are instances of some neuron model and where connections indicate coupling between any two neurons. When discussing network topologies it is expedient to look at standard network characteristics from graph theory (cf. [55, 56]):

- The *mean degree* is the average number of in-going connections (the same number is obtained if one takes the average over all out-going connections).

- The *in-/out-degree-distribution* is the distribution of the number of in-/outgoing connections.

- The *clustering coefficient* describes the probability that two vertices which are connected two a third vertex are themselves connected.

- The *mean path length* is the average over all distances between every two nodes in the graph, where distance is the length of the smallest chain that connects two vertices.

- The *highest betweenness centrality* is the maximum of the number of shortest paths that go through a connection.

These measures allow one to characterize different structural aspects of networks, but it is still matter of ongoing discussion whether they can be related to functional aspects of networks.

Small-world and Scale-free Networks

Although regular lattices and random graphs have been studied for many years, they differ from many natural networks when describing their properties using the above-mentioned network characteristics. In 1998, Watts and Strogatz proposed a new kind of network that seems better suited for a characterization of natural networks, the so-called small-world network [57]. In such a network there are many *local* connections and only a few *long-range* connections. Watts and Strogatz showed that the brain of the nematode worm *C. elegans*, which is the first animal brain completely mapped, does fit into their concept of small-worldness. Hilgetag and colleagues observed this type of network in the brains of the macaque monkey and the cat [58, 59]. Buzsáki and colleagues proposed that the brain's small-world topology constitutes a trade-off between connection of distant parts and wiring length, dictated by evolutionary constraints, such as conservation of space, material, and energy [60]. Small-world graphs can thus be considered an efficient organization principle for larger neuronal networks like the cerebral cortex.

A small-world configuration is characterized by a high clustering coefficient (as in a regular lattice) and by a low mean path length (as in a random graph). To build such a network, one can start with a regular lattice and then rewire each connection with a certain *rewiring probability*. The constructed network topologies continuously interpolate between lattices and random networks. Moreover, they all have the same mean degree.

Another property of real world networks that has received much attention during recent years is the degree distribution. For regular lattices the degree distribution is sharply peaked, and for random graphs it follows a Poisson distribution. Many natural networks, however, show a distribution that follows a potential law at the right tail. This indicates that vertices with many connections (so called *hubs*) are encountered more often than expected. Networks with this property are called *scale-free networks*. Their occurrence can be demonstrated by simulations of evolving networks where new connections are added in such a way that vertices which already have many connections are preferentially linked [61]. Hubs are thought to play pivotal roles in the coordination of information flow and can thus be regarded as an attractive concept in the understanding of the functionality of brain structures with a high number of in-/outgoing connections, such as hippocampus or association cortices (for an overview see, e.g., [62, 63]).

The Influence of Network Topology on Synchronizability

In 1998, Pecora and Carroll [64] derived a link between the ability of a network to synchronize and the underlying network topology. Recently, Nishikawa and colleagues generalized the original findings and concluded that a network is best suited for synchronization if every node has the same sum of in-going couplings [65]. This is in agreement with earlier simulation studies [66, 67] where the fact that homogeneous degree distributions favor synchronization was received with surprise. A number of research groups have addressed the question whether other network characteristics also reflect synchronizability of a network, such as mean path length [68, 69], highest betweenness centrality [70], or correlation between in- and out-degrees [71].

For networks of small-world type there is evidence from simulation studies that synchronization is enhanced for higher rewiring probabilities [72, 73], and in [74, 75] the authors hypothesized that synchronization in small-world networks might also have implications for epilepsy. It is currently matter of debate whether the epileptic brain can indeed be regarded as more *random* in a functional sense (i.e., characterized by a higher rewiring probability) and because of this, is more susceptible to seizures than a healthy brain [76].

Modeling Epileptiform Phenomena with Neural Networks

Seizures are thought to arise from an imbalance of excitation and inhibition. The first attempts to model the influence of synaptic properties were presented by Traub and collaborators who simulated the activity of small parts of the CA3 region of hippocampus using networks of interconnected neurons [77–79]. Seizure generation and seizure propagation in hippocampal and other cortical networks was also addressed in several computational network models composed of more or less detailed neuron models [80–85]. More recently, the influence of the network topology on seizure generation has been investigated. Studying small-world networks of excitatory neurons and using several types of model neurons, Netoff and colleagues provided a possible explanation for the fact that, in a hippocampal slice model of epilepsy, the CA3 region exhibits bursts, while the CA1 region exhibits seizure-like activity [74]. For an increasing number of long-range connections the dynamics of their simulated neural network changed from normal to seizure and to a bursting state. The authors speculated that the less connected CA1 region could be in a state similar to the observed seizure state while the more connected CA3 region is in the bursting state. Dyhrfjeld-Johnsen and colleagues investigated seizure-induced structural changes in the dentate gyrus as can be observed in temporal lobe epilepsy [86]. In their simulations with a realistic model (rat dentate gyrus with one billion neurons, and no more than three synapses between any two neurons – suggestive of a small-world architecture) the authors observed that the local axonal sprouting outweighs the loss of long distance hilar cells and, surprisingly, hyperexcitability was not decreased due to structural changes. Feldt and colleagues investigated synchronization phenomena between two small-world networks of integrate-and-fire neurons [87]. Their simulations allowed them to reproduce the observed preictal drop in phase synchrony on the EEG from patients suffering from focal epilepsy [88, 89].

The aforementioned network models are able to generate activity patterns that closely mimic epileptic activity recorded *in vitro*, as well as complex interactions patterns of EEG activity seen in patients. These models have been particularly useful in uncovering the roles of various cell types and connections in ictogenesis. Together with recent reports that aimed to identify *functional* network properties in ictal EEG recordings from epilepsy patients (see [76] for an overview), these studies might help us to gain deeper insights into the complex interplay between structure and function in epileptogenesis.

3.4
Neural Mass Models of the EEG

The simulation of a neural network consisting of a large number of model neurons, as described in the previous section, requires huge computational effort (see e.g. [6]). Alternatively, it is possible to model *populations of neurons* with similar properties as a single unit. These models are called *neural mass models*. Since the EEG reflects macroscopic electrical activity of neural populations, it may not be necessary to model each single neuron in detail but instead to model the statistical distributions of membrane potentials of many neurons and their dynamical changes as a set of differential equations. This simplification is similar to the one achieved by the Hodgkin–Huxley approach, where the statistical properties of a bulk of ion channels are modeled instead of the opening and closing of each single ion channel. It is reasonable to model populations of neurons with similar properties as a single unit, e.g., inhibitory and excitatory neurons in a cortical column. In neural mass models, however, the spatial resolution is dismissed and therefore some phenomena, e.g., waves, cannot be sufficiently described using this formalism (see e.g. [90]).

It is generally accepted that the nervous system is organized in columns. This had already been established in the 1950s through anatomical and electrophysiological investigations of neocortex in primates [91]. A column is a densely connected set of neurons that exhibit a functional coherence throughout the neuron populations. Depending on the region of the cortex it has a cylindrical shape (diameter $=$ $300-600\,\mu$m) that spans the whole cortical thickness. The functional coherence of neurons in a column permits the description of its dynamics as a single unit. Experimental investigations of the dynamical behavior of such functional units and their interpretation toward modeling were summarized by Freeman [92, 93].

Due to the fundamental functional difference in excitatory and inhibitory neurons, it is useful to model excitatory and inhibitory subpopulations of neurons. The dynamical behavior of a larger neuron population arises mainly from interactions between subpopulations. The macroscopic electrical activity of neural populations, which is measured on the EEG, is said to arise from an average of post-synaptic membrane potential changes [94]. The reduction of the degrees of freedom that arises from comprising neurons to populations can partly be compensated for by an introduction of undirected influences on the population modeled by Gaussian white noise.

The passive behavior of the averaged membrane potential was accessed experimentally through the so-called open-loop response, where the electrical response of a neural mass in deep anesthesia to a delta-like stimulus was analyzed [92]. It is now widely accepted that the response of the averaged membrane potential can be described by

$$h_{inh/exc}(t) \propto t e^{-a_{inh/exc} t}, \tag{3.6}$$

where the time constants $a_{inh/exc}$ can be understood as lumped properties of the dendritic network and the membranes. In contrast, some authors [83, 95] have modeled the response of the membrane potential using an approach similar to [96]. The interaction of a neural population with itself was modeled using statistical averages. Excitatory and inhibitory subpopulations, e.g., excitatory pyramidal cells and inhibitory stellate cells [97], can be modeled separately, and then connected with a coupling strength representing average axonal densities. The conversion from the average membrane potential to a mean firing rate can be achieved by assuming simple threshold models. When averaging over distributions of thresholds in a population one usually obtains a sigmoidal function. For this purpose Gaussian distributions are usually assumed (see, however, [16]), which results in a mean firing rate of

$$S(V) \propto \frac{1}{1 + e^{-r(V - v_0)}}, \tag{3.7}$$

where v_0 is a mean threshold, r is a slope parameter, and V is the averaged cell potential. Considering several population models of this kind it was possible to introduce a physiological motivated coupling mechanism to study, for example, the interaction of columns from cortex and regions of the thalamus (thalamus to cortex [97–99], cortex to cortex [100, 101], hippocampus to hippocampus [102]).

Disregarding the inexactness of estimated physiological parameters and the simplified assumptions on the underlying neuron models, it is yet arguable whether this approach to neural mass systems is relevant since, for example, topological properties of the coupling (cf. Section 3.3) cannot be easily taken into account. However, there have been several attempts to associate the dynamics of the model of single or multiple columns to certain pathophysiological and physiological phenomena observed on the EEG. Normal background EEG was represented by a steady state of the model, where its dynamics is dominated by the influence of dynamical noise. In certain parameter ranges some models showed epileptiform dynamics, which were deterministically dominated. The analysis of these signals, which are comparable to those observed during epileptic seizure, is especially prominent in the literature [99, 101–104].

Because of the relatively simple structure of the neural mass models it was possible to treat the stable states analytically via linear response theory [93, 97, 98, 105].

In this way power spectral densities of the linearized model could be obtained and compared to those from EEG time-series. By such a comparison Robinson and colleagues constrained physiologically motivated model parameters – e.g., soma potential rise and decay rates – and compared these estimates with comparable quantities derived from independent studies [98]. This approach may lead to a patient-based classification of different types of epilepsies through measurements of the individual background EEG, but a statistical evaluation remains to be carried out.

For certain parameter ranges some models exhibit epileptiform dynamics. This could be explored in detail using methods of bifurcation analysis. Breakspear and colleagues related different types of seizure to the dynamics of their model of the cortico-thalamic system [99]. Other theoretical works [83, 102] treated the matter of cortical synchronization phenomena, and it was observed that multi-column models can exhibit states of intermittency and generalized synchronization. Others investigated techniques of brain electrical stimulation in the framework of bifurcation control with their spatial-temporal model of cortical electric activity [104, 106, 107]. The method of bifurcation control has become relevant to the treatment of patients suffering from Parkinson's disease and is currently under investigation in the framework of epilepsy [108].

Based on models of the transition between normal and epileptic activity proposed by Lopes da Silva and colleagues [109], Suffczyinski and collaborators analyzed a model that exhibits the coexistence of a steady interictal-like and an oscillatory ictal-like state [95, 110]. In this model the stochastic influence induces transitions between these dynamical states. The authors showed that this type of dynamics leads to a Poisson distribution of interictal and ictal states (see also [111]), which deviated from their findings observed in certain biological systems. They argued that a randomly fluctuating parameter responsible for the state changes would lead to similar results. These findings indicate that either the approach of neural mass systems cannot explain the phenomenon of the interictal-to-ictal transition or that the time-dependence of model parameters should be described by a non-random process which follows some underlying dynamics that can be estimated from EEG data. The authors proposed two ways to address this issue.

Extending a neurophysiologically relevant model initially proposed in 1974 by Lopes da Silva and colleagues [97], Wendling and collaborators exemplified that the time-dependence of a bivariate nonlinear measure applied to the EEG recorded during a seizure can probably be explained by changes in the coupling strength between two model columns [101]. Extending their model by including a physiologically relevant fast inhibitory feedback loop, the same authors [112] showed that the transition between interictal to fast ictal activity is explained, in the model, by the impairment of dendritic inhibition (see also [113] and references therein). In a recent study [114], Wendling and colleagues proposed a more direct approach to identify the three main parameters of their model of hippocampus EEG activity (related to excitation, slow dendritic inhibition and fast somatic inhibition): the best fitting model parameters at certain instances during seizure onset were identified by minimizing a spectral distance between the model time-series and segments of the intracranially recorded EEG (see also [115]). The authors demonstrated that their model generates very realistic signals for automatically identified model parameters. Moreover, their findings indicate that the interictal-to-ictal transition cannot be simply explained by an increase in excitation and a decrease in inhibition but rather by a variety of complicated time-varying ensemble interactions between pyramidal cells and interneurons with slow and fast $GABA_A$ kinetics.

The studies described above are far from covering the topic *EEG modeling* completely, but they clearly illustrate that macroscopic statistical and dynamical phenomena seen on the EEG from epilepsy patients can be reproduced with high accuracy, despite the simplified assumptions on the underlying neuron models and network topology.

3.5
Conclusion

In this chapter we have tried to give an overview of the field of computational neuroscience with special emphasis on ictogenesis. In a work of this scope it is inevitable that some contributions may be over or under emphasized, depending upon the points to be made in the text. We discussed modeling approaches on different scales, from single neurons to neural networks and to neural mass models of the EEG using different levels of detail. Computational neuroscience is a relatively young but fast growing branch of science. The rapid development of refined or new models on different scales can, of course, be related to the growing availability of fast digital computers. On the other hand, it also indicates an increasing awareness of, and confidence in, theoretical approaches based on computational models in order to advance our understanding of the complex mechanisms underlying epileptogenesis and ictogenesis.

The models currently available provide valuable insights into how the modification of system parameters, or intrinsic fluctuations, may lead to the transition between apparently normal activity and epileptic seizures. Findings obtained from modeling studies thus corroborate experimental results and can help to make predictions concerning clinically relevant questions such as seizure predictability and control. Further improvements demand the integration of experimental and theoretical approaches, which requires interdisciplinary research and collaborations (cf. [116]).

Despite significant advances, the questions of detail and scale still remain. Too detailed models may be too complex to be treated analytically and may thus be difficult to interpret. Too general models may not capture the essence of phenomena observed experimentally. Modeling on one scale may help to understand experimental observation on this neurophysiological level but it may be inadequate for another scale. Bridging microscopic and macroscopic levels of description remains a challenge for computational neuroscience.

References

1 C. Koch. *Biophysics of Computation: Information Processing in Single Neurons*. Computational Neuroscience. Oxford University Press, New York, Oxford (**1999**).

2 B. Hille. *Ion Channels of Excitable Membranes*. Sinauer Associates, Inc., Sunderland, MA, U.S.A., third edition (**2001**).

3 E. M. Izhikevich. *Dynamical Systems in Neuroscience: The Geometry of Excitability and Bursting*. The MIT Press, Cambridge, MA (**2007**).

4 F. Rieke, D. Warland, R. de Ruyter van Steveninck and W. Bialek. *Spikes: Exploring the Neural Code*. MIT Press, Cambridge, MA (**1997**).

5 P. Suffczynski, F. Wendling, J. J. Bellanger and F. H. Lopes Da Silva. Some insights into computational models of (patho)physiological brain activity. *Proc. IEEE*, **94**, 784 (**2006**).

6 H. Markram. The blue brain project. *Nat. Rev. Neurosci.*, **7**, 153–60 (**2006**).

7 A. L. Hodgkin and A. F. Huxley. Currents carried by sodium and potassium ions through the membrane of the giant axon of *Loligo*. *J. Physiol.*, **116**, 449–72 (**1952**).

8 A. L. Hodgkin and A. F. Huxley. The components of membrane conductance in the giant axon of *Loligo*. *J. Physiol.*, **116**, 473–96 (**1952**).

9 A. L. Hodgkin and A. F. Huxley. The dual effect of membrane potential on sodium conductance in the axon of *Loligo*. *J. Physiol.*, **116**, 497–506 (**1952**).

10 A. L. Hodgkin and A. F. Huxley. A quantitative description of membrane current and its application to conduction and excitation in nerve. *J. Physiol.*, **117**, 500–44 (**1952**).

11 E. M. Izhikevich. *Bifurcations in Brain Dynamics*. PhD thesis, Michigan State University (**1996**).

12 Y. Kuramoto. *Chemical Oscillations, Waves and Turbulence*. Springer Verlag, Berlin (**1984**).

13 W. S. McCulloch and W. Pitts. A logical calculus of the idea immanent in nervous activity. *Bull. Math. Biophys.*, **5**, 115–33 (**1943**).

14 W. Gerstner. Time structure of the activity in neural network models. *Phys. Rev. E*, **51**, 738–58 (**1995**).

15 W. Gerstner and W. M. Kistler. *Spiking Neuron Models. Single Neurons, Populations, Plasticity*. Cambridge University Press, Cambridge, UK (**2002**).

16 H. R. Wilson and J. D. Cowan. Excitatory and inhibitory interactions in localized populations of model neurons. *Biophys. J.*, **12**, 1–24 (**1972**).

17 J. Cronin. *Mathematical Aspects of Hodgkin–Huxley Neural Theory*. Cambridge University Press, Cambridge, U.K. (**1987**).

18 N. T. Carnevale and M. L. Hines. *The NEURON Book*. Cambridge University Press, Cambridge, UK (**2006**).

19 J. M. Bower and D. Beeman. *The Book of GENESIS: Exploring Realistic Neural Models with the GEneral NEural SImulation System*. Springer, Berlin (**1998**).

20 E. N. Warman, D. Durand and G. L. F. Yuen. Reconstruction of hippocampal CA1 pyramidal cell electrophysiology by computer simulation. *J. Neurophysiol.*, **71**, 2033 (**1994**).

21 W. C. Stacey and D. M. Durand. Stochastic resonance improves signal detection in hippocampal CA1 neurons. *J. Neurophysiol.*, **83**, 1394–402 (**2000**).

22 W. C. Stacey and D. M. Durand. Synaptic noise improves detection of subthreshold signals in hippocampal CA1 neurons. *J. Neurophysiol.*, **86**, 1104–12 (**2001**).

23 W. C. Stacey and D. M. Durand. Noise and coupling affect signal detection and bursting in simulated physiological neural network. *J. Neurophysiol.*, **88**, 2598–611 (**2002**).

24 L. Gammaitoni, P. Hänggi, P. Jung and F. Marchesoni. Stochastic resonance. *Rev. Mod. Phys.*, **70**, 223–87 (**1998**).

25 F. Moss, L. M. Ward and W. G. Sannita. Stochastic resonance and sensory information processing: a tutorial and review of application. *Clin. Neurophysiol.*, **115**, 267–81 (**2004**).

26 H. Eichenbaum. A cortical – hippocampal system for declarative memory. *Nat. Rev. Neurosci.*, **1**, 41–50 (**2000**).

27 D. A. McCormick and D. Contreras. On the cellular and network bases of epileptic seizures. *Annu. Rev. Physiol.*, **63**, 815–46 (**2001**).

28 J. G. R. Jefferys. Models and mechanisms of experimental epilepsies. *Epilepsia*, **44**(Suppl. 12), 44–50 (**2003**).

29 E. R. G. Sanabria, H. Su and Y. Yaari. Initiation of network bursts by Ca^{2+}-depedendent intrinsic bursting in the rat pilocarpine model of temporal lobe epilepsy. *J. Physiol.*, **532**, 205–16 (**2001**).

30 Y. Yaari and H. Beck. 'Epileptic neurons' in temporal lobe epilepsy. *Brain Pathol.*, **12**, 234–9 (**2002**).

31 E. M. Izhikevich. Neural excitability, spiking, and bursting. *Int. J. Bifurcat. Chaos*, **10**, 1171–266 (**2000**).

32 J. Tropp Sneider, J. J. Chrobak, M.C. Quirk, J. A. Oler and E. J. Markus. Differential behavioral state-dependence in the burst properties of CA3 and CA1 neurons. *Neurosci.*, **141**, 1665–77 (**2006**).

33 J. E. Lisman. Bursts as a unit of neural information: making unreliable synapses reliable. *Trends Neurosci.*, **20**, 38–43 (**1997**).

34 R. D. Traub. Neocortical pyramidal cells: a model with dendritic calcium conductance reproduces repetitive firing and epileptic behavior. *Brain Res.*, **173**, 243–57 (**1979**).

35 R. D. Traub Simulation of intrinsic bursting in CA3 hippocampal neurons. *Neuroscience*, **7**, 1233–42 (**1982**).

36 A. D. Heilman and J. Quattrochi. Computational models of epileptiform activity in single neurons. *Biosystems*, **78**, 1–21 (**2004**).

37 D. Golomb, C. Yue and Y. Yaari. Contribution of persistent Na^+ current and M-type K^+ current

to somatic bursting in CA1 pyramidal cells: combined experimental and modeling study. *J. Neurophysiol.*, **96**, 1912–26 (**2006**).

38 E.-H. Park and D. M. Durand. Role of potassium lateral diffusion in nonsynaptic epilepsy: A computational study. *J. Theor. Biol.*, **238**, 666–82 (**2006**).

39 D. K. Binder and C. Steinhäuser. Functional changes in astroglial cells in epilepsy. *Glia*, **54**, 358–68 (**2006**).

40 S. Nadkarni and P. Jung. Spontaneous oscillations of dressed neurons: A new mechanism for epilepsy? *Phys. Rev. Lett.*, **91**, 268101 (**2003**).

41 J. Rinzel and Y. X. Li. Equations for InsP3 receptor-mediated $[Ca^{2+}]$-oscillations derived from a detailed kinetic model: a Hodgkin–Huxley like formalism. *J. Theor. Biol.*, **166**, 461–73 (**1994**).

42 A. Aaraque, V. Parpura, R. Sanzgiri and P. Haydon. Glutamate-dependent astrocyte modulation of synaptic transmission between cultured hippocampal neurons. *Eur. Neurosci.*, **10**, 2129–42 (**1998**).

43 A. Cornell and A. Williamson. *Biology and Pathology of Astrocyte–Neuron Interactions*, pp 51–65. Plenum Press, New York (**1993**).

44 H. Kager, W. J. Wadman and G. G. Somjen. Seizure-like afterdischarges simulated in a model neuron. *J Comput. Neurosci.*, **22**, 105–28 (**2007**).

45 F. Bartolomei, F. Wendling, J. J. Bellanger, J. Règis and P. Chauvel. Neural networks involving the medial temporal structures in temporal lobe epilepsy. *Clin. Neurophysiol.*, **112**, 1746–60 (**2001**).

46 S. S. Spencer. Neural networks in human epilepsy: Evidence of and implications for treatment. *Epilepsia*, **43**, 219–27 (**2002**).

47 M. Guye, J. Règis, M. Tamura, F. Wendling, A. McGonial, P. Chauvel and F. Bartolomei. The role of corticothalamic coupling in human temporal lobe epilepsy. *Brain*, **129**, 1917–28 (**2006**).

48 K. Schindler, H. Leung, K. Lehnertz and C. E. Elger. How generalised

are secondarily 'generalised' tonic-clonic seizures? *J. Neurol. Neurosurg. Psychiatry*, **78**, 993–6 (**2007**).

49 K. Schindler, H. Leung, C. E. Elger and K. Lehnertz. Assessing seizure dynamics by analysing the correlation structure of multichannel intracranial EEG. *Brain*, **130**, 65–77 (**2007**).

50 K. Schindler, C. E. Elger and K. Lehnertz. Increasing synchronization may promote seizure termination: Evidence from status epilepticus. *Clin. Neurophysiol.*, **118**, 1955–68 (**2007**).

51 C. S. Peskin. Mathematical aspects of heart physiology. Courant Institute of Mathematical Sciences, pp 268–78 (**1975**).

52 T. I. Netoff, M. I. Banks, A. D. Dorval, C. D. Acker, J. S. Haas, N. Kopell and J. A. White. Synchronization in hybrid neuronal networks of the hippocampal formation. *J. Neurophysiol.*, **93**, 1197–208 (**2005**).

53 R. Brette, M. Rudolph and T. Carnevale *et al.* Simulation of networks of spiking neurons: A review of tools and strategies. *J. Comput. Neurosci.*, **23**, 349–98 (**2007**).

54 P. Hammarlund and Ö. Ekeberg. Large neural network simulations on multiple hardware platforms. *J. Comp. Neurosci.*, **5**, 443–59 (**1998**).

55 R. Albert and A.-L. Barabási. Statistical mechanics of complex networks. *Rev. Mod. Phys.*, **74**, 47–97 (**2002**).

56 M. E. J. Newman. The structure and function of complex networks. *SIAM Rev.*, **45**, 167–256 (**2003**).

57 D. J. Watts and S. H. Strogatz. Collective dynamics of 'small-world' networks. *Nature*, **393**, 440–2 (**1998**).

58 C.C. Hilgetag, G.A. Burns, M.A. O'Neill, J.W. Scannell and M.P. Young. Anatomical connectivity defines the organization of clusters of cortical areas in the macaque monkey and the cat. *Philos. Trans. Roy. Soc. Lond. B. Biol. Sci.*, **355**, 91–110 (**2000**).

59 K. E. Stephan, C. C. Hilgetag, G. A. Burns, M. A. O'Neill, M. P. Young and R. Kötter. Computational analysis of functional connectivity between areas of primate cerebral cortex. *Philos. Trans. R. Soc. Lond. B. Biol. Sci.*, **355**, 111–26 (**2000**).

60 G. Buzsáki. Large-scale recording of neuronal ensembles. *Nat. Neurosci.*, **7**, 446–51 (**2004**).

61 A.-L. Barabási and R. Albert. Emergence of scaling in random networks. *Science*, **286**, 509–12 (**1999**).

62 C. J. Honey, R. Kötter, M. Breakspear and O. Sporns. Network structure of cerebral cortex shapes functional connectivity on multiple time scales. *Proc. Natl. Acad. Sci. USA*, **104**, 10240–5 (**2007**).

63 O. Sporns, C. J. Honey and R. Kötter. Identification and classification of hubs in brain networks. *PLoS ONE*, **2**, e1049 (**2007**).

64 L. M. Pecora and T. L. Carroll. Master stability functions for synchronized coupled systems. *Phys. Rev. Lett.*, **80**, 2109–12 (**1998**).

65 T. Nishikawa and A. E. Motter. Synchronization is optimal in nondiagonalizable networks. *Phys. Rev. E*, **73**, 065106 (**2006**).

66 T. Nishikawa, A. E. Motter, Y.-C. Lai and F. C. Hoppensteadt. Heterogeneity in oscillator networks: Are smaller worlds easier to synchronize? *Phys. Rev. Lett.*, **91**, 014101 (**2003**).

67 A. E. Motter, C. Zhou and J. Kurths. Network synchronisation, diffusion and the paradox of heterogeneity. *Phys. Rev. E*, **71**, 016116 (**2005**).

68 I. V. Belykh, V. N. Belykh and M. Hasler. Connection graph stability method for synchronized coupled chaotic systems. *Physica D*, **195**, 159–87 (**2004**).

69 I. V. Belykh and M. Hasler. Synchronization and graph topology. *Int. J. Bifurcat. Chaos*, **15**, 3423–33 (**2005**).

70 H. Hong, B. J. Kim, M. Y. Choi and H. Park. Factors that predict better synchronizability on complex networks. *Phys. Rev. E*, **69**, 067105 (**2004**).

71 M. Chavez, D.-U. Hwang, A. Amann and S. Boccaletti. Synchronizing weighted complex networks. *Chaos*, **16**, 015106 (**2006**).

72 M. Barahona and L. M. Pecora. Synchronization in small-world systems. *Phys. Rev. Lett.*, **89**, 054101 (**2002**).

73 H. Hong and M. Y. Choi. Synchronization on small-world networks. *Phys. Rev. E*, **65**, 026139 (**2002**).

74 T. I. Netoff, R. Clewley, S. Arno, T. Keck and J. A. White. Epilepsy in small-world networks. *J. Neurosci.*, **24**, 8075–83 (**2004**).

75 B. Percha, R. Dzakpasu and M. Zochowski. Transition from local to global phase synchrony in small world neural network and its possible implications for epilepsy. *Phys. Rev. E*, **72**, 031909 (**2005**).

76 J. C. Reijneveld, S. C. Ponten, H. W. Berendse and C. J. Stam. The application of graph theoretical analysis to complex networks in the brain. *Clin. Neurophysiol.*, **118**, 2317–31 (**2007**).

77 R. D. Traub, R. Miles and R. K. Wong. Model of the origin of rhythmic population oscillations in the hippocampal slice. *Science*, **243**, 1319–25 (**1989**).

78 R. D. Traub and R. Dingledine. Model of synchronized epileptiform bursts induced by high potassium in CA3 region of rat hippocampal slice. Role of spontaneous EPSPs in initiation. *J. Neurophysiol.*, **64**, 1009–18 (**1990**).

79 R. D. Traub, R. Miles and G. Buzsáki. Computer simulation of carbachol-driven rhythmic population oscillations in the CA3 region of the in vitro rat hippocampus. *J. Physiol.*, **451**, 653–72 (**1992**).

80 P. J. Franaszczuk, P. Kudela and G. K. Bergey. External excitatory stimuli can terminate bursting in neural network models. *Epilepsy Res.*, **53**, 65–80 (**2003**).

81 P. Kudela, P. J. Franaszczuk and G. K. Bergey. Changing excitation and inhibition in simulated neural networks: effects on induced bursting behavior. *Biol. Cybern.*, **88**, 276–85 (**2003**).

82 K. H. Yang, P. J. Franaszczuk and G. K. Bergey. Inhibition modifies the effects of slow calcium-activated potassium channels on epileptiform activity in a neuronal network model. *Biol. Cybern.*, **92**, 71–81 (**2005**).

83 M. Breakspear, J. R. Terry and K. J. Friston. Modulation of excitatory synaptic coupling facilitates synchronization and complex dynamics in a biophysical model of neural dynamics. *Netw. Comput. Neural Syst.*, **14**, 703–32 (**2003**).

84 M. A. Kramer, A. J. Szeri, J. W. Sleigh and H. E. Kirsch. Mechanism of seizure propagation in a cortical model. *J. Comput. Neurosci.*, **22**, 63–80 (**2007**).

85 D. Takeshita, Y. D. Sato and S. Bahar. Transitions between multistable states as a model of epileptic seizure dynamics. *Phys. Rev. E*, **75**, 051925 (**2007**).

86 J. Dyhrfjeld-Johnsen, V. Santhakumar, R. J. Morgan, R. Huerta, L. Tsimring and I. Soltesz. Topological determinants of epileptogenesis in large-scale structural and functional models of the dentate gyrus derived from experimental data. *J. Neurophysiol.*, **97**, 1566–87 (**2007**).

87 S. Feldt, H. Osterhage, F. Mormann, K. Lehnertz and M. Zochowski. Internetwork and intranetwork communications during bursting dynamics: application to seizure prediction. *Phys. Rev. E*, **76**, 021920 (**2007**).

88 F. Mormann, R. Andrzejak, T. Kreuz, C. Rieke, P. David, C. E. Elger and K. Lehnertz. Automated detection of a preseizure state based on a decrease in synchronization in intracranial electroencephalogram recordings from epilepsy patients. *Phys. Rev. E*, **67**, 021912 (**2003**).

89 F. Mormann, T. Kreuz, R. G. Andrzejak, P. David, K. Lehnertz and C. E. Elger. Epileptic seizures are preceded by a decrease in synchronization. *Epilepsy Res.*, **53**, 173–85 (**2003**).

90 A. Omurtag, B.W. Knight and L. Sirovich. On the simulation of large populations of neurons. *J. Comp. Neurosci.*, **8**, 51–63 (**2000**).

91 V. B. Mountcastle. The columnar organization of the neocortex. *Brain*, **120**, 701–22, April (**1997**).

92 W. J. Freeman. Linear analysis of the dynamics of neural masses. *Annu. Rev. Biophys. Bioeng.*, **1**, 225–56 (**1972**).

93 W. J. Freeman. *Mass Action in the Nervous System*. Academic Press, New York (**1975**).

94 N. Schaul. The fundamental neural mechanisms of electroencephalography. *Electroencephalogr. Clin. Neurophysiol.*, **106**, 101–7 (**1998**).

95 P. Suffczynski, F. H. Lopes da Silva, J. Parra, D. Velis and S. Kalitzin. Epileptic transitions: Model predictions and experimental validation. *J. Clin. Neurophysiol.*, **22**, 288–99 (**2005**).

96 C. Morris and H. Lecar. Voltage oscillations in the barnacle giant muscle fiber. *Biophys. J.*, **193**, 193–213 (**1981**).

97 F. H. Lopes da Silva, A. Hoeks, H. Smits and L. H. Zetterberg. Model of brain rhythmic activity. *Kybernetik*, **15**, 27–37 (**1973**).

98 P. A. Robinson, C. J. Rennie, D. L. Rowe, S. C. O'Connor and E. Gordon. Multiscale brain modelling. *Philos. Trans. Roy. Soc. B*, **360**, 1043–50 (**2005**).

99 M. Breakspear, J. A. Roberts, J. R. Terry, S. Rodrigues, N. Mahant and P. A. Robinson. A unified explanation of primary generalized seizures through nonlinear brain modeling and bifurcation analysis. *Cereb. Cortex*, **16**, 296–313 (**2006**).

100 B. H. Jansen and V. G. Rit. Electroencephalogram and visual evoked potential generation in a mathematical model of coupled cortical columns. *Biol. Cybern.*, **73**, 357–66 (**1995**).

101 F. Wendling, J. J. Bellanger, F. Bartolomei and P. Chauvel. Relevance of nonlinear lumped-parameter models in the analysis of depth-EEG epileptic signals. *Biol. Cybern.*, **83**, 367–78 (**2000**).

102 R. Larter, B. Speelman and R. M. Worth. A coupled ordinary differential equation lattice model for the simulation of epileptic seizures. *Chaos*, **9**, 795–804 (**1999**).

103 L. H. Zetterberg, L. Kristiansson and K. Mossberg. Performance of a model for a local neuron population. *Biol. Cybern.*, **31**, 15–26 (**1978**).

104 M. A. Kramer, B. A. Lopour, H. E. Kirsch and A. J. Szeri. Bifurcation control of seizing human cortex. *Phys. Rev. E*, **73**, 041928 (**2006**).

105 J. W. Kim and P. A. Robinson. Compact dynamical model of brain activity. *Phys. Rev. E*, **75**, 031907 (**2007**).

106 A. G. Balanov, V. Beato, N. B. Janson, H. Engel and E. Schöll. Delayed feedback control of noise-induced patterns in excitable media. *Phys. Rev. E*, **74**, 016214 (**2006**).

107 B. Hauschildt, N. B. Janson, A. Balanov and E. Schöll. Noise-induced cooperative dynamics and its control in coupled neuron models. *Phys. Rev. E*, **74**, 051906 (**2006**).

108 C. Hauptmann and P. A. Tass. Therapeutic rewiring by means of desynchronizing brain stimulation. *Biosystems*, **89**, 173–81 (**2007**).

109 F. Lopes da Silva, W. Blanes, S. N. Kalitzin J. Parra, P. Suffczynski and D. N. Velis. Epilepsies as dynamical diseases of brain systems: basic models of the transition between normal and epileptic activity. *Epilepsia*, **44**(Suppl. 12), 72–83 (**2003**).

110 P. Suffczynski, S. Kalizuin and F. H. Lopes da Silva. Dynamics of nonconvulsive epileptic phenomena modeled by a bistable neuronal network. *Neurosci.*, **126**, 467–84 (**2004**).

111 S. Sunderam, I. Osorio, M. G. Frei and J. F. Watkins, *III*. Stochastic modeling and prediction of experimental seizures in Sprague–Dawley rats. *J. Clin. Neurophysiol.*, **18**, 275–82 (**2001**).

112 F. Wendling, F. Bartolomei, J. J. Bellanger and P. Chauvel. Epileptic fast activity can be explained by a model of impaired GABAergic dendritic inhibition. *Eur. J. Neurosci.*, **15**, 1499–508 (**2002**).

113 M. A. Whittington, R. D. Traub, N. Kopell, B. Ermentrout and E. H. Buhl. Inhibition-based rhythms: experimental and mathematical observations on network dynamics.

Int. J. Psychophysiol., **38**, 315–36 (**2000**).

114 F. Wendling, A. Hernandez, J. J. Bellanger, P. Chauvel and F. Bartolomei. Interictal to ictal transitions in human temporal lobe epilepsy: insights from a computational model of intracerebral EEG. *J. Clin. Neurophysiol.*, **22**, 343–56 (**2005**).

115 J. Kybic, O. Faugeras, M. Clerc and T. Papadopoulo. Neural mass model parameter identification for MEG/EEG. In A. Manduca and P. Hu Xiaoping, editors, *Proceedings of SPIE: Medical Imaging 2007. Physiology, Function, and Structure from Medical Imaging*, volume 6511, pages 1–9. SPIE (**2007**).

116 J. M. Bower and C. Koch. Experimentalists and modelers: can we all just get along? *Trends Neurosci.*, **15**, 458–61 (**1992**).

4
Effective and Anatomical Connectivity in a Rat Model of Spontaneous Limbic Seizure

Paul R. Carney, Alex Cadotte, Thomas B. DeMarse, Baba Vemuri, Thomas H. Mareci, William Ditto

4.1
Introduction

Temporal lobe epilepsy is one of the most common forms of partial-onset epilepsy. Animal models of temporal lobe epilepsy offer the opportunity for controlled *in vivo* studies of ictogenesis. Animal models in combination with advanced *in vivo* multi-scale microarray electrodes acquisition of ongoing continued neurophysiological activity, allows for characterization of limbic system interactions that precede seizure onset. Advanced high-field imaging tools and methods of data analysis allow for the simultaneous evaluation *in vivo* of white matter association tracts both within hippocampal circuitry and associate limbic system tracts. The hippocampus contains distinct layers; densely packed neuron cell body layers such as the stratum pyramidale are surrounded by layers of neuropil, such as stratum radiatum, where relatively few neurons cell bodies are interspersed with glia and a complex interdigitation of dendrites and axonal projections. The hippocampus also contains several well-described neuronal circuits like the trisynaptic intrahippocampal pathway (Figure 4.1), which is linked to seizure onset, and this provides an interesting structure for fundamental investigations into ictogenesis in temporal lobe epilepsy. The trisynaptic pathway contains several coherent neuronal pathways such as Schaffer collaterals, perforant and mossy fiber, that are significantly altered in temporal lobe epilepsy.

The experimental and computational studies briefly presented here are consistent with the view that the hippocampus circuitry is composed of functionally specialized local populations of neurons that are interacting dynamically along reentrant anatomical loops and pathways, within, and between, hipocampi. These large-scale patterns of temporal activity generated by the dynamics of neuronal interactions across the brain are often referred to as functional connectivity. By its nature, functional connectivity involves statistical relationships between potentially large numbers of segregated elements. Clearly, a system's dynamics must strongly depend on the underlying structure of the network. In the case of the brain, this structure is equivalent to its neuroanatomy. Here we present novel results which

Seizure Prediction in Epilepsy. Edited by Björn Schelter, Jens Timmer and Andreas Schulze-Bonhage
Copyright © 2008 WILEY-VCH Verlag GmbH & Co. KGaA, Weinheim
ISBN: 978-3-527-40756-9

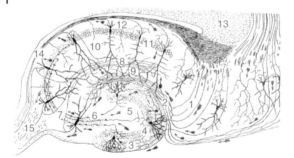

Fig. 4.1 Hippocampal anatomy and connectivity depicted by Cajal (1911) [1] based on the silver chromate method, which only stains a limited number of neurons. Here, Cajal used arrows to demonstrate the fiber connectivity of the trisynaptic pathway in the hippocampus. In this circuit, entorhinal cortex neurons innervate dentate granule neuron dendrites via the perforant pathway. Granule cell axons, called mossy fibers, then project to CA3 pyramidal neurons. A diffuse projection of CA3 axons forms the Schaffer collateral system innervating the CA1 strata radiatum and oriens. Finally, axons from CA1 and CA3 project via the alveus to the fimbria. The orientation and localization of these pathways define anatomical layers within the hippocampus [1 = ubiculum, 2 = perforant pathway axons, 3 = molecular layer, 4 = granule cell layer, 5 = hilum, 6 = mossy fiber axons, 7 = CA3 pyramidal neurons, 8 = Schaffer collateral axons, 9 = stratum lacunosum–moleculare, 10 = stratum radiatum, 11 = CA1 pyramidal neurons, 12 = stratum oriens, 13 = dorsal hippocampal commissure, 14 = alveus, 15 = fimbria].

describe the neuroanatomical patterns of the hippocampus of local circuits and pathways linking distinct subfields with the hippocampus. It is postulated that both anatomical and functional tools and measures of connectivity may give novel insights to further our understanding of ictogenesis. In order to investigate the relationship between anatomy connectivity and functional connectivity, we need to develop tools (acquisition and hardware) and measures to characterize both the structure of networks as well as the dynamics of their activity.

4.2
Granger Causality

The original concept of Granger causality has played a significant role in the field of economics since the 1960s. Now it is increasingly used in neuroscience, especially in the past 15 years, to understand the dynamic interactions of brain circuitry. Granger Causality (GC) provides a strong mathematical basis used to determine the causal influence and direction of neural interactions. The basic idea can be traced back to Wiener [2]. He proposed that, for two simultaneously measured time series, one series can be called causal to the other if we can better predict the second series by incorporating past knowledge of the first one. This concept was later adopted and formalized by Granger in 1969 [3] for linear regression models of stochastic processes. Specifically, if the variance of the prediction error for the second time series at the present time is reduced by including past measurements from the first time series in a linear regression model, then the first time series can be said to have a causal (directional or driving) influence on the second time series. The advantage over cross-correlation metrics is that GC is more robust with respect to changes in overall rate of activity, provides a measure of the 'causal' influence, is directional, and often requires less data for analysis.

Autoregressive (AR) models [4] are at the core of parametric Granger causality methods. GC methods make use of the variance of prediction errors from various combinations of AR models to determine causal relationships. Pairwise Granger Causality (PGC) is used to investigate the directional interactions between two time series. Two AR models are required to determine each direction of influence. Consider two time series $X(t)$ and $Y(t)$ where separate AR models are created to predict current value of the X series from m (a previously determined AR model order) previous X values and:

$$X(t) = \sum_{j=1}^{m} b_{XX}(j)X(t-j) + \varepsilon(t), \quad \Sigma_1 = \text{var}(\varepsilon(t)) \tag{4.1}$$

$$Y(t) = \sum_{j=1}^{m} b_{YY}(j)Y(t-j) + \gamma(t), \quad \Gamma_1 = \text{var}(\gamma(t)) \tag{4.2}$$

The variance of the error series $\varepsilon(t)$, Σ_1, is a gauge of the linear prediction accuracy of $X(t)$ as Γ_1 is for $Y(t)$. Now consider a bivariate multivariate autoregressive (MVAR) model, $W(t)$, where both $X(t)$ and $Y(t)$ are calculated from the previous values of the X and Y time series:

$$W(t) = \begin{pmatrix} X(t) = \sum_{j=1}^{m} a_{XX}(j)X(t-j) + \sum_{j=1}^{m} a_{XY}(j)Y(t-j) + \eta(t) \\ Y(t) = \sum_{j=1}^{m} a_{YX}(j)X(t-j) + \sum_{j=1}^{m} a_{YY}(j)Y(t-j) + \gamma(t) \end{pmatrix} \tag{4.3}$$

The variance of the new error series is a gauge of the prediction accuracy of the new expanded predictor.

$$\Sigma_W = \begin{pmatrix} \Sigma_2 & \Upsilon_2 \\ \Upsilon_2 & \Gamma_2 \end{pmatrix} = \begin{pmatrix} \text{var}(\eta(t)) & \text{cov}(\eta(t), \gamma(t)) \\ \text{cov}(\eta(t), \gamma(t)) & \text{var}(\gamma(t)) \end{pmatrix} \tag{4.4}$$

Based on Wiener's idea, Granger [3] formulated that if the X prediction is improved by incorporating past knowledge of the Y series, the Y time series can then be said to have a Granger-causal influence on the X time series. This is the basis for the time domain version of Granger Causality where the variance of the linear prediction error of X alone, Σ_1, is compared to the variance of the linear prediction error of X including Y, Σ_2:

$$F_{Y \to X} = \ln \frac{\Sigma_1}{\Sigma_2} \tag{4.5}$$

Note that when $\Sigma_1 = \Sigma_2$ (i.e., the linear prediction error is not improved by including Y) this relationship will yield a PGC value of zero. Driving in the opposite direction is addressed by simply reversing the roles of the two time series. It is clear from this definition that timing plays an essential role in directional causal influences. Spectral and conditional methods have also been developed but will not be explained here. For a more detailed exposition of these additional methods, AR modeling, and PGC, please refer to Mingzhou Ding's recent chapter [5] on Granger Causality.

Because of the strengths mentioned earlier, Granger Causality is now increasingly being applied in various neuroscience paradigms including *in vivo* plasticity [6, 7], functional connectivity using fMRI [8, 9], human sleep analysis [10] and connectivity within complex neural systems [11, 12]. Causal methods have also been applied in the study of epilepsy to examine seizure initiation and propagation to identify the seize focus in human patients using data from EEG and ECoG recordings [5, 13, 14]. In the following example, we use PGC to explore the dynamic interactions between different brain areas during seizure recorded from microelectrode arrays implanted *in vivo* into the hippocampus of rats.

4.2.1
Analysis of Temporal Lobe Seizures

The continuous *in vivo* microarray electrode recordings present a unique opportunity to explore the *in vivo* dynamics and effective connectivity relationships of seizures from an animal model of temporal lobe epilepsy (TLE). High-quality multichannel data was continuously recorded from the rats, including multiple seizure events from several animals. The 32 channel microwire electrodes sampled at 25 kHz (band pass (0.5–12) kHz) recorded from these animals in hippocampus, afford the opportunity to examine over 992 ($n!/(n - k)!$; $n = 32$ channels, $k = 2$ for bivariate model) different directional hippocampal interactions on multiple time scales using Granger Causality. The anatomy and the physiology of the hippocampus is well established [15]. However, descriptions of the effective connectivity of the hippocampus from implanted microwire arrays are rare in general, and nonexistent for seizure behavior. The goal of this analysis is to provide a better description of the dynamics and causal relationships underlying seizure activity and how they differ from those observed prior to seizure.

The TLE animal model used in this study has been in use for the last two decades [16]. In the weeks that follow this stimulation, the rat progressively demonstrates recurrent spontaneous hippocampal seizures. The temporal and electrographic progression of rats treated with this model closely resembles that of human patients that have chronic mesial temporal seizures. The surgical and electrophysiological techniques used by the Evolution Into Epilepsy (EIE) project team at the University of Florida are described below.

The rats were prepared according to procedures approved by the University of Florida IACUC. The rats are adult male Sprague–Daly rats that are 50 days old. The surgical procedure consists of a craniotomy from 1.7 mm lateral to 3.5 mm

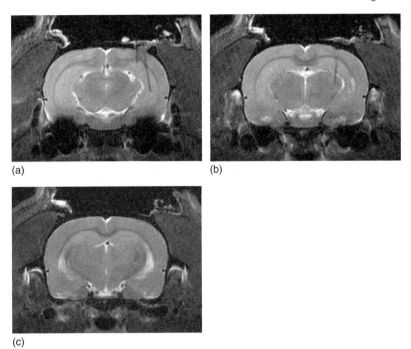

(a)　　　　　　　　　　(b)

(c)

Fig. 4.2 MR images of rat brain *in vivo* seven days after simulation. In (a) the path of one simulating electrode and the partial path of the other simulating electrode are shown on the right side of the brain image. In (b) the path of the other stimulating electrode is shown on the right side of the brain image. In (c), an increase in T2-weighted image contrast is visible bilaterally in regions around the ventricles.

lateral of the bregma so that electrodes could be implanted into areas due to their prominent role in epileptic discharges [17]. A 2×8 rectangular array of electrodes was implanted bilaterally in both hemispheres so that they cover the CA1/CA2 area to the DG. The electrode placement is shown in Figure 4.2 with the surgery shown in the left panel of Figure 4.2. The spacing between electrodes is 400 mm on the short axis and 200 mm on the long axis at a depth of 4 mm from the surface of the brain. The length for each electrode is designed such that the tip of each electrode is within the brain region of interest. The location of the end of each array closest to the centerline of the brain is 4 mm caudal to the bregma and 1.7 µm lateral. A single Teflon coated bipolar twist electrode was placed in the posterior ventral hippocampus in the right hemisphere of the animal, posterior to the electrode array for stimulation into status epilepticus. The rats were allowed to recover for one week post surgery before stimulation to seizure. A 10 second train of 50 Hz 1 ms bipolar 240 µA pulses was delivered every 12 seconds for 60 minutes to induce status epilepticus in the rats. 32 channels of simultaneous recordings were taken from the rats 24 hours a day for 76 days while the rat was engaged in sleep and normal exploratory behavior in a 30 cm diameter cage. After the experiment was completed the rats were sacrificed and high-resolution magnetic resonance imaging (MRI) [18] was used to verify the position of the electrodes, as shown in the lower right panel of Figure 4.2(a,b).

Seizure detection was conducted daily by manual inspection of the raw traces and verification against video records by an expert. The seizure grade was catalogued and the start time of the seizure was noted. A 10-minute window surrounding the seizure was then extracted for all 32 channels from the raw recording and low pass filtered to 1 kHz. This frequency was chosen because it maintains most of the features in the data while allowing for a reasonable model order for autoregressive modeling required for parametric PGC. A 60-second window was then chosen that encompassed the seizure. The stationarity of the time series data was addressed by normalizing the mean and variance. The Bayesian information criterion (BIC) was used to optimize the AR model order, which was found to be $m = 25$. Pairwise Granger Causality was then calculated in a one second moving window with 50 % overlap for all possible electrode pairings (992 directional combinations) for the entire 60-second window for 15 seizures. The analysis produces a single time domain causal value for each electrode pairing for each window resulting in a $32 \times 32 \times 119$ matrix describing the causal interactions over time. These matrices were translated into movies to better visualize changes in the effective connectivity between the observed hippocampal regions over the time course of the seizures.

4.2.2
Results

Qualitative visual analysis of the raw waveform, behavioral videos, and PGC movies reveal a common progression of events during all 15 seizures that were analyzed (Figure 4.3). To classify this progression and to allow better statistical quantification, five distinct stages of causal interaction were identified. These five stages each have their own unique combination of electrographic, behavioral, and causal interactions. Panel A shows the results of Granger analysis before the onset of a single seizure for one animal in which some causal activity occurred within the CA1 region of the left hemisphere. The seizure began in Stage 1 (shown in Panel B) in which a mixture of intrahemisphere activity was dominated by causal activity within the L-CA1 region and also from the L-CA1 region into to the L-DG. This activity gradually accumulates, sometimes includes interictal spiking, building up to the beginning of seizure. Behaviorally, Stage 1 often includes the beginning of the tonic phase of the seizure, where the animal remains motionless and blinks its eyes.

Stage 2, shown in Panel C, is the most striking and is marked by a high-magnitude mono-directional transfer from the left hemisphere into the right hemisphere as well as substantially increased intra-hemisphere activity. These directional transfers often reverberate sequentially from one hemisphere to the other. After several directional transfers, the seizure moves into Stage 3 activity, marked by mainly intra-hemisphere activity shown in Panel D. Unlike Stage 1,

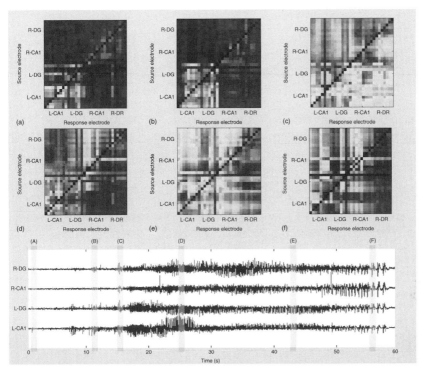

Fig. 4.3 Progression of Granger causality analyses through a single seizure event. The bottom electrode voltage traces are from four of the 32 electrodes from each of the four major hippocampal regions including the L-CA1, L-DG, R-CA1, and R-DG. The Granger analyses in panels A through F correspond to the red shaded windows of data highlighted over the electrode traces (A through F, respectively). In each Granger causality plot the driving brain area (source) is on the y-axis and the target area (response) is on the x-axis. For example, the driving influence from the R-DG to the L-CA1 in panel D is located in the upper left corner of the plot. The magnitude of the Granger causality interaction is represented by color ranging through blue (near zero), light blue, green, yellow, to red (highly causal). The interictal activity about 15 seconds prior to seizure in panel A shows a strong interaction within the L-CA1. Stage 1 in panel B shows this relationship increasing and spreading into the L-DG as well. Stage 1 is the beginning of the tonic behavioral state for the rat that persists until the Stage 4 transition. This driving influence from the CA1 to the DG is abnormal and persists until the Stage 4 transition. Panel C shows a directional transfer in Stage 2 from the left hemisphere to the right hemisphere. These transfers reverberate directionally several times across hemispheres. Panel D shows intrahemisphere activity common to Stage 3. Panel E shows across the board synchronization for the entire array for the 'transition stage' or Stage 4. Stage 4 is called the transition stage because behaviorally the rat moves from tonic to clonic activity and the primary driving relationship transitions from CA1 driving DG to DG driving CA1. Panel F shows activity from Stage 5 after the transition. In this stage the rat exhibits clonic behavior marked by rhythmic shaking of the limbs. (Please find a color version of this figure on the color plates.)

however, the causality patterns are more dynamic, of larger magnitude, and less confined to a single area or pathway. Occasional weak cross-hemisphere directed transfers also sometimes occur in Stage 3.

Stage 4, or the transition stage (Panel E), is usually a brief and highly causal epoch where all areas both intra and cross hemisphere are bidirectionally causal. This stage is usually characterized by a saturation of causal effect with nearly the entire plot shows strong causal interactions. Behaviorally, Stage 4 marks the transition from the tonic stage where the rat is motionless and appears tense to the clonic stage of the seizure where the rats appendages jerk rapidly and rhythmically. Additionally, the driving influence from the CA1 to the DG seen during all of the prior seizure stages flips to the DG driving the CA1. Following Stage 4, the pattern smoothly transitions into Stage 5 (shown in Figure 4.3 F), which is made up of heavy spiking activity with a causality pattern and with a very strong driving influence from the DG to the CA1. This driving pattern is similar to normal driving patterns observed in interictal activity after and far removed from ictal events suggesting a transition from abnormal to normal driving patterns.

4.2.3
Discussion

Currently, synchronization is thought to be the dominant dynamic to occur across the limbic system during epileptic seizures. Synchronization is defined as adjustment of rhythms of oscillating objects due to their interaction [19]. Synchronization across a diffuse excitable network is thought to be a possible mechanism that initiates seizure in the limbic system [20]. These diffuse models of connectivity often included universal bidirectional connectivity. In contrast, others suggest that a large amount of asynchrony is necessary to maintain seizure events [21]. The results from this analysis do indeed find synchronization within the hippocampus prior to, during, and after the ictal event. *The detailed interactions illustrated in these results have never previously been described in the study of TLE.* However, Granger Causality analysis shows that these interactions are highly directional and follow a common progression.

One of the distinct features in Stage 1, seen in nearly all seizures, is an increase in local interactions within the CA1 area that tend build over time and eventually directionally influence DG within that hemisphere. The large-magnitude directional cross-hemisphere reverberations of Stage 2 immediately follow this activity pattern. This suggests a hierarchical progression of causation within the CA1 that spreads to the DG and finally crosses over into the neighboring hemisphere.

By comparison, a coherence analysis would erroneously suggest that these directional transfers were bidirectional phenomena because this method is non-directional. Non-directional measures would also not be able to detect the reversal in directional connectivity between the DG and CA1 during the Stage 4 transition. The entire seizure would seem to be dominated by bidirectional synchronization across the entire network using non-directional measures. However, Granger causality analysis provides a very different story. The dominant directional influences in the beginning stage of the seizure are abnormal, with the CA1 areas showing directional causation into DG. This is the opposite of interictal patterns far removed from ictal events that often show directional influences from the DG to CA1. Additionally, this abnormal pattern is time-locked behaviorally with the tonic portion of the seizure. Interestingly, the flip of the directional connectivity to connectivity resembling baseline connectivity occurs during the transition from the tonic to clonic behavior. The implications of this directional flip are not completely clear.

One of the possible explanations of the driving influence of the CA1 into DG during the early phase of the seizure, highlights one of the weaknesses of Granger causality-based methods. It is possible that the entorhinal cortex may be differentially driving both of these areas creating a false directed influence from CA1 to the DG, instead of the CA1 to DG influence being mediated through the entorhinal cortex. The entorhinal cortex has been implicated as a possible initiator of ictal events [15]. For example, it has been suggested that disinhibited neurons in the entorhinal cortex may directly drive pyramidal cells in CA1 [15] in pilocarpine treated network interactions. The entorhinal cortex also anatomically connects to the DG. The alternative is that the entorhinal cortex may mediate the activity from the CA1 to the DG during this epoch. Both of these situations are supported by the literature and may even be happening simultaneously. It is unlikely, based on the anatomical organization of the hippocampus that the effects of CA1 are mediated through CA2 and CA3 to the DG. Further experimental analysis using conditional Granger causality with electrode coverage within the entorhinal cortex will be necessary to determine which of these connectivity patterns is correct.

Even without these details, the results also suggest that a Granger causality metric may be useful for seizure detection. The Stage 4 transition from abnormal driving (CA1 to DG) to a normal driving pattern (DG to CA1) suggests that, at some point prior to seizure, the driving pattern became abnormal. The abnormal driving influence has been seen in the pre-ictal analysis at least a minute before seizure. An natural question to ask would be 'when does this driving influence become abnormal?' Once identified, the transition from normal to abnormal driving influences could potentially be used as an indicator of an impending seizure. However, the role of the entorhinal cortex in this connectivity should be determined before this relationship is used for seizure detection. For example, if the entorhinal cortex turns out to be the source of the abnormal driving influence it may better aid detection if electrodes were placed in this area. Additionally, detection of this abnormal driving influence may increase the sensitivity of current seizure-detection methods. This could be useful for detecting epochs during epileptigenesis that are not easily detected using behavioral or current electrographic seizure-detection methods.

In summary, using Granger causality to reveal effective connectivity led to a new understanding of the interactions within the hippocampus during seizure. This understanding provides the opportunity to make new predictions about seizure detection and suggests new experiments that should expand understanding of TLE. However, we have only used the most basic of GC-based tools in this analyisis. Future work will make use of conditional and spectral GC analysis to paint a clearer picture of effective connectivity during seizure. We anticipate that Granger causality will become an important tool used in the study of epilepsy.

4.3
Structural Visualization with Magnetic Resonance

We have developed an MR imaging protocol to examine structural changes in the rat brain during the epileptogenic period following injury. This protocol consists of examining the rat brain *in vivo* at 11.1 Tesla by measuring pre-injury control images, post-electrode-implant images, then images at 3, 5, 7, 10, 20, 40 and 60 days following injury. Following sacrifice after day 60, the brain is fixed and excised for further study. Very high resolution MR images of the intact excised brain are measured at 17.6 Tesla. Then the brain is destructively processed for histological analysis. In this MR imaging protocol, we measure anatomical images, quantify MR relaxation times, and measure diffusion-weighted images. An example of our *in vivo* MR imaging results at 11.1 Tesla is shown in Figure 4.2. In these images, the Teflon-coated bipolar simulation electrodes are visualized in parts (a) and (b) within a rat brain seven days following stimulation. In addition, a bilateral increase in T2-weighted image contrast is visible in part (c) in regions around the ventricles indicating areas of pathology. An example of the very high resolution images we obtain at 17.6 Tesla is shown in Figure 4.4. This gradient echo three-dimensional image has a resolution of 75 mm × 75 mm × 75 mm. In (a), the path of the recording electrodes in indicated in the white ellipse and the tip of other recording electrodes in shown in (b). However, the most informative structural information is provided by the diffusion-weighted images which can be modeled as either diffusion tensor images or as images of water displacement probability maps.

4.3.1
Diffusion Tensor Imaging

Diffusion tensor imaging (DTI) is a non-invasive imaging technique that allows the measurement of water molecular diffusion through tissue *in vivo*, and indicates the local fiber direction within the tissue [22]. In this method, the directional features and rate of water diffusion are modeled as a positive-definite rank-2 tensor which allows the inference of connectivity patterns prevalent in tissue and track changes in this connectivity over time. This rank-2 tensor model of diffusion requires only the measurement of seven diffusion-weighted images with clinically feasible diffusion gradient strengths. This approach enables

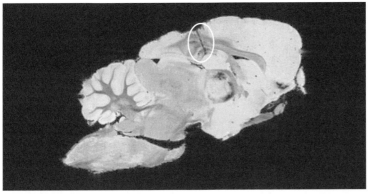

(a)

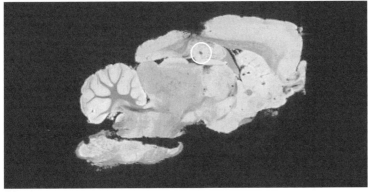

(b)

Fig. 4.4 MR images of excised fixed rat brain. In (a) the path of recording electrodes is shown within the white ellipse, and in (b) the end of other electrodes, terminating in the hippocampus, is shown within the white circle.

simple estimation of diffusion anisotropy, through the calculation of orientation-independent-parameter fractional anisotropy, and infers fiber orientation by the principal eigenvector of the diffusion tensor [23, 24]. Despite its modest requirements, the results achieved using DTI have been very successful in regions of the brain and spinal cord with substantial white matter coherence and have enabled the mapping of many anatomical connections in the central nervous system [25–27].

4.3.2
High Angular Resolution Diffusion Imaging

The major drawback of DTI is that it can only reveal a single fiber orientation in each voxel and fails in voxels with orientational heterogeneity, which makes DTI inappropriate for the study of detailed connectivity relationships in the regions of the brain containing crossing or bifurcating fiber bundles [24, 28, 29]. The limited ability of DTI to visualize regions of complex tissue structure has prompted interest in the development of both improved image acquisition strategies and more sophisticated reconstruction methods. An improved image acquisition scheme is available in an emerging approach, called high angular resolution diffusion imaging (HARDI) MRI [30, 31], in which apparent diffusion coefficients are measured along a large number of directions, and distributed uniformly on a unit hemisphere in the diffusion wave vector space. By sampling the diffusion signal with high angular resolution, the complexity of tissue structure can be visualized in each image voxel by determining the displacement probability density function for water diffusion in the tissue.

Several approaches to modeling the HARDI-derived displacement probability function have been proposed by our group and others [32–35]. We recently completed developmental work on methods to acquire HARDI data and processing algorithms to model the diffusion displacement probability [33–37] in order to visualize detailed fibrous structure in complex tissue regions in excised rat brains [37] and human brains *in vivo*. Our first development was motivated by earlier work from Tuch et al. [31, 38] who introduced HARDI and the diffusion orientation distribution function. In our work, we expanded upon the concept of the rank-2 tensor model of diffusion [22, 39] with the introduction of higher rank tensor [33] in order to expand upon the concept of diffusion in MR [40] and provide a physical basis for the visualization of complex structures by diffusion MRI. As part of this work, we developed a measure of displacement probability anisotropy [36], which provides a measure of tissue anisotropy in complex tissue regions, as does the DTI-based factional anisotropy in simple tissue regions.

As shown in Figure 4.5, recent investigations in our laboratory regarding the white matter structural changes during the 'latent period' of epileptogenesis have revealed that MR HARDI methodology can identify functional changes in limbic system connectivity in advance of the onset of epilepsy [35, 37]. These images of an excised fixed rat brain, sacrificed ≈60 days following injury, were measured at 17.6 Tesla. Figure 4.5 shows the displacement probabilities in a control, (a) and an epileptic rat, (b). The hippocampus and entorhinal cortex are shown in expanded views with the regions in the red boxes. The displacement probability map in each voxel depicts the orientations of the highly anisotropic and coherent fibers. Note voxels with crossing orientations located in the dentate gyrus (dg) and entorhinal cortex (ec). The region superior to CA1 represent the stratum lacunosum-moleculare and statum radiatum. In the control hippocampus, the molecular layer and stratum radiatum fiber orientations paralleled the apical dendrites of granule cells and pyramidal neurons, respectively. In the epileptic hippocampus, the CA1 subfield pyramidal cell layer is notably lost relative to the control. Structures of the dentate gyrus are also altered with more evidence of crossing fibers. We are continuing to develop HARDI acquisition and analysis methods to apply these methods to study evolving brain pathology in the living subject.

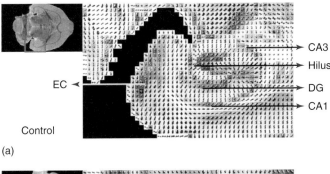

Control

(a)

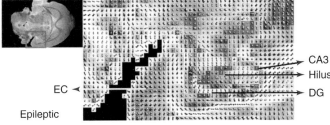

EC ◄

Epileptic

(b)

Fig. 4.5 Probability maps of coronal images of a control and an epileptic hippocampus. The upper left corner shows the corresponding reference images where the rectangle regions enclose the hippocampus. In the control hippocampus, the molecular layer and stratum radiatum fiber orientations paralleled the apical dendrites of granule cells and pyramidal neurons, respectively; whereas in the stratum lacunosum, molecular orientations paralleled Schaffer collaterals from CA1 neurons. In the epileptic hippocampus, the overall architecture is notably altered; the CA1 subfield is lost, while an increase in crossing fibers can be seen in the hilus and dentate gyrus (dg). Increased crossing fibers can also be seen in the entorhinal cortex (ec). Fiber density within the statum lacunosum moleculare and statum radiale is also notably reduced, although fiber orientation remains unaltered (from NeuroImage 37 (2007) 164–176, used with permission). (Please find a color version of this figure on the color plates.)

4.4
Acknowledgments

This research was supported by the National Institutes of Biomedical Imaging and Bioengineering (NIBIB) through Collaborative Research in Computational Neuroscience (CRCNS) Grant Numbers R01 EB004752 and EB007082, the Wilder Center of Excellence for Epilepsy Research, and the Children's Miracle Network. PRC was partially supported through the Wilder Center of Excellence for Epilepsy Research Endowment funds. WLD was partially supported through the J. Crayton Pruitt Family Endowment funds. SST was partially funded by a Fellowship Grant from the Epilepsy Foundation of America. The analysis work in this project was sponsored through a grant from the office of Naval research (Grant Number N00014-02-1-1019). We would like to thank Morgan Guoan who performed all the surgeries for electrode implantations, Dr. Wendy Norman who provided her kind assistance in scanning the EEG/video data to identify the seizures and creating the control data sets for the statistical analysis, Lan Hoang-Minh, Stephen Myers, and Hector Sepulveda for assistance with data collection, and Linda Dance for developing the data acquisition software. We would also like to thank Dr. Mingzhou Ding and Dr. Yonghong Chen for useful discussion and support in the application of Granger Causality methods. MRI data were obtained at the Advanced Magnetic Resonance Imaging and Spectroscopy (AMRIS) facility in the McKnight Brain Institute of the University of Florida.

References

1 S. Cajal, *Histologie Du Systeme Nerveux De L'Homme Et Des Vertebrates* (A. Maloine, Paris, **1911**).

2 N. Wiener, *The Theory of Prediction*, vol. Series 1. B. EF (**1956**).

3 C. W. J. Granger, *Econometrica* **3**, 424,438 (**1969**).

4 C. Chatfield, *The Analysis of Time Series: An Introduction* (Chapman & Hall, **1996**).

5 M. Ding, Y. Chen and S. L. Bressler, Granger Causality: Basic Theory and Application to Neuroscience. In B. Schelter, J. Timmer and M. Winterhalder, *Handbook of Time Series Analysis* (John Wiley & Sons Ltd, **2007**).

6 L. Zhu, Y.-C. Lai, F. C. Hoppensteadt and J. He, *Neural Comput.* **15**, 2359 (**2003**).

7 L. Zhu, Y.-C. Lai, F. C. Hoppensteadt and J. He, *Math. Biosci. Eng.* **2**, 1 (**2005**).

8 R. Goebel, A. Roebroeck, D.-S. Kim and E. Formisano, *Magn. Reson. Imaging* **21**, 1251 (**2003**).

9 J. R. Sato, E. A. Junior, D. Y. Takahashi, M. de Maria Felix, M. J. Brammer and P. A. Morettin, *Neuroimage* **31**, 187 (**2006**).

10 M. Kamiński, M. Ding, W. A. Truccolo and S. L. Bressler, *Biol. Cybern.* **85**, 145 (**2001**).

11 A. K. Seth, *Network* **16**, 35 (**2005**).

12 A. K. Seth and G. M. Edelman, *Neural Comput.* **19**, 910 (**2007**).

13 P. J. Franaszczuk, K. J. Blinowska and M. Kowalczyk, *Biol. Cybern.* **51**, 239 (**1985**).

14 P. Franaszczuk, G. Bergey and M. Kamiński, *Electroencephalogr. Clin. Neurophysiol.* **91**, 413 (**1994**).

15 M. Avoli, M. D'Antuono, J. Louvel, R. Köhling, G. Biagini, R. Pumain, G. D'Arcangelo and V. Tancredi, *Prog. Neurobiol.* **68**, 167 (**2002**).

16 E. W. Lothman, E. H. Bertram, J. Kapur and J. L. Stringer, *Epilepsy Res.* **6**, 110 (**1990**).

17 K. Kaneda, Y. Fujiwara-Tsukamoto, Y. Isomura and M. Takada, *Neurosci. Res.* **52**, 83 (**2005**).

18 G. T. Buracas, A. M. Zador, M. R. DeWeese and T. D. Albright, *Neuron* **20**, 959 (**1998**).

19 Rosenblum, *Synchronization* (Scholarpedia, http://www.scholarpedia.org/article/Synchronization, **2007**).

20 E. H. Bertram, D. X. Zhang, P. Mangan, N. Fountain and D. Rempe, *Epilepsy Res.* **32**, 194 (**1998**).

21 T. I. Netoff and S. J. Schiff, *J. Neurosci.* **22**, 7297 (**2002**).

22 P. J. Basser, J. Mattiello and D. LeBihan, *J. Magn. Reson. B* **103**, 247 (**1994**).

23 P. J. Basser and C. Pierpaoli, *J. Magn. Reson. B* **111**, 209 (**1996**).

24 P. J. Basser, S. Pajevic, C. Pierpaoli, J. Duda and A. Aldroubi, *Magn. Reson. Med.* **44**, 625 (**2000**).

25 T. E. Conturo, N. F. Lori, T. S. Cull, E. Akbudak, A. Z. Snyder, J. S. Shimony, R. C. McKinstry, H. Burton and M. E. Raichle, *Proc. Natl. Acad. Sci., U S A* **96**, 10422 (**1999**).

26 S. Mori, B. J. Crain, V. P. Chacko
    and P. C. van Zijl, *Ann. Neurol.* **45**,
    265 (**1999**).

27 P. J. Basser, *NMR Biomed.* **8**, 333
    (**1995**).

28 M. R. Wiegell, H. B. Larsson and
    V. J. Wedeen, *Radiology* **217**, 897
    (**2000**).

29 E. A. H. von dem Hagen and R. M.
    Henkelman, *Magn. Reson. Med.* **48**,
    454 (**2002**).

30 D. S. Tuch, R. M. Weisskoff, J. W.
    Belliveau and V. J. Wedeen (Philadel-
    phia, PA, **1999**), Proc. of the 7th
    Annual Meeting of ISMRM, p. 321.

31 D. S. Tuch, T. G. Reese, M. R.
    Wiegell, N. Makris, J. W. Belliveau
    and V. J. Wedeen, *Magn. Reson. Med.*
    48, 577 (**2002**).

32 V. J. Wedeen, T. G. Reese, V. J.
    Napadow and R. J. Gilbert,
    Biophys. J. **80**, 1024 (**2001**).

33 E. Ozarslan and T. H. Mareci,
    Magn. Reson. Med. **50**, 955 (**2003**).

34 E. Ozarslan, T. M. Shepherd, B. C.
    Vemuri, S. J. Blackband and T. H.
    Mareci, *Neuroimage* **31**, 1086 (**2006**).

35 B. Jian, B. C. Vemuri, E. Ozarslan,
    P. R. Carney and T. H. Mareci,
    Neuroimage **37**, 164 (**2007**).

36 E. Ozarslan, B. C. Vemuri and T. H.
    Mareci, *Magn. Reson. Med.* **53**, 866
    (**2005**).

37 T. M. Shepherd, E. Ozarslan,
    M. A. King, T. H. Mareci and S. J.
    Blackband, *Neuroimage* **32**, 1499
    (**2006**).

38 D. S. Tuch, T. G. Reese, M. R.
    Wiegell and V. J. Wedeen, *Neuron* **40**,
    885 (**2003**).

39 E. Stejskal, *J. Chem. Phys.* **43**, 3597
    (**1956**).

40 H. Torrey, *Phys. Rev.* **104**, 563
    (**1956**).

5
Network Models of Epileptiform Activity: Explorations in Seizure Evolution and Alteration

Pawel Kudela, William S. Anderson, Piotr J. Franaszczuk, Gregory K. Bergey

5.1
Introduction

The EEG during a seizure (ictal EEG) has specific electrophysiologic characteristics of the underlying changes in spatially and temporally ordered brain network synchrony. Recent advances in methods of signal analysis [1] have provided an improved accuracy in describing the frequency components of the ictal EEG. Time-frequency analyses of intracranial EEG (ICEEG) recordings from patients with mesial temporal lobe epilepsy reveal common elements in evolution of this seizure type. These elements are represented by rapidly evolving dynamic changes, including a period of organized rhythmic activity that undergoes a monotonic decline in frequency before transitioning to a period of intermittent bursting prior to seizure termination [2].

The application of time-frequency analysis to the ICEEG reveals these ictal characteristics, but raises many new questions that remain unanswered. The identified phases in the ictal ICEEG indicate dynamic changes in the brain neuronal synchrony and currently it is not known why seizures of temporal lobe onset undergo these changes. The molecular, cellular, or functional mechanisms that may cause or contribute to the ictal evolution remain poorly understood. Identifying potential mechanisms underlying the dynamic changes seen in ictal ICEEG could yield important insights into understanding seizure evolution and termination. If these mechanisms were better understood, this could lead to development of new treatment options for patients with epilepsy.

This chapter starts with a brief overview of seizure dynamics and evolution followed by the description of network models used to simulate the evolution of seizures of temporal lobe onset. The various outputs from the network model simulations are analyzed and compared directly with ictal EEG signals or with results of time-frequency analyses of the ictal EEG. Finally, the attempt to simulate the interruption of seizure evolution by external electrical stimulation acting on a modeled network is described. The disruption studies are performed in networks of different size and complexity varying from two synaptically connected small neuronal networks to a large realistic network with cortical architecture.

Seizure Prediction in Epilepsy. Edited by Björn Schelter, Jens Timmer and Andreas Schulze-Bonhage
Copyright © 2008 WILEY-VCH Verlag GmbH & Co. KGaA, Weinheim
ISBN: 978-3-527-40756-9

5.2
Time-frequency Analyses of Seizure Dynamics and Evolution

Time-frequency analyses of ICEEG can provide precise information about the occurrence of specific frequency changes in these signals. These analyses were used to reveal specific seizure characteristics, and to study seizure dynamics and evolution [2, 3]. Figure 5.1 shows a Matching Pursuit (MP) time-frequency energy distribution of the ictal ICEEG signal recorded from depth electrodes in patients undergoing pre-resection evaluation. Such studies demonstrated after the seizure initiation, a period of relatively low complexity ICEEG signal, which is dominated by one predominant rhythm. This rhythm peaks at 8 Hz and undergoes a monotonic decline in frequency with the subsequent transition of the ICEEG into a high complexity signal before seizure termination. A similar evolution pattern is frequently observed in mesial temporal seizures. Pacia and Ebersole [4] similarly identified a pattern of frequency evolution (5–9 Hz) in hippocampal onset seizures

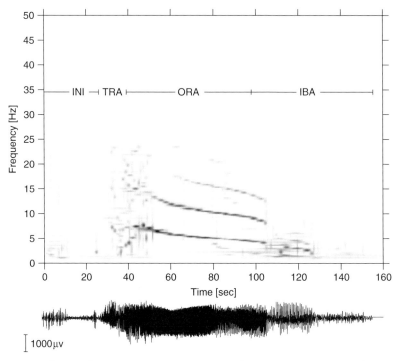

Fig. 5.1 The time-frequency energy distribution of the entire seizure recorded by the deepest depth electrode contact, located near the region of seizure onset. The ICEEG recorded from contact is shown below the plot. The left vertical axis shows the frequency in Hz. The vertical calibration bar is for the illustrated ICEEG. The effective sampling rate was 100 Hz. The first 296 atoms with the highest energy are shown. The periods of seizure initiation (INI + TRA), organized rhythmic activity (ORA), and intermittent bursting activity (IBA) are marked. (Figure reprinted from [2] with permission from Elsevier.)

in scalp recordings. Quiroga et al. [5] reported a mean value of 7.9 Hz frequency after seizure onset and a mean of 1.4 Hz at the end of the seizure in tonic clonic seizures. These examples indicate that there is a shift from high into low frequencies as the seizure of temporal lobe onset progresses.

5.3
Model Assumptions and Modeling Approach

Since the low frequencies might be the signature of late phase or seizure termination, it is important to recognize the mechanisms that are responsible for this frequency shift and others that are implicated in seizure evolution. Before presenting details of the network model, it is necessary to introduce the model assumptions along with some general comments on the modeling approach. The occurrence of a predominant rhythm in the EEG is most likely a consequence of the activity of some population of neurons, which are recurrently activated, and fire trains of action potentials (APs) at exact intervals matching the frequency of the predominant rhythm. The decline of the frequency of the predominant rhythm possibly reflects the involvement of some cell-specific mechanisms, which might occur at the molecular, membrane or the synaptic level. The above two assumptions serve as working hypotheses that will be tested in the network model. In order to build an appropriate network model that explains the rhythm, and its decline, the following two questions need to be formulated. First, what is the source of the periodic excitation and refractoriness of neurons in the rhythmic repetitive neuronal firing? Second, what is the mechanism that modulates periods of recurrent neuronal firing and refractoriness?

The present model will attempt to simulate epileptiform activity in a disinhibited neuronal network, where continuous neuronal activity arises from mutual neuronal excitation in the network. Our strategy is to add the transient refractoriness to neurons in order to periodically stop continuous activity. With regard to the first question, the neuronal refractoriness might arise from any outward current with activation/inactivation dynamics matching the appropriate time scale of the observed frequencies. We considered slow hyperpolarizing Ca^{2+} dependent K^+ current (AHP) as the source of neuronal refractoriness, because this current underlies neuronal bursting and adaptation behaviors and is regulated by slowly changing $[Ca^{2+}]_i$. It should be also emphasized that Ca^{2+}-dependent K^+ currents have for a long time been implicated in epileptogenesis [6–9]. Regarding the second question, we considered intracellular calcium clearance mechanisms as a factor that can modulate periods of neuronal inactivity. This hypothesis is supported by two observations: 1) that epileptiform activity is associated with enhanced Ca^{2+} influx and 2) the evidence of impaired ability of neurons to remove calcium in conditions of high $[Ca^{2+}]_i$ in experimental models of epilepsy [10–12].

These hypotheses are tested in a simple neuronal network model consisting of n by n excitatory and m by m inhibitory neurons uniformly distributed in a

2 dimensional plane. The typical size of the network varies from 1600 to 65 000 neurons and consists of 10% inhibitory neurons. Inhibitory neurons make synaptic contacts locally with neighboring excitatory and inhibitory neurons. Excitatory neurons synapse with neighboring inhibitory neurons and can make long range connections with remote excitatory neurons but the probability of connections decreases exponentially with distance. Neurons are simulated as single compartment using a conductance-based model (Equation (5.1) in the Appendix). The active membrane model includes sodium, calcium, and potassium currents. Sodium and potassium are Hodgkin–Huxley type currents, while the calcium current is modeled with the Goldman–Hodgkin–Katz equation. These networks are activated by external APs delivered at random intervals derived from a Poisson distribution to selected excitatory neurons in the center of the network array. The procedure of induction of bursts in neurons includes the gradual reduction of the excitatory drive to inhibitory neurons in the network. Typically, the inhibitory drive removal starts from the 5th second of simulation and at the 10th second the inhibitory drive is entirely removed. In these networks, each neuron has a minimum of two excitatory synapses on input from randomly selected presynaptic neurons and the strength of the excitatory synapses is adjusted to the level allowing burst spreading from pre- to postsynaptic neurons throughout the network.

5.4
Recurrent Neuronal Bursting and Mechanism of Burst Frequency Decline

In this network, bursts of APs arise simultaneously in a population of excitatory neurons after removal of the inhibitory drive. This is started by random neuronal firing that occurs in neurons located in the center of the network, with subsequent spread of the firing along recurrent excitatory synaptic pathways. The synchrony is characterized by simultaneous occurrence of AP bursts or AP trains in a population of excitatory neurons. During bursts that last 50–100 ms, synchronization implies that bursts in any two neurons occur within tens of milliseconds of one another. Within a burst, the simultaneous occurrence of APs in two or more neurons is not important but may occur as well.

In the first simulation a very simple model of calcium dynamics was used. $[Ca^{2+}]_i$ in neurons was simulated in a thin shell under the membrane. Ca^{2+} flow is proportional to the I_{Ca} current amplitude and the Ca^{2+} clearance mechanism is represented by single removal rate parameter R. Figure 5.2 shows the total number of action potentials fired by a population of excitatory neurons in a network as a function of time after removal of the inhibitory drive from the network. The 8 Hz periodic bursting in this network was evoked by APs delivered at random intervals and a high rate to neurons located in the center of the network array. Within each peak, excitatory neurons fire 100 ms long trains of APs. Each single AP is associated with activation of I_{Ca} and Ca^{2+} influx, which is next followed by the increase of the I_{AHP} current. Inactivation of I_{AHP} is a slowly changing $[Ca^{2+}]_i$-dependent process,

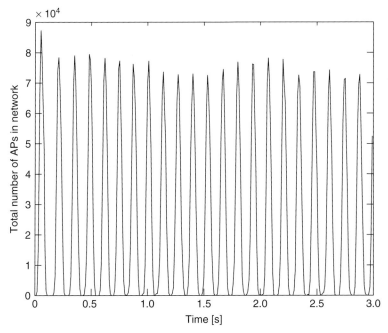

Fig. 5.2 The total number of action potentials generated in the network of 51 840 excitatory neurons vs time after the removal of the inhibitory drive from the network. Periodic activity in the network array is triggered by action potentials continuously delivered at random intervals (from Poisson process with $\lambda = 0.03$) to 324 excitatory neurons in the center of the network array. Continuous activation evokes periodic activity with a frequency of 7.5 Hz in the network. The calcium removal rate $R = 0.018 \times 10^3$ s^{-1}. (Figure reprinted from [13] with permission from Elsevier.)

so this current is sustained between APs. The amplitude of I_{AHP} increases with increasing number of APs, and after several APs, the I_{AHP} current hyperpolarizes the membrane sufficiently to prevent neuronal response to excitation. Therefore each peak in Figure 5.2 is followed by a period of neuronal inactivity resulting from summation of I_{AHP}. The length of that period depends on how fast $[Ca^{2+}]_i$ returns to the baseline, which in turn depends on the rate of Ca^{2+} clearance. In consequence, interburst intervals are regulated by the Ca^{2+} clearance rate (R) in neurons.

In later simulations the rate of Ca^{2+} clearance R was altered in order to vary the frequency of recurrent periodic neuronal bursting in the network. The Ca^{2+} clearance rate R decreased exponentially and changed dynamically during simulations. This decreasing R can be interpreted as an alteration in the ability of neurons to remove calcium after an enhanced Ca^{2+} influx consistent with epileptiform activity reported in experimental models of epilepsy [10–12]. The frequencies of bursting in 81 adjacent neurons remote from the site of network activation are plotted in Figure 5.3(a). Points were obtained from the instantaneous burst-to-burst intervals measured in these neurons. In a 60 s long simulation,

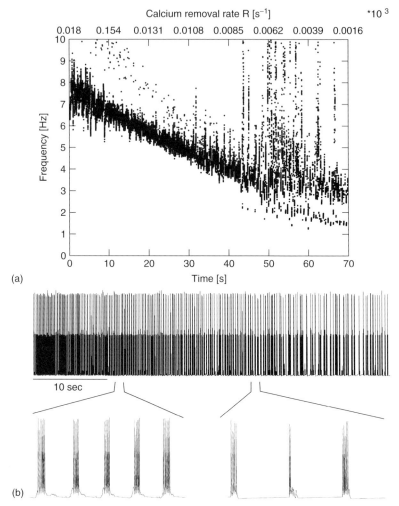

(a)

(b)

Fig. 5.3 (a) Decreasing the calcium removal rate R during the simulation decreases the frequency of bursts in neurons. Each of 26, 263 points represents the instantaneous frequency of bursts, measured as the reciprocal of the current period of bursting, in excitatory neurons ($n = 81$) in a remote area from the site of activation. The vertical axis indicates the frequency of bursts. The bottom horizontal axis indicates time and the top axis indicates the calcium removal rate R, which decreased exponentially from $0.018 \times 10^3 \, s^{-1}$ ($t = 0 \, s$) to $0.0016 \times 10^3 \, s^{-1}$ ($t = 70 \, s$). (b) Change in neuronal bursting patterns caused by change in the calcium removal rate R. The top trace shows the membrane potential of one neuron located near the border of the network array (remote from the site of network activation). Total time shown on trace is 45 s. The two lower traces show an expanded view of two 1-s periods from the top trace at the 12th and at the 28th s of simulation. The calcium removal rate R has decreased exponentially from $0.089 \times 10^3 \, s^{-1}$ ($t = 0 \, s$) to $0.0019 \times 10^3 \, s^{-1}$ ($t = 45 \, s$) during the period shown in the top trace. (Figures adapted from [13].)

the frequency of bursts decreased approximately 2.5 times (from 8 Hz to 3 Hz) while the Ca^{2+} clearance rate R decreased at the same time tenfold (from 0.018 to 0.0018). Figure 5.3(b) shows the activity pattern from a single neuron, with expanded views of two 1s long periods, approximately at the 19th and 30th second of the simulation. When the Ca^{2+} removal rate R decreases, the burst-to-burst intervals increase and the frequency of burst occurrence per second decreases. The comparison of ictal ICEEG signal from a depth electrode and the simulated local field potential (LFP) is shown in Figure 5.4. This simulated LFP signal represents an average membrane potential in the proximity of the hypothetical electrode and was calculated from the membrane voltage of 81 adjacent neurons. The corresponding time-frequency energy distributions are also illustrated for both signals. The comparison of the predominant frequency decline observed in the ictal IEEG and the burst frequency decline obtained in two network models is shown in Figure 5.5. The first model was simulated with an exponential and the second with a linear Ca^{2+} removal rate R decrease. The plotted points represent centers of the atoms in the corresponding time-frequency energy distributions. In both simulations, patterns of changes in the frequency follow the pattern of frequency changes observed in the signal from a depth electrode. In the model with an exponential decrease in R the pattern is slightly more consistent than in the model with a linear decline in R.

5.5
Network Models of Epileptiform Activity Disruption by External Stimulation

These network models can reproduce typical characteristics of seizure dynamics and evolution. This includes the signal similarities and the time frequency characteristics of that signal that are consistent with characteristics of the ictal ICEEG. The network model studies revealed potential factors and mechanisms involved in seizure dynamics and evolution. This knowledge would be particularly useful when one considers new treatment options based on the idea of modification of seizure evolution. Because of the nature of seizures, these efforts could point toward methods that would have acute abortive effects on seizures after onset. The paradigm of external electrical stimulation can potentially offer such options. Driven by evidence from human and animal studies, the possibility of seizure control by external electrical stimulation is gaining considerable attention. The actual therapeutic effect of electrical stimulation is however poorly understood. Theoretical computational studies performed in the context of neuronal networks again can help identify mechanisms leading to suppression or disruption of repetitive neuronal firing. Examples of network models of disruption of activity in neurons by externally administered stimuli are described below. The first two network models focused solely on the mechanism of action rather than providing the exact description of electrode-neuron interactions. The latter are described in the model of a network with cortical architecture, where neurons are stimulated by a realistic electrode using realistic stimulation parameters.

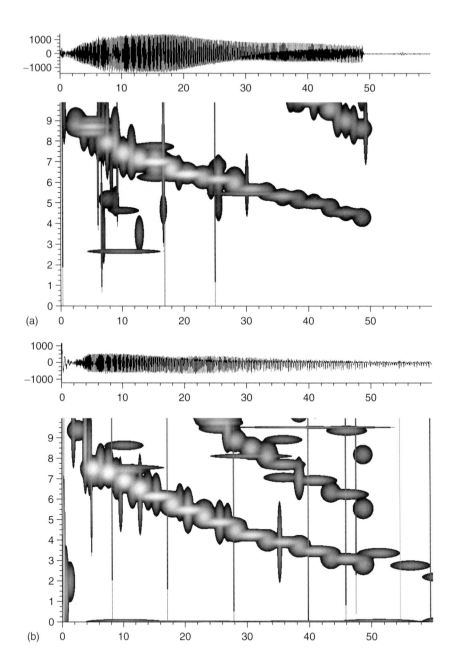

5.6
Chain Network Model Studies

The mechanism of recurrent bursting termination was studied in two synaptically connected neuronal networks [14]. Each network consisted of less than a hundred excitatory neurons that were randomly synaptically connected. A few neurons from network A make synaptic contact with a few neurons in network B. In the same way, neurons in network B make contact with neurons in network A via a synaptic feedback loop illustrated in Figure 5.6 by the dashed line. The active membrane model used in these neuron simulations included a slow Ca^{2+}-dependent K^+ current (I_{AHP}), which hyperpolarizes the neuronal membrane after each AP. This current is responsible for the spike frequency adaptation in neurons in the presence of prolonged stimulation.

Neurons in network A activated by an external stimulation or a current pulse injection respond with APs spreading over the population of neurons instantly and bursts of APs occurring simultaneously in all neurons. These bursts stop simultaneously in all neurons after approximately one hundred milliseconds, as a result of summation of the I_{AHP} current in the neurons. After each burst the neuronal membrane remains slightly hyperpolarized for several hundred milliseconds until the I_{AHP} current inactivates. These post-bursting periods are characterized by decreased excitability of neurons. Activity in network A causes activation of neurons in network B via the existing synaptic connections and similar bursts occur in neurons in network B. Next, neurons in network A are again activated via the excitatory feedback loop so the bursting activity reappears in network A. In order to maintain such a continuous network interaction and to have stable periodic bursting in both networks, the feedback loop delay must be larger than the time of recovery of neurons from hyperpolarization caused by the I_{AHP} current. Typically, these delays must be set on the order of hundreds of milliseconds. The continuous recurrent bursting can be stopped by the same stimulus that was used to start this activity if the second stimulation of neurons in network A occurs before these neurons receive excitation from B. Additional stimulation would extend the refractory period, and network A would not be able

Fig. 5.4 Time-frequency decompositions of the dominant rhythm in simulated LFPs computed for an array of 81 locally connected neurons (b) and the recorded signal from the depth electrode contact nearest the mesial temporal region of seizure onset in a human during the period of organized rhythmic activity in (a). Each panel shows the trace of signal and the energy map of time-frequency atoms obtained from matching pursuit decomposition of the band-filtered (2–10 Hz) signal to separate the predominant frequency. Only 40 atoms with the highest energy are shown. The horizontal axes show time in seconds, the vertical axis shows relative amplitude in arbitrary units for signal plots, and frequency in Hz for energy maps. The gray areas in the energy plots indicate the areas of concentration of energy for Gabor atoms used for decomposition. The pattern of changes of the dominant frequency in the simulated LFP (b) is consistent with the pattern of changes in the dominant frequency observed during the organized rhythmic activity period in the recorded signal from the human with the seizure (reprinted from [19] with permission © 2004 IEEE).

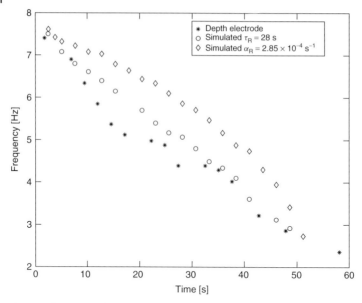

Fig. 5.5 Illustration of the decline in the dominant frequency in the signal from the depth electrode and simulated LFPs. Points represent the centers of atoms from time-frequency energy distributions with the highest energy. Asterisks (∗) represent atoms from the decomposition of the recorded signal, open circles (○) represent atoms obtained from simulations with an exponential decrease in calcium removal rate $R = 0.018 \times 10^3 s^{-1} \times \exp(-t/28\,s)$; and open diamonds (◇) represent atoms from simulations with a linear decline of R (e.g., $R = 0.018 \times 10^3 s^{-1} \times (1 - 2.85 \times 10^{-4} s^{-1} \times t)$). (Figure reprinted from [13] with permission from Elsevier.)

to fully respond to excitation received from network B. Also network B would not be able to fully respond to excitation from network A (after the second stimulation) because neurons in network B are still in the process of recovery from afterhyperpolarization. This mechanism is shown in Figure 5.6.

Typically, synaptic transmission delays are shorter than 2 ms. A few hundred millisecond delayed feedback loop in the above two-network model was used in order to reduce the complexity of the model. This allows us to investigate the possibility of recurrent activity interruption and to demonstrate possible mechanisms of stopping such recurrent activity. In situ, such a feedback loop can exist from several interconnected networks. Figure 5.7 illustrates an example of a hypothesized loop consisting of 16 networks (called subnetworks hereafter). Usually stimulation of one selected subnetwork induces traveling wave activity in the network chain loop. The same stimulation that was used to start the activity can be later used to stop it, but it requires the precise selection of a time window for stimulation. This is illustrated in Figure 5.7, where the stimulus was applied to the selected subnetwork before the traveling wave passes. Interestingly, when the number of subnetworks in the loop increases, it is possible to induce several waves of activity in a single network chain loop. Simultaneous termination in all

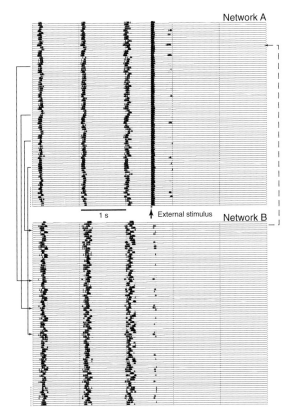

Network A

1 s   ↑ External stimulus   Network B

Fig. 5.6 Illustration of recurrent and periodic bursting in a two-network model. Bursting activity was initiated earlier by the current pulse injection ($I = 10\,\mu A\,cm^{-2}$, $t = 50\,ms$) to neurons in network A (not shown). When the same external current is applied to all neurons in network A at a specified time before excitability has returned to baseline in network A, the bursting of the network ceases. The four selected neurons from network B have inputs from the four selected neurons in network A. A feedback loop is modeled as a single connection with a delay of 800 ms. (Figure reprinted from [14] with permission from Elsevier.)

subnetworks is also possible but requires multipoint stimulation along with careful selection of time windows for both stimuli [15].

5.7
Networks with Realistic Cortical Architecture

These chain network model studies suggest that the effectiveness of stimulation depends on the underlying network activity phase and the timing of stimulation. Although these network models may account for the actual mechanism underlying the interruption of recurrent network activity, they cannot realistically describe the actual neuronal interactions in the brain. In order to gain a better understanding

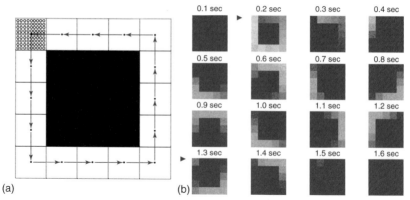

(a) (b)

Fig. 5.7 Schematic diagram (a) of a loop of 16 randomly connected local networks. Each small square represents 81 locally connected neurons (shown only in one square and indicated by circles). Each neuron in a subnetwork has two randomly assigned inputs from neurons inside the subnetwork and one randomly assigned input from a neuron in the preceding subnetwork. Arrows show the pattern of connections between subnetworks. (b) Recurrent bursting in a loop shown in the diagram. The activity in the loop is evoked by applying an external current to all neurons in one subnetwork indicated by the arrow. The time is shown at the top of each frame. The first stimulus which initiates activity is applied at 0.2 s. The second at 1.3 s ceases recurrent bursting. The gray scale indicates the relative intensity of bursting in the subnetworks, black represents lack of bursting. (Figure reprinted from [15] with permission from Elsevier.)

of the mechanisms of suppression and to find effective stimulation parameters, more advanced models are needed. The effects of electrical stimulation on the suppression of neuronal activity was simulated in networks with a realistic cortical architecture. Neurons in these networks are arranged in vertical minicolumns and in four horizontally oriented layers. Each minicolumn consists of 12 pyramidal neurons (equally distributed between layers II/III, V, and VI), one stellate neuron in layer IV, and three inhibitory neurons (basket and double bouquet in layer II/III, and chandelier in layer IV). Stellate and pyramidal neurons in layers II/III and IV are simulated as regular spiking neurons with spike frequency adaptation. Pyramidal neurons in layer V are simulated as intrinsically bursting neurons. Inhibitory neurons are simulated as fast spiking neurons, except chandelier neurons, which exhibit regular spiking. The characteristics of the neurons as well as the connectivity patterns were drawn from a variety of histological data sources. The connectivity pattern inside a minicolumn is based on a model of visual cortex [16]. A diagram of the intrinsic connectivity within a given minicolumn is illustrated in Figure 5.8. The extra-columnar wiring (between remote minicolumns) was also derived from several histological studies. The complete list of references can be found elsewhere [17]. Typically, pyramidal neurons from layer II/III make isotropic connections within 300 μm radius to other pyramidal neurons in layer II/III with a probability of connection of 5% and to layer IV stellate neurons with probability 5% and to layer V with probability 9%. Pyramidal neurons in layer II/III project also to surrounding basket and chandelier neurons and the number of connections was

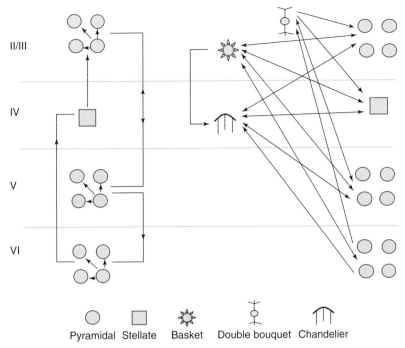

○	☐	☀	☡	⋔
Pyramidal	Stellate	Basket	Double bouquet	Chandelier

Fig. 5.8 Schematic diagram of the types of neurons as well as their intrinsic connectivity within a minicolumn. Each minicolumn consists of 16 neurons including 12 pyramidal, one stellate and three inhibitory. The connections between excitatory and inhibitory neurons are on the right. Excitatory–excitatory connections are on the left. The connectivity pattern is based on a model of the visual cortex [16].

set to 5% of the total number of excitatory to excitatory connections. Connection ranges and connectivity probabilities between other neurons within and between layers can be found in [17]. The center to center distance between minicolumns is assumed to be 25 μm. The number of simulated minicolumns is 32 by 32 and includes 16 384 neurons. This corresponds to an 800 μm × 800 μm square piece of gray matter. No attempt was made to simulate white matter connections.

Networks are activated by APs delivered at random intervals to the 16 most central pyramidal neurons in layer II/III. It involves random current injection to these neurons at a very low probability. This provides robust ongoing activity and reinitiates activity after stimulation. In order to have various patterns of activity in the network, we introduced five different levels (B to F) of network connectivity. These connectivity patterns were derived from the initial connectivity pattern (A) by introducing multiplicative factors (ranging from 0.25 to 1) that modify the number of excitatory and inhibitory connections independently. The stimulation procedure includes the calculation of a voltage field distribution (Equation (5.2)) near the circular disk electrode of radius a, with current amplitude I applied to that disk in homogenous medium of σ conductivity. ρ is the radial distance from the center of the disk in the X Y plane, and z is the Z coordinate (distance below the disk).

The applied current I has the form of a square wave (positive followed by negative lobe, 200 μs each) delivered at a frequency from 60 to 200 Hz for up to 100 ms total stimulation time. The induction of action potentials was assumed to take place in the vertically oriented initial segments of the neurons, with the length of the initial segment assumed to be $\delta = 30\,\mu m$. The threshold was calculated according to Equation (5.3), which is a second difference of the voltage drop along the neuronal initial segment. The threshold value for the induction of an action potential is 3 mV, and each time the threshold function exceeds that value, a single current injection pulse is delivered to a neuron. We also simulated the induction of APs in the nodes of Ranvier in horizontally oriented axon branches. This mechanism, however, cannot be satisfactorily modeled at this neuronal network scale. A typical internodal spacing for a 5 μm diameter axon is 500 μm. The size of the simulated network in our model is 800 μm by 800 μm and the majority of neurons make connections within less than 300 μm. These computational constraints limit the total number of APs generated in axons that occur after stimulation in the network. The mechanism of induction of APs in axons, however, can be well modeled in larger networks.

Figure 5.9 illustrates the instantaneous activity of all neurons in the simulated network in time (before, during, and after stimulation). The electrode has a 1 mm radius and was positioned in the upper left corner of the network, with the center of the electrode overlapping with the corner of the network. 200 Hz stimulation occurs from 0.5 to 0.6 s and is indicated as a step. After the stimulation, there is a suppression of activity in the network lasting approximately 0.5 s.

A set of similar simulations of monopolar stimulation with application of biphasic frequency in the range 60–200 Hz was also performed. The electrode was positioned in the upper left corner of the network. Five simulations were performed each for a different network connectivity pattern labeled from B to F. Stimulations occurred from 0.5 to 0.6 s. Figure 5.10(a) shows the number of APs fired by pyramidal neurons in layer II/III before and after 200 Hz stimulation. Applied stimulation causes a temporary suppression of activity in the network, which reappears from random APs constantly delivered to the network.

The time delay to return of activity, which is defined as the time from the end of stimulation to the first peak in activity in the network after stimulation is plotted in Figure 5.10(b). The delay time to return to activity is plotted as a function of frequency for all tested connectivity patterns (B–F). There is a plateau effect, with longer return to activity times at higher stimulation frequency.

The position of the electrode and its effect on the delay time to return of activity in a network was examined. This included moving the electrode in the parallel and perpendicular directions to the network plane as illustrated in Figure 5.11. The electrode was positioned 155 μm above the hypothetical surface of the network. Plot (a) illustrates the return to activity time as a function of electrode position relative to the center of the network. The abscissa indicates the distance from the center of the electrode to the center of the network. The ordinate indicates the delay time to return to activity. Five simulations were performed, each for a different connectivity pattern labeled from B to F using 200 Hz stimulation. There is a peak

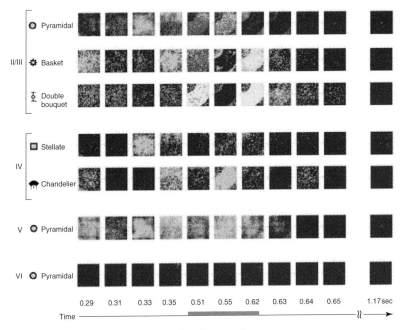

Fig. 5.9 Snapshots of network activity for all types of neurons as a function of time. In this particular simulation a network with connectivity pattern C was used. Stimulation was applied from 0.5 to 0.6 s and momentary suppression of activity occurs from 0.65 s to 1.17 s.

in the delay time to return of activity for all patterns of connectivity as the center of the electrode is 0.75 mm–0.5 mm from the center of the network.

In the case of one pattern of connectivity B the electrode was moved away from the center of the network in the opposite direction in order to validate the results. There is a symmetric peak corresponding to the position of the electrode on the opposite side. Simulation of the effect of moving the electrode perpendicular to the network plane is illustrated in Figure 5.11(b). The vertical line indicates the hypothetical surface of the cortex. When the electrode is 200 μm or more above the surface of the network, the effect of suppression of activity in a network disappears entirely. This is a reflection of the fact that the threshold values for the initial segment activation, as well as the Z coordinates of the initial segments within a given layer, were the same for all of the neurons. Introducing a 1 mV threshold value dispersion smoothed this effect and the sharp cutoff in the time delay to return to activity in Figure 5.11 is replaced by a tail.

Simulations of bipolar stimulation were also performed, including dipole stimulation with two electrodes of 250 μm radius each and separated by 1.6 mm, center to center. The same stimulation protocol as in the monopolar studies was used but with opposite polarity voltage applied between the electrodes. The entire network was divided into four equal sections and two patches on opposite corners of the

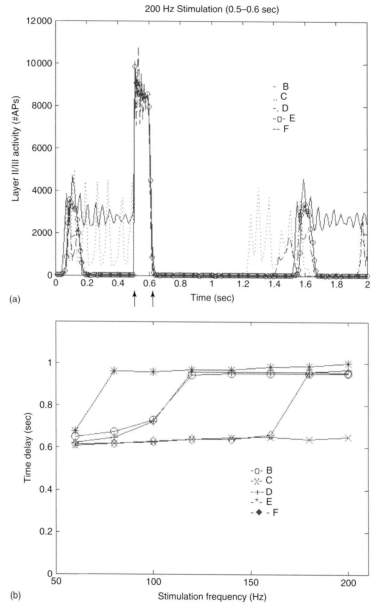

Fig. 5.10 The effect of stimulation on neuronal activity as a function of frequency of stimulation for five different connectivity pattern (B–F). Plot (a) illustrates the number of APs fired in time by pyramidal neurons in layer II/III. 200 Hz stimulation occurs from 0.5 to 0.6 s (arrows). The time delay to return of activity is defined as the time after 0.6 s at which the first peak in the layer II/III network activity occurs. Plot (b) shows delay time to return of activity (layer II/III pyramidal cells) as a function of frequency for the tested connectivity patterns (B–F). (Figure reprinted from [17] with permission from Springer Science and Business Media.)

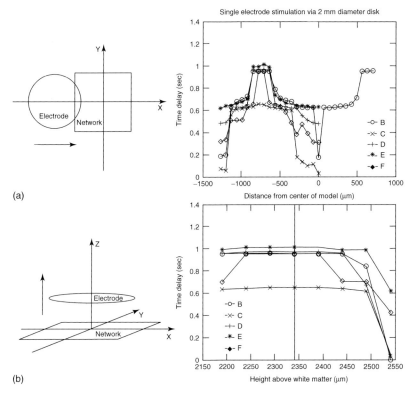

Fig. 5.11 (a) Stimulation effects as a function of electrode position parallel to the network surface. The time to return of activity (layer II/III pyramidal cells) is plotted as a function of the stimulation electrode distance from the center of the network for five types of connectivity patterns. The pattern B data was continued through the center of the model to show the approximate dimension of the active central ring of stimulation under the electrode. (b) Stimulation effects as a function of height above the network for the tested connectivity patterns (B–F) (layer II/III pyramidal cells). The single vertical line marks the hypothetical cortical surface. (Figure reprinted from [17] with permission from Springer Science and Business Media.)

network were preferentially connected to each other mainly between layer II/III pyramidal cells. The connected patches areas are marked in gray in Figure 5.12(a). This was done to introduce some anisotropy to the network, and may reflect long-range patch connections (not via white matter) that are observed in the mammalian cortex. These studies were performed in the context of how suppression of activity depends on dipole electrode orientation relative to the patch connection orientation. The dipole was rotated with the center of rotation overlapping the center of the network. The stimulation was applied between 0.5 to 0.6 ms of simulation. The longest time delay to return of activity was observed in the network when the dipole axis was parallel to the connected patch axis. Figure 5.12(b) illustrates the time delay to return of activity in the network as a function of the dipole rotation angle. The time delay to return of activity as a function of the electrode separation is plotted

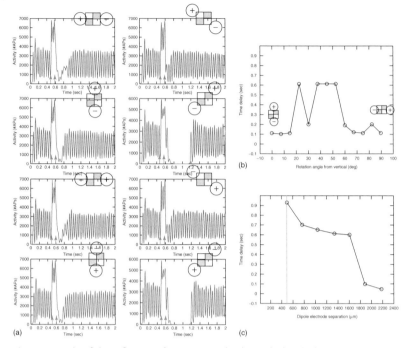

Fig. 5.12 Study of the influence of stimulation on longer range patch connections in the network. Two opposite corners of the network, specifically between layer II/III pyramidal cells in each patch, are preferentially connected. Stimulation is via the dipole electrode arrangement, and the plots of layer II/III pyramidal cell activity are presented as a function of the dipole orientation. Stronger stimulation induced effects are observed with the dipole axis parallel to the connected patches. The connected patches are shaded gray in the schematic, and the two dipole disks are represented as circles around the periphery of the model. The arrows mark the onset and cessation of stimulation (connectivity pattern C). The dipole marked with the '+' sign undergoes the positive lobe of stimulation first, the '−' electrode is switched in polarity. The dipole is rotated in the clockwise direction around the model. (b) The time delay to return of activity as a function of dipole rotation angle taken between the '0' orientation presented in the inset on the left side of the figure, and the '90' orientation shown on the right side. (c) The time delay dependence on the separation of the two electrodes, with the dipole system oriented along the connected patches. A vertical line marks the point of contact of the two electrodes. (Figure reprinted from [17] with permission from Springer Science and Business Media.)

in Figure 5.12(c). The vertical line marks the point of contact of the two electrodes. A set of simulations were performed in order to determine the importance of the timing of stimulation. These simulations show that the delay time to return of activity depends on the timing of stimulation onset with respect to the phase of network activity. A brief pulse of 200 Hz stimulation was applied with the onset varying from −60 ms to +30 ms with respect to a local peak in neuronal activity, Figure 5.13(a). The dependence of the delay time to return of activity on the timing of the stimulus is illustrated in Figure 5.13(b). A longer delay time to return of

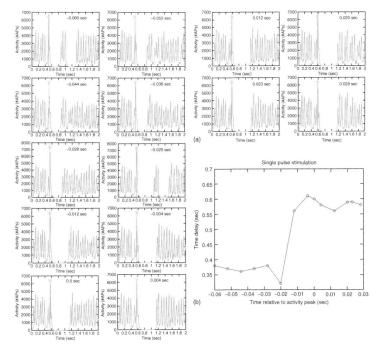

Fig. 5.13 Study of timing of single-pulse stimulation with respect to the underlying network rhythmicity, as demonstrated by the connectivity pattern C. The time to the return of layer II/III pyramidal cell activity as a function of delay time relative to the network activity peak is shown in (a). A vertical line and arrow mark the stimulation point. The plots of network activity for each stimulation time are presented in (b), demonstrating the phase-sensitive alterations in the post-stimulation network activity (Figure reprinted from [17] with permission from Springer Science and Business Media.)

activity was observed when the stimulus was delivered in about the same time or after the occurrence of the local peak in neuronal activity. A shorter delay time was observed when the onset of stimulation preceded the local peak in activity (−15 ms or more). In addition to shorter delay time, some alterations in spatiotemporal pattern of network activity were observed when the stimulus was delivered before the local peak in neuronal activity (−60 ms, −20 ms in Figure 5.13(a)).

5.8
Conclusions

Alternative therapies for medically refractory epilepsies have gained considerable attention. Identifying mechanisms underlying seizure evolution and dynamics could yield novel seizure-aborting therapies. The network model of seizure dynamics and evolution aims to identify potential mechanisms associated with seizure termination. Typical characteristics of epileptic ICEEG include a monotonic decline in frequency of the predominant rhythm prior to seizure termination.

The predominant frequency decline can be explained in this network model by the alteration in neuronal membrane excitability, which is maintained by a slow and calcium-dependent potassium current. In the model, the rate of Ca^{2+} clearance in neurons steadily decreases. This is reflected in a decline in frequency of the synchronous neuronal bursting in the rhythmic network activity pattern and in simulated LFPs. This mechanism in part may account for the seizure abortive effect of electrical stimulation. Indeed, in addition to AP induction in axons and the initial segment of stimulated neurons, the stimulation may momentarily increase the calcium influx into neurons and hence acutely alter the membrane excitability. This mechanism has been shown to be important in the termination of recurrent bursting in models of synaptically connected networks. In the network loop model, the timing of the stimulus that will terminate bursting is determined for the given loop length. Once determined, the window for stimulation after an individual burst is very critical. Stimuli that are applied outside of the relatively narrow window fail to terminate bursting permanently. This is due to the underlying transient alteration in neuronal excitability by applied stimulation. Similarly, in a cortical network model, applied stimulation resulted in momentary suppression of rhythmic activity. The efficacy of the stimulation depends on stimulation parameters, including the stimulation frequency and on the position and the orientation of the electrode as well. The timing of the stimulation with respect to the underlying rhythmic activity also demonstrated a phase dependent sensitivity.

5.9
Acknowledgment

This work was supported by NIH grants NS 51382, NS 38958 and by the Epilepsy Foundation.

5.10
Appendix

The neuronal membrane model incorporates two inward currents: I_{Na} and I_{Ca}, three outward potassium currents: the delayed rectifier I_K, the Ca-dependent $I_{K(Ca),AHP}$ and the transient I_A current, and a leak current I_L. I_{syn} represents the synaptic current and is described by a double exponential function [14, 15].

$$C_m \frac{dV}{dt} = -I_{Na} - I_{Ca} - I_K - I_{K(Ca)} - I_A - I_L - I_{syn} \tag{5.1}$$

The potential field from a single metallic disk of radius a, with a current I applied to it, measured in an isotropic medium of conductivity σ is described by the equation:

$$\Phi(\rho, z) = \left(\frac{I}{4\pi \sigma a} \right) \arcsin \frac{2a}{[z^2 + (\rho - a)^2]^{1/2} + [z^2 + (\rho + a)^2]^{1/2}} \tag{5.2}$$

$$\sigma = 0.3 \text{ S}^{-1}\text{m}$$

where z and ρ are Z coordinate and a radial distance from the center of the electrode in the X Y plane, respectively. The tissue inhomogeneities and CSF loss effects are ignored. A derivation of this equation can be found in [18].

The action potential production is assumed to take place at the axon initial segment, with the size of initial segment $\delta = 30\,\mu m$. The threshold function is:

$$\Delta^2 V = V(z_{AIS} + \delta) + V(z_{AIS} - \delta) - 2V(z_{AIS}) \tag{5.3}$$

where z_{AIS} is the z coordinate of the axon initial segment. The threshold value of the electric field gradient for an action potential induction in the initial segment is $3\,mV$. This value was taken from the literature and the complete list of references can be found in [17].

References

1 S. Mallat and Z. Zhang, *Signal Processing, IEEE Transactions on* (see also *Acoustics, Speech, and Signal Processing*), *IEEE Transactions on* **41**, 3397 (**1993**).

2 P. J. Franaszczuk, G. K. Bergey, P. J. Durka and H. M. Eisenberg, *Electroencephalogr. Clin. Neurophysiol.* **106**, 513 (**1998**).

3 C. C. Jouny, P. J. Franaszczuk and G. K. Bergey, *Clin. Neurophysiol.* **114**, 426 (**2003**).

4 S. V. Pacia and J. S. Ebersole, *Epilepsia* **38**, 642 (**1997**).

5 R. Q. Quiroga, H. Garcia and A. Rabinowicz, *Electromyogr. Clin. Neurophysiol.* **42**, 323 (**2002**).

6 B. E. Alger and R. A. Nicoll, *Science* **210**, 1122 (**1980**).

7 S. Verma-Ahuja, M. S. Evans and T. L. Pencek, *Epilepsy Res.* **22**, 137 (**1995**).

8 C. Silva-Barrat and J. Champagnat, *Neurosci. Lett.* **189**, 105 (**1995**).

9 N. L. Chamberlin and R. Dingledine, *Brain Res.* **492**, 337 (**1989**).

10 S. Pal, S. Sombati, D. D. Limbrick and R. J. DeLorenzo, *Brain Res.* **851**, 20 (**1999**).

11 S. Pal, D. Sun, D. Limbrick, A. Rafiq and R. J. DeLorenzo, *Cell Calcium* **30**, 285 (**2001**).

12 M. Raza, S. Pal, A. Rafiq and R. J. DeLorenzo, *Brain Res.* **903**, 1 (**2001**).

13 P. Kudela, P. J. Franaszczuk and G. K. Bergey, *Epilepsy Res.* **57**, 95 (**2003**).

14 P. J. Franaszczuk, P. Kudela and G. K. Bergey, *Epilepsy Res.* **53**, 65 (**2003**).

15 P. Kudela, P. J. Franaszczuk and G. K. Bergey, *Neurocomputing* **44**, 897 (**2002**).

16 R. J. Douglas and K. A. C. Martin, *Ann. Rev. Neurosci.* **27**, 419 (**2004**).

17 W. S. Anderson, P. Kudela, J. Cho, G. K. Bergey and P. J. Franaszczuk, *Biol. Cybern.* **97**, 173 (**2007**).

18 J. D. Wiley and J. G. Webster, *IEEE Trans. Biomed Eng.* **29**, 381 (**1982**).

19 P. Kudela, P. J. Franaszczuk and G. K. Bergey, *Conf. Proc. IEEE Eng. Med. Biol. Soc.* **1**, 715 (**2004**).

6
Recurrent Cortical Network Activity and Modulation of Synaptic Transmission

Yousheng Shu

6.1
Introduction

The brain is able to generate a variety of electrical activities that relate to different behaviors, and during these cortical activities cortical neurons are constantly bombarded with synaptic inputs that arise from other cortical cells. An individual cortical pyramidal neuron receives thousands of synaptic inputs, the majority of which come from its neighboring excitatory neurons and each cortical neuron innervates a large group of other neurons, typically within a local cortical area, by forming a few synapses with each of them. Therefore a local cortical network is built as a combination of convergent and divergent connectivity [1–3].

The excitatory pyramidal neuron not only sends output to other pyramidal neurons through its recurrent axon collaterals, but also innervates GABAergic inhibitory interneurons, which compose approximately one-fifth of the total number of cortical neurons. In return, the inhibitory interneuron sends its densely branched axons to innervate local pyramidal neurons, which mainly form synaptic contacts onto their perisomatic and distal dendritic regions. Therefore, the wiring between excitatory and inhibitory neurons provides the morphological basis for the operation of the cerebral cortex [4, 5].

It has been proposed that the recurrent excitation and inhibition in the cortex is well balanced, but there are few experimental studies carried out to probe this fundamental property of the cortical network. Computational studies have predicted that the recurrent excitation and inhibition are proportional to each other [6], large changes in excitation may result in proportional changes of feedback inhibition to keep the network relatively stable. A stable and balanced network is required for normal functioning of the cortex, while disruption of the balance between excitation and inhibition may result in malfunctioning of the cortex, such as epileptic seizures, schizophrenia and anxiety [7–11].

The arrival of postsynaptic potentials in cortical pyramidal neurons during the generation of recurrent network activity causes fluctuation of the membrane potentials in the cell body and generation of action potentials in the axon if the firing threshold is reached. Traditionally, the all-or-none action potentials are believed

Seizure Prediction in Epilepsy. Edited by Björn Schelter, Jens Timmer and Andreas Schulze-Bonhage
Copyright © 2008 WILEY-VCH Verlag GmbH & Co. KGaA, Weinheim
ISBN: 978-3-527-40756-9

to be the only form of communication between the neuronal somata and axon terminals; the traveling of subthreshold voltage changes in the axon is largely ignored. Here we demonstrate that the spread of the somatic membrane potential changes along the axon could reach the presynaptic terminals and modulate postsynaptic responses evoked by presynaptic action potentials [12, 13]. These findings may shed light on the mechanisms for normal cortical processing and abnormal pathological conditions, including epilepsy.

6.2
The Ability of the Cortical Network to Generate Recurrent Activity

During slow wave sleep, or under certain types of anesthesia, the cerebral cortex is able to generate periodic recurrent activities; the slow oscillations [14, 15]. Two distinguished states could be detected by extracellular recordings: one is the active state with relatively high frequency firing, termed the Up state; the other is the quiet state with very low activity termed the Down state. The two states occur intermittently at a frequency of less than 1 Hz. Combination of anesthetics, such as urethane followed by injection of ketamine and xylozine, could promote the recurrent pattern of the two states, but the duration of Up and Down states depends on the depth of anesthesia. Deep anesthesia results in a shorter Up and longer Down state, while light anesthesia is associated with a longer Up and shorter Down state. Since the slow oscillation is also robust during natural sleep [15], this indicates a normal physiological phenomena of the cortex, instead of the abnormal consequence of anesthesia.

Intracellular studies *in vivo* demonstrated that the membrane potential of a cortical neuron was depolarized by 10–20 mV during the Up state. The depolarization is mediated by the arrival of barrages of excitatory and inhibitory postsynaptic potentials [14, 16–19] and maintains for approximately 0.5–1.5 seconds and results in the generation of action potentials in pyramidal neurons at 5–20 Hz. During the intermittent Down state, the membrane potential returns to its resting level because no, or sparse, postsynaptic potentials are received by a given cortical neuron. The transition between Up and Down states is relatively rapid, it takes about 100–150 ms from the Down to the Up state, and 150–200 ms from the Up to the Down state.

Not only could the cortical slow oscillation occur in the intact whole brain, but also it could be spontaneously generated by cortical slices [20–22], indicating that the ability to generate slow oscillation is a property of a local network independent of subcortical areas and the large scale of the cortex. The experiments demonstrated that slow oscillation was less frequent in slices maintained in traditional artificial cerebrospinal fluid (ACSF), but the slices could generate robust Up and Down states if they were incubated in a modified ACSF in which the composition of the ionic medium was changed to more closely match that found *in situ*. This *in vitro* preparation allowed one to explore the mechanisms underlying the generation of slow oscillations by easily manipulating the extracellular ionic environment, delivering neurotransmitter or ion channel-related drugs, and simultaneously probing the interaction between different neuronal compartments, such as dendrite, soma and even axon [13, 21, 22].

How could the cortical network generate the slow oscillations? We propose that the deep layer intrinsic bursting neurons may activate their neighbor neurons through recurrent excitation and cause the transition from a Down state to an Up state, while the buildup of the outward conductances during continuous activation of cortical neurons, such as the Ca^{2+} and Na^+-activated potassium currents, may contribute to the cessation of the Up state [20]. Although there are other hypotheses available to explain the mechanisms underlying the generation of spontaneous slow oscillations, very few studies have been carried out to clarify the situation.

6.3
Cortical Network Activity as Propagating Electrical Waves

The slow oscillation generated *in vivo* and *in vitro* is a global activity. In the anesthetized animal, simultaneous recording from multiple cortical sites indicated that the slow oscillation is a propagating wave that travels at a speed of approximately $100 \, mm \, s^{-1}$ [23]. In cortical slices, the activities are propagating at a considerably lower velocity (approximately $10 \, mm \, s^{-1}$), presumably owing to the isolation of the small network from the whole brain [20].

The slice preparation allowed us to find out where the slow oscillation first starts. We arranged a one-dimensional electrode array (16 multiple unit recording electrodes) vertically on a cortical slice from the pia to the white matter. The recordings revealed that the activity starts first at layer V, and then propagates to layer VI and superficial layers. Cross-correlation between the activity in layer V and that of other layers showed that the activity in layer V leads the activities in other layers. To answer the question whether or not there is a specific location in layer V cross-cortical columns that could always generate the network activity first, we performed similar experiments by arranging the 16 electrodes only in layer V. The results showed that the activity could be generated first at any location a cross the electrodes, although the probability of first generating an Up state at some spots is higher than other locations. These multiple electrode recordings revealed that, among the six cortical layers, layer V generates the slow oscillation first, but no specific location in layer V always generates network activity first.

Local excitation by puffing glutamate could initiate the Up states. Again, with the 16 electrodes, we recorded the spontaneous activities in layer V; brief injection of glutamate to one of the electrode locations could reliably trigger an Up state if the injection took place after a delay that was long enough from the cessation of the previous Up state. Increasing this delay resulted in an increase in the duration of the evoked Up states. As soon as a new wave of activity was initiated by the injection of glutamate, the activity propagated across the electrodes. Interestingly, if we puffed glutamate at the two ends of the electrode array simultaneously, the evoked activities propagated across the electrodes and collapsed in the middle, indicating that the refractory mechanisms controlled the timing of the next round of activity[1].

6.4
Balance of Excitation and Inhibition during Cortical Network Activity

During the Up state, the cortical neuron was bombarded with barrages of postsynaptic potentials including excitatory and inhibitory PSPs; as a result, the membrane potential was depolarized and kept relatively stable for 0.5–2.5 seconds [15, 24]. Theoretical studies have predicted a strong relation between the amplitude of recurrent excitation and inhibition in cortical networks, resulting in a proportionality of theses two feedback signals [6]. If this is the case, during the Up state, the postsynaptic currents should have a stable reversal potential somewhere between −75 mV (the reversal potential for Cl⁻-mediated IPSPs) and 0 mV (the reversal potential for AMPA receptor-mediated EPSPs).

Indeed, voltage-clamp experiments in the cortical pyramidal neuron [22] showed that the Up state currents reversed at approximately −30 mV, and at this membrane potential the current associated with the Up state was surprisingly stable (Figure 6.1) during the entire period of time. The recordings were done with sharp electrodes filled with CsAc and QX314 to reduce the voltage-gated potassium and sodium currents. To reveal the dynamic changes of both excitatory and inhibitory conductances and the relationship between them, we held the membrane potential of cortical neurons at different levels ranging from −80 to +30 mV and monitored the Up state currents at each membrane potential level. Calculation of the changes in total membrane conductance revealed a steady increase followed by a decrease. Excitatory and inhibitory conductances could be derived from the total conductances because we knew the reversal potentials for EPSCs and IPSCs and the I-V relation at each time point during the Up state. Plotting the inhibitory conductance as a function of excitatory conductance revealed their relative linear relationship, indicating that excitation and inhibition are precisely proportional and increase or decrease in a balanced manner (Figure 6.1). This finding provides direct experimental data supporting the theoretical assumption; the balance of feedback excitation and inhibition during recurrent cortical activity.

6.5
Initiation and Termination of Cortical Network Activity by Electrical Shock

To investigate whether or not Up states could be initiated and terminated through the activation of synaptic pathways, we delivered electrical shock to activate neuronal elements in layer V (close to the recording electrode), layer II/III, or in the border

---

1) Shu and McCormick: unpublished observations

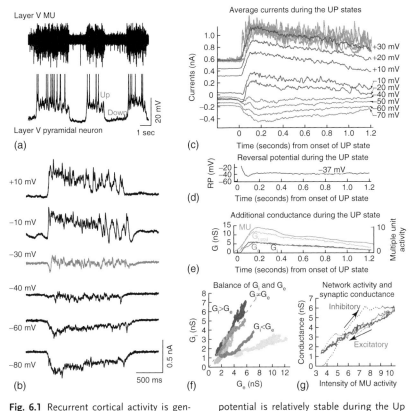

Fig. 6.1 Recurrent cortical activity is generated by a balance of excitation and inhibition. (a) The prefrontal cortical slice maintained *in vitro* spontaneously generates slow oscillations. Simultaneous extracellular multiple-unit (MU) recording and intracellular recording from a layer V pyramidal neuron reveal the two intermittent states; the Up and Down state. The action potentials are truncated. (b) Voltage-clamp recording demonstrates that the Up state currents have a reversal potential around −30 mV. (c) Averaged currents during the Up state at holding potentials from −70 to +30 mV. Several raw current traces at +30 mV are shown with the average for comparison. The neuron was recorded with sharp electrode filled with CsAc and QX314 to minimize the contribution from K^+ and Na^+ currents. (d) Reversal potential is relatively stable during the Up state. (e) Calculation of the total conductance (G_{total}), excitatory (G_e) and inhibitory conductance (G_i). Note the relationship between the intensity of multiple unit activity and the changes in these conductances. (f) Linear relationship between G_e and G_i indicates the proportionality between excitation and inhibition. (g) Relationship between the average intensity of MU activity and the amplitude of G_e and G_i. In this neuron, inhibitory conductance lagged the excitatory conductance as the network transitioned into the Up state. Following the onset of the Up state the G_e and G_i were proportional and correlated strongly with the intensity of MU activity recorded locally. (Figure reproduced from [22].) (Please find a color version of this figure on the color plates.)

of white matter and layer VI right below the recording electrode. Single electrical stimulation resulted in initiation of an Up state whose duration depended on the intensity of the stimulus. Increasing the intensity of the stimulation decreased the duration of the Up state, such that a strong electrical shock that was about four times the threshold could only evoke a brief burst of activity (approximately 20 ms). Intracellular recordings revealed that strong stimulation resulted in big and prolonged IPSPs or IPSCs, such that the evoked activity could not sustain. Interestingly, the same electrical shock that applied to initiate an Up state could also cause the cessation of the network activity depending on the time since the onset of activity and the strength of the stimuli. The intensity of the stimuli is also critical for the efficiency of termination of the network activity. Relatively weak stimuli (20–80 μA) could not stop the activity, although it was able to initiate the Up states reliably. With modest intensity (80–200 μA), the same stimuli could cause a transition from the Up state to the Down state, but only after a delay, the duration of which depended on the stimulus intensity. Strong stimuli (200–500 μA) could stop the UP states efficiently at an inter-stimulus interval as short as 100 ms [22].

Why could the delivery of the same stimuli result in both initiation and cessation of the Up states? The neuronal substrates in the network may contribute differently to the termination of the network activity; therefore, we performed intracellular recording from excitatory and inhibitory neurons to monitor their firing behavior in response to the starting and stopping stimuli.

The excitatory pyramidal neurons either discharged a single action potential or failed to fire in response to the starting and stopping stimuli, but the stopping stimuli more reliably evoked an action potential across trials, and the latency jitter of these action potentials is significantly less than the responses evoked by the starting stimuli. Therefore, the excitatory neurons were more synchronous after the stopping stimuli; consequently the increased synchronization of pyramidal neurons may switch the network activity to a refractory period.

On the other hand, the GABAergic inhibitory fast-spiking neuron could generate more than one action potential in response to both starting and stopping stimuli. In comparison with the response evoked by the starting stimuli, the stopping stimuli triggered more action potentials and fired at higher frequencies. These effects appear to result largely from the depolarization associated with the Up state, because close examination of the PSP barrages evoked by the start and stop stimuli, after hyperpolarization to prevent the generation of action potentials, failed to reveal marked differences in the amplitude time course of these events. These results suggest that the increase in responsiveness of inhibitory neurons may result in more inhibition and cause the cessation of the Up state [22].

Inconsistent with the current-clamp results, the voltage-clamp recording experiments revealed that the PSCs evoked by the stopping stimulus had a more hyperpolarized and prolonged reversal potential than that induced by the starting stimulus. Holding the membrane potential at 0 mV to isolate IPSCs, the recordings directly demonstrated that the stopping stimulus evoked an IPSC that was significantly larger than that evoked by the starting stimulus [22].

6.6
Epileptiform Activity Results from Imbalance of Excitation and Inhibition

The balance of excitation and inhibition is critical for the generation of normal Up states, while an imbalanced network resulted in malfunctioning of the cortical network including the generation of epileptiform activity. Animal models with either an increase in excitation or a depression of inhibition by pharmacological manipulations demonstrated experimental epileptic seizures. Similarly, *in vitro* experiments revealed that cortical slices bathed with an ACSF without addition of Mg^{2+}, which is a voltage-dependent blocker of NMDA receptors, could spontaneously generate seizure-like activity; removal of inhibition by bath application of GABAA receptor antagonists, picrotoxin or bicuculline, could also cause the generation of epileptiform activity [25–27].

6.7
Conduction of Action Potentials in the Axon during Normal and Epileptiform Activity

During slow oscillation and epileptiform activity, cortical neurons were bombarded with synaptic potentials and discharged action potentials when the summated PSPs reached the firing threshold. The all-or-none action potentials functioned as digital signals that traveled along the axon and passed information to the postsynaptic cells. The fidelity of axonal conduction of action potentials is critical for cortical processing and may control the pathological changes of the cortex during seizure-like activities.

Using our newly developed axonal whole-cell recording combined with simultaneous somatic recording [13], we calculated the distance between the soma and the initiation site of the action potential by evoking somatic and axonal action potentials by current injection in the soma and axon. The results revealed that the action potential was always generated in the axon initial segment, and the initiation site was about 46 μm away from the soma. This calculation was based on the assumption that the velocities of spike propagation from the initiation site to the somatic and axonal recording site are equal. However, under physiological and experimental conditions, spikes propagated from the initiation site down the axon should be faster than back propagation to the soma. In considering the difference of the conduction velocity, computer models using NEURON simulation environment [28] revealed that the action potentials are actually initiated in the axon about 38 μm away from the soma.

Although there are lines of evidence showing that the action potentials are generated in the axon initial segment under the relatively quiescent conditions of the *in vitro* slices, the site of fast spike initiation under conditions when the cortical network is active is still relatively unexplored. With the cortical slices that spontaneously generate slow oscillation, we were able to investigate the initiation site of action potentials evoked by the arrival of barrages of PSPs. Our recordings showed that the axonal action potential always preceded the somatic action potential if the axonal recording site was less than 90 μm from the soma, indicating that the initiation site was closer to the axonal recording site. Additional evidence demonstrated that the initiation site was in the axonal initial segment because the somatic action potential preceded the axonal one when the recording site was far down the axon (more than 90 μm) [13].

During the epileptiform activity, cortical neurons are dramatically depolarized and discharge bursts of potentials with an instantaneous firing frequency of more than 200 Hz. Simultaneous dendritic and somatic whole-cell recording revealed that action potentials generated during the epileptic bursts were initiated near the somatic compartment and back-propagated to the dendritic trees, while simultaneous somatic and axonal recording clearly demonstrated that the spike initiation site was in the axon initial segment. The depolarization during the epileptic burst was so large that somatic sodium current was inactivated quickly after the onset of the burst, therefore the height of somatic action potentials decreased progressively with the increase in depolarization. Surprisingly, our axonal recording revealed the initiation and conduction of burst-related action potentials that were not detected in the soma. These results provide evidence that under normal (for example, the slow oscillation) and abnormal (for example, the epileptiform activity) conditions, fast action potentials are preferentially initiated in the axon initial segment and propagate reliably down the axon [13].

6.8
Traveling of Subthreshold Potentials in the Axon

During the generation of slow oscillation and epileptic seizures, the changes in the somatic membrane potential are up to 20 mV and 50 mV, respectively, which result from the integration of synaptic inputs, the activity of intrinsic ion channels and the accumulation of extracellular K^+. The traveling of these somatic voltage changes along the axon is largely ignored, although there is a huge body of studies investigating the axonal mechanisms of action potentials since the discovery of saltatory conduction by Hodgkin and Huxley [29]. Our knowledge about the function of axons in the central nervous system mainly comes from the extracellular recordings, and whole-cell recordings from special axon terminals such as the Calyx of Held in the brain stem [30] and the mossy fiber boutons in the hippocampus [31]. The axonal tight-seal whole cell recording from the axon bleb (3–5 μm in diameter) formed during dissection of the cortical slices provides direct access to the cell membrane of the axon, and solves the problem of inaccessibility to the patch pipette of the thin axon [13].

With the axonal whole cell recording (Figure 6.2), surprisingly, we found that the subthreshold membrane potential changes that associated with the Up states in the soma of layer V pyramidal neuron could spread far down the axon with a length constant of 417 μm. That is, somatic voltage changes decreased to 37% when spread to this axonal site. The axonal form of synaptic barrages was a close copy of that recorded in the soma, with some attenuation at higher frequencies in the axon. Consistently, DC current pulse injection (−0.2 nA, 500 ms) through the somatic recording pipette, results in membrane potential changes in the soma and axon. Calculation of the length constant revealed a slightly longer distance (455 μm) than the traveling of barrages of PSPs, presumably due to the effect of the low pass filter of the axonal cable [12].

Similarly, during the generation of epileptiform activity, the enormous depolarization could propagate long distances along the main axon. Simultaneous recording from the soma and axon revealed that somatic membrane potential changes are bigger than those recorded in the axon, and this might be the reason why the height of somatic action potentials were significantly reduced but axonal action potentials were relatively well preserved [13].

6.9
Modulation of Intracortical Synaptic Transmission by Presynaptic Somatic Membrane Potential

The spread of a subthreshold potential along the axon may influence the excitability of the axon and participate in modulating synaptic transmission. According to the length constant, the subthreshold potential could travel along a large portion of the axon, but we don't know whether or not there are presynaptic terminals distributed within this portion of the axon. To examine how many presynaptic boutons that are distributed in the axonal compartments are close to the soma, we labeled some layer V pyramidal neurons in cortical slices by injecting biocytin followed by DAB staining, and used neurolucida to detect the location of presynaptic boutons. The results revealed that approximately 155 putative boutons were distributed along the axon collaterals within the first 500 μm from soma, and about 270 putative boutons within the first 1 mm [12]. These values are probably a significant underestimate of local synaptic connectivity, owing to the cutting of axon branches and limitations of axonal staining using the slice technique.

Do the changes of membrane potential in the axon terminals influence synaptic transmission? To answer this question, we performed pair or triple recordings from layer V pyramidal neurons that have excitatory synaptic connections (Figure 6.2). The somatic membrane potential of the presynaptic cell was depolarized and hyperpolarized to mimic the Up and Down states, or kept at various membrane potential levels through constant current injection, an action potential was initiated by a brief (1–2 ms) positive current pulse injection every second, and the postsynaptic responses were averaged corresponding to each membrane potential level of the presynaptic cell. Our recordings revealed that, indeed, the averaged amplitude of EPSPs correlated well with presynaptic membrane potential (Figure 6.2), depolarization in the presynaptic soma resulted in facilitation of postsynaptic responses at a rate of approximately 30% per 10 mV, indicating the synaptic transmission

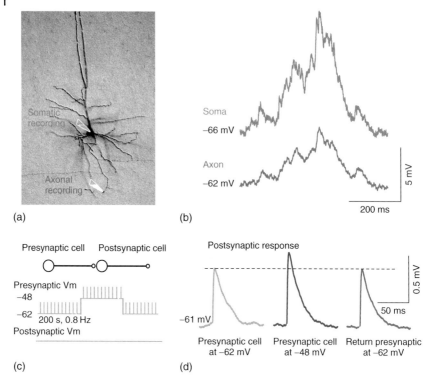

Fig. 6.2 Membrane potential fluctuations in the soma could propagate down the axon and modulate synaptic transmission. (a) Simultaneous whole-cell recording from the soma and axon cut end of a layer V pyramidal neuron. The morphology of the neuron was obtained by DAB staining after recording. The axonal recording site was 266 µm away from the soma. (b) During the generation of network activity, voltage fluctuations were recorded at both somata and axon terminal bleb. (c) Protocols for pair recording. Action potentials were initiated in the presynaptic cell at a rate of 0.8 Hz by injection of brief (1 ms) current pulses. The resting and depolarized periods were 100–200 s each. (d) Average EPSPs at the two presynaptic somatic membrane potentials. Note the facilitation induced by presynaptic somatic depolarization. (Reproduced from [12].)

between cortical pyramidal neurons not only through action potential-mediated digital communication but also through membrane potential-mediated analog communication [12]. Similarly, in some other central synapses, simultaneous recording from a presynaptic terminal and a postsynaptic cell revealed this hybrid form of synaptic transmission [32]. It was found previously that the mixture of digital and analog modes of communication occurs in some invertebrate synaptic connections and is characterized by a change in the amplitude of the synaptic responses evoked in the postsynaptic cell by a change in the membrane potential of the presynaptic neuron [33–37]. In these cases, it is believed that synaptic release sites are sufficiently electrotonically close to the presynaptic soma to be affected by changes in somatic membrane potential.

Since the huge depolarization associated with epileptic seizures could spread efficiently down the axon [12, 13]; therefore, we reasoned that this depolarization in the soma of a pyramidal neuron should modulate the synaptic transmission induced by the action potentials that reached the axon terminals during the burst. It is difficult, however, to probe the synaptic changes during epileptiform activities due to the large fluctuations in the membrane potential and membrane conductances in the postsynaptic cells.

6.10
Mechanisms Underlying EPSP Facilitation Induced by Somatic Depolarization

How could the depolarization of the axon terminals of pyramidal neurons cause enhancement of postsynaptic responses? One possible mechanism is that depolarization in the terminal may result in small increases in intracellular calcium concentration [38]. To test this possibility, we recorded from neurons using pipettes that contained the calcium chelators BAPTA at a concentration of $25\,\mu M$ and either $10\,mM$ EGTA or $1\,mM$ EGTA. Higher concentrations of EGTA significantly reduced the number of pairs that showed somatic depolarization-induced facilitation of EPSPs, indicating that Ca^{2+} is a mediator for the expression of EPSP facilitation.

An Increase in intracellular calcium concentration may result from the activation of calcium channels induced by subthreshold depolarization of the membrane potential. Patch-clamp recording from the terminal of Calyx of Held revealed that modest depolarization could activate certain calcium channels and consequently caused an increase in background Ca^{2+} concentration and resulted in the enhancement of postsynaptic responses [30].

Broadening of action potentials [36, 39] is another mechanism for an increase in the calcium concentration in the axonal terminal. Whole cell recording from soma and axon simultaneously demonstrated that somatic depolarization from -85 to $-45\,mV$ did not significantly change the width of action potential evoked in the soma by a brief current pulse, while the duration of the axonal action potential increased dramatically with somatic depolarization. Although the amplitude of the action potentials decreased, the integrated area under the axonal action potentials increased with the depolarization in the soma [12].

It has been shown that the cumulative inactivation of an A-type potassium current could prolong the action potentials in the presynaptic terminal and cause enhancement of the postsynaptic responses [31]. Axonal whole cell recording revealed that a selectively expressed potassium current was responsible for the broadening of axonal action potentials. This type of potassium current was sensitive to a low concentration of 4-AP and α-dendrotoxin. Modest depolarization of the axon could activate this current rapidly but inactivated very slowly over a time period of six seconds. The inactivation time period is consistent with that of the broadening of axonal action potentials and EPSP facilitation induced by presynaptic somatic depolarization [40]. Future experiments would focus on the contribution of axonal ion channels to the pathogenesis of epileptic seizures.

6.11
Summary

1. The cortical recurrent network activity is generated through a balance of excitation and inhibition.

2. Recurrent network activity was generated first in the layer V and propagated vertically to the superficial layers and horizontally across cortical columns.

3. Activation of synaptic elements by electrical stimulation could start and stop the recurrent network activity. Increased synchronization of the excitatory pyramidal neurons and increased excitability of the inhibitory interneurons may cause the network activity to stop.

4. Somatic membrane potential changes associated with the UP states and epileptiform activities could spread along the axon and reach the presynaptic terminals. Depolarization of the axon terminals could facilitate the postsynaptic responses induced by presynaptic action potentials.

5. Broadening of action potentials induced by somatic depolarization may cause more calcium influx in the axon terminal, and participate in modulating synaptic transmission.

6.12
Acknowledgments

I am grateful to Dr. David A. McCormick for his full support and encouragement and also to Andrea Hasenstaub, Alvaro Duque, Jing Yang, Yuguo Yu, and Bilal Haider for their contribution to these studies. The work was supported by the NIH, the 973 Program of China (2006CB806600) and the Shanghai Commission of Science and Technology Grant (06dj14010).

References

1 E. L. White, *Cortical circuits* (Boston: Birkhauser, **1989**).

2 M. Abeles, *Corticonics. Neural Circuits of the Cerebral Cortex* (Cambridge University Press, **1991**).

3 V. Braitenburg and A. Shuz, *Cortex: Statistics and Geometry of Neuronal Connectivity* (Springer, **1998**).

4 A. Peters and E. G. Jones, *Cerebral Cortex, Vol. 1. Cellular Components of the Cerebral Cortex* (**1984**).

5 A. Gupta, Y. Wang and H. Markram, *Science* **287**, 273 (**2000**).

6 A. Compte, M. V. Sanchez-Vives, D. A. McCormick and X.-J. Wang, *J. Neurophysiol.* **89**, 2707 (**2003**).

7 P. Marco, R. G. Sola, P. Pulido,
   M. T. Alijarde, A. Sanchez,
   S. R. y Cajal and J. DeFelipe,
   Brain 119 (Pt 4), 1327 (1996).

8 R. Cossart, C. Dinocourt, J. C.
   Hirsch, A. Merchan-Perez,
   J. DeFelipe, Y. Ben-Ari, M. Esclapez
   and C. Bernard, *Nat. Neurosci.* 4, 52
   (2001).

9 E. M. Powell, D. B. Campbell, G. D.
   Stanwood, C. Davis, J. L. Noebels
   and P. Levitt, *J. Neurosci.* 23, 622
   (2003).

10 I. Cobos, M. E. Calcagnotto, A. J.
   Vilaythong, M. T. Thwin, J. L.
   Noebels, S. C. Baraban and J. L.
   Rubenstein, *Nat. Neurosci.* 8, 1059
   (2005).

11 D. A. Lewis, T. Hashimoto and
   D. W. Volk, *Nat. Rev. Neurosci.* 6, 312
   (2005).

12 Y. Shu, A. Hasenstaub, A. Duque,
   Y. Yu and D. A. McCormick, *Nature*
   441, 761 (2006).

13 Y. Shu, A. Duque, Y. Yu, B. Haider
   and D. A. McCormick,
   J. Neurophysiol. 97, 746 (2007).

14 M. Steriade, A. Nunez and
   F. Amzica, *J. Neurosci.* 13, 3252
   (1993).

15 M. Steriade, I. Timofeev and
   F. Grenier, *J. Neurophysiol.* 85, 1969
   (2001).

16 R. Metherate and J. H. Ashe,
   J. Neurosci. 13, 5312 (1993).

17 M. Steriade, F. Amzica and
   A. Nunez, *J. Neurophysiol.* 70, 1385
   (1993).

18 M. Steriade, A. Nunez and
   F. Amzica, *J. Neurosci.* 13, 3266
   (1993).

19 I. Lampl, I. Reichova and D. Ferster,
   Neuron 22, 361 (1999).

20 M. V. Sanchez-Vives and D. A.
   McCormick, *Nat. Neurosci.* 3, 1027
   (2000).

21 Y. Shu, A. Hasenstaub, M. Badoual,
   T. Bal and D. A. McCormick,
   J. Neurosci. 23, 10388 (2003).

22 Y. Shu, A. Hasenstaub and D. A.
   McCormick, *Nature* 423, 288 (2003).

23 F. Amzica and M. Steriade,
   J. Neurophysiol. 73, 20 (1995).

24 E. A. Stern, A. E. Kincaid and C. J.
   Wilson, *J. Neurophysiol.* 77, 1697
   (1997).

25 D. Johnston and T. H. Brown, *Science*
   211, 294 (1981).

26 J. J. Hablitz, *J. Neurophysiol.* 58, 1052
   (1987).

27 R. D. Chervin, P. A. Pierce and B. W.
   Connors, *J. Neurophysiol.* 60, 1695
   (1988).

28 M. L. Hines and N. T. Carnevale,
   Neural Comput. 9, 1179 (1997).

29 A. L. Hodgkin and A. F. Huxley,
   J. Physiol. 117, 500 (1952).

30 G. B. Awatramani, G. D. Price and
   L. O. Trussell, *Neuron* 48, 109 (2005).

31 J. R. Geiger and P. Jonas, *Neuron* 28,
   927 (2000).

32 H. Alle and J. R. Geiger, *Science* 311,
   1290 (2006).

33 T. Shimahara and L. Tauc, *J. Physiol.*
   247, 299 (1975).

34 J. Nicholls and B. G. Wallace,
   J. Physiol. 281, 171 (1978).

35 J. Nicholls and B. G. Wallace,
   J. Physiol. 281, 157 (1978).

36 E. Shapiro, V. F. Castellucci
   and E. R. Kandel, *Proc. Natl.
   Acad. Sci. U S A* 77, 629 (1980).

37 P. Meyrand, J. M. Weimann and
   E. Marder, *J. Neurosci.* 12, 2803
   (1992).

38 A. I. Ivanov and R. L. Calabrese,
   J. Neurosci. 23, 1206 (2003).

39 J. H. Byrne and E. R. Kandel,
   J. Neurosci. 16, 425 (1996).

40 Y. Shu, Y. Yu, J. Yang and D. A.
   McCormick, *Proc. Natl. Acad.
   Sci. U S A* 104, 11453 (2007).

7

Epilepsy as a Disease of the Dynamics of Neuronal Networks – Models and Predictions

Fernando Lopes da Silva

7.1
Introduction

In this chapter I consider some general principles of how the transition to an epileptic seizure (ictus) takes place from a relatively normal state (interictal) of brain activity. It is well known that epileptic patients do not seize all the time, so that there are periods where the patient is free of seizures, which fortunately are very long, as a rule, and relative short episodes where seizures occur. Therefore a theory that has the objective of explaining why and how seizures occur should also explain why the same brain can demonstrate long periods without seizures.

The main question that is the focus of the present discussion is whether precursors of an ictal manifestation, that may be considered to be characteristic of a *proictal state* may be identified reliably. We prefer to introduce the term proictal rather than to use the more common term preictal; indeed preictal denotes just the activity immediately preceding the seizure, which is, of course, always present, whether or not it may be causally related to the transition from an apparently normal EEG signal to a seizure; proictal implies that the activity reflects a state of excitability of the underlying neuronal networks that may lead to a seizure.

A discussion about general principles of brain dynamics is relevant in this context, since it is difficult to develop and interpret EEG data obtained in epileptic patients and very often analyzed by sophisticated analytical methods, without having a general theoretical framework of the dynamics of brain activity that may account both for normal and epileptic neuronal activities. In this context, we assume that in the epileptic brain, some neuronal networks can display different kinds of dynamical states because they possess an abnormal set of control parameters. In other words, they may have bi(multi) stable properties. This means that, in addition to a normal steady state, they also have an abnormal one characterized by widespread synchronous activity, and that the transition between these two states may occur more or less abruptly.

This accounts for the two main characteristics of epilepsy: (a) that an epileptic brain can function apparently normally between seizures (i.e., during the interictal state); and (b) that the seizures occur in a paroxysmal way, thereby impairing brain

Seizure Prediction in Epilepsy. Edited by Björn Schelter, Jens Timmer and Andreas Schulze-Bonhage
Copyright © 2008 WILEY-VCH Verlag GmbH & Co. KGaA, Weinheim
ISBN: 978-3-527-40756-9

functioning to a lesser or greater extent. In this sense, epileptic disorders may be considered especial cases of the large class of dynamical diseases, meaning those pathophysiologic states characterized by the occurrence of abnormal dynamics, a theoretical concept proposed by [1] that we [2–4], and others [5], have used in the context of epilepsy.

The theory of nonlinear dynamics offers the possibility of understanding, in formal terms, how the occurrence of the manifestations of dynamical diseases takes place. In the case of epilepsy, the basic question is how changes in the dynamics of a neuronal network may occur, such that paroxysmal widespread synchronous oscillations abruptly emerge. It is often difficult to study these changes in the human brain due to the inherent limitations of obtaining detailed neuronal data from recordings in the human brain. Therefore, it is important to investigate these phenomena in experimental animals that provide reliable models of some types of epilepsy. Thus, in this chapter, we focus also on some experimental observations obtained in epileptic animals, that shed light on the main question of how dynamical changes leading to seizures can occur in well defined neuronal networks *in vivo*.

We consider two cases which are paradigmatic for two different kinds of routes to seizures:

- Case 1: The occurrence of Spike-and-Wave discharges in the thalamo-cortical system, typical of absence seizures.

- Case 2: The occurrence of seizure activity in the hippocampus.

7.2
Experimental Observations – Case 1: The WAG/Rij Rat as a Genetic Model for Absence Epilepsy

The WAG/Rij rat model is a well established animal model of absence epilepsy [6]. Similar to the EEG characteristics of human absences, Spike and Wave Discharges (SWDs) occur in the EEG of the WAG/Rij rat as the animal displays freezing behavior for a few seconds, usually accompanied by vibrissal twitches [7].

Typical SWDs start and end abruptly. We found [8, 9] that SWDs are initiated in the facial somatosensory cortex and propagate to other cortical areas (Figure 7.1) and to the thalamus. This was confirmed in another experimental rat model of absences: the Genetic Absence Epilepsy Rats of Strasbourg (GAERS) where [10] it was found that the display of SWDs in the local ElectroCorticoGram (ECoG) coincides with rhythmic membrane depolarizations superimposed on a tonic hyperpolarization of layer IV neurons, that occurs abruptly (Figure 7.2). Furthermore it was demonstrated in a detailed microelectrophysiolgical study [11] also in GAERS that layer VI corticothalamic neurons act as 'drivers' in the generation of SWDs in the somatosensory thalamocortical system. These observations reveal also that SWDs occur abruptly in this animal model.

A general conclusion that may be drawn from these experimental studies is that the SWDs, characteristic of absence seizures, tend to occur without

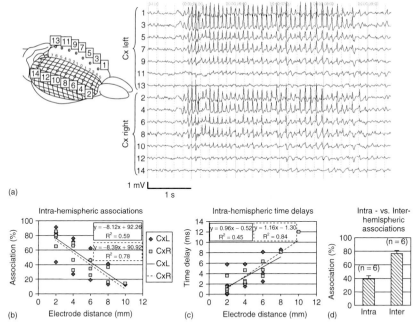

Fig. 7.1 Cortico–Cortical Associations: bilaterally symmetric sites. Recordings from the cortex of a WAG/Rij rat during an absence seizure showing (a) an episode of Spike-and-Wave Discharges (SWDs). Plots showing (b) the intra-hemispheric nonlinear associations and (c) corresponding time delays. In (d) it is shown that the average of intra-hemispheric associations (average of 6 rats) is smaller than the average interhemispheric association between homologous cortical sites. (Adapted from [8].)

a clear proictal electrophysiological ECoGraphic pattern. In order to understand in more general terms how this abrupt change of neuronal dynamics takes place a computational model has been constructed and proved to be useful.

7.3
Computational Model of the Thalamo–Cortical Neuronal Networks

With this objective we constructed a computational model [12] of absence epilepsy that included features of a distributed neuronal network with respect to different synaptic currents and membrane properties (for example Ca^{2+} channels, burst firing mechanisms) and which was designed as a circuit consisting of interconnected neuronal populations. That is, we did not simulate the explicit behavior of individual neurons but rather modeled the activity of populations of interacting neurons integrating neuronal and network properties. In this way we assumed that such a computational model can take into account the most relevant (patho)physiological experimental findings to simulate the dynamics of brain activity as transitions to absence seizures occur.

B

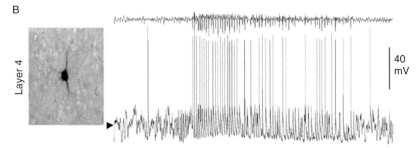

Fig. 7.2 Interictal–ictal transition. Recording in a GAER rat of a SWD in the local ECoG and the corresponding intracellular recording of a layer IV neuron. Note that the two phenomena coincide. The latter displays rhythmic membrane depolarizations superimposed on a tonic hyperpolarization of this layer IV neuron. (Adapted from [10].)

A number of detailed, distributed models of thalamic and thalamo–cortical networks were recently developed (e.g., [13–16]). These models have given insight into some basic neuronal mechanisms of SW discharges, but do not address specifically the most essential issue of this type of epileptic activity: that a given thalamocortical loop can display both kinds of activity – normal brain activity and epileptic seizures – without specific adjustment of parameters being expressly made.

Thus, the main aim of the computational model study was to find out the mechanisms responsible for transitions from normal activity to paroxysmal SWDs. Without entering into detail (see [12, 17, 18] for details of the model studies) we may state that the model studies (Figure 7.3) revealed that: (i) SWDs result from dynamical bifurcations that occur in a bistable neuronal network; (ii) the durations of paroxysmal and normal epochs have exponential distributions, indicating that transitions between these two stable states occur randomly over time with constant probabilities; (iii) the probabilistic nature of the onset of paroxysmal activity implies that it is not possible to predict its occurrence; (iv) the bistable nature of the dynamical system allows that an ictal state may be aborted by a single counter-stimulus.

7.4
Model Predictions

The main prediction of the model is that the transition from normal brain activity to an absence seizure characterized by SWDs occurs randomly (Poisson process); this implies that the distribution of the intervals between seizures should be exponential. This prediction was tested by calculating the distributions of durations of seizures and of inter-paroxysm intervals as described by [17]. The prediction could be confirmed both in the genetic rat models and in some human cases, by the fact that these distributions in many cases are exponential (Figure 7.4), although not in all cases. This led us to perform a more extensive investigation of these

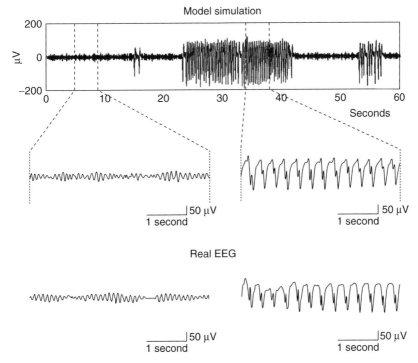

Fig. 7.3 Example of a bifurcation between two states: 'normal' and 'seizure' (absence type), both in the computational model and in EEG real signals.

distributions in several conditions, both in human and animal models, and to apply a more general statistical model to describe the distributions. This model can be described by a gamma distribution:

$$y = C x^{\alpha-1} e^{-x/\beta},$$

where α and β are distribution parameters and C is a normalization constant. Gamma distributions are flexible in terms of their overall shape. The shape is determined by the shape parameter, α; for $\alpha < 1$, the distribution has the maximum at the origin and is monotonically decreasing, for $\alpha = 1$ the distribution has an exponential shape, and for $\alpha > 1$, the distribution has zero at the origin and maximum at non-zero values. As derived analytically an exponential distribution of intervals between events corresponds to a Poisson process, in which events occur along time with constant probability of occurrence. In [17] the results of this investigation are described in detail, but here we summarize the main conclusions as follows: (1) the distribution of shape parameter α, can be close to one in both ictal and interictal recordings as already indicated above; (2) the dynamical processes during ictal epochs are, in general, different from those during the interictal states; (3) a majority of ictal epochs have an α parameter larger than one. This suggests

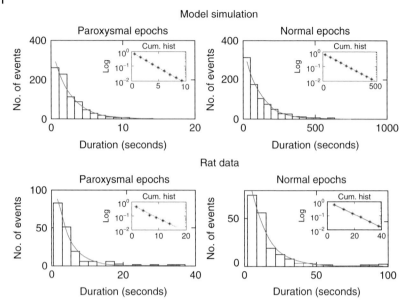

Fig. 7.4 Distributions of epochs duration. Distributions of the durations of ictal and interictal epochs of the data obtained from the computational model and from WAG/Rij rats (under treatment with Vigabatrin).

that deterministic time-dependent mechanisms are involved in seizure termination and the probability of terminating an ictal state increases with time spent in that state. On the contrary, interictal epochs are described predominantly with an α parameter smaller than one. This suggests that the longer the system remains in a seizure-free state, the higher the chance it shall remain seizure-free in the immediate future. This kind of dynamics results in a grouping of seizures, i.e., in the appearance of clusters of ictal episodes separated by long interictal periods. Seizure clustering has been reported in other studies. In [19] it was reported that, in about half of epileptic patients with different seizure types, the occurrence of seizures was indistinguishable from a Poisson process, while other patients showed seizure clustering.

Thus the studies using the computational model led to the conclusion that deterministic time-dependent mechanisms have to be assumed, with respect to seizure termination. This implies that we have to consider which neuronal parameters may change as a seizure progresses. We may hypothesize that candidates for such a 'use-dependent parameter' might be the extracellular accumulation of K^+ in glial cells affecting the excitability of neurons, an increase of intra-cellular Ca^{2+} leading to an increase of cAMP that can cause an increase of the Ih current, or use-dependent changes in the dynamics of GABA receptors, or a combination of all these factors.

In conclusion, the model has not only showed that the transitions to absence seizures can occur randomly, but also that additional physiological mechanisms have to be tested to explain some aspects of specific statistical distributions, particularly of the durations of seizures.

7.5
Experimental Observations – Case 2: Hippocampal Seizures

A recent *in vitro* study in hippocampal slices revealed some interesting findings that shed light on the question that we are dealing with here; namely, whether or not in this kind of epileptic seizures a proictal state can be identified, in contrast to what was observed with respect to absence seizures (Figures 7.2, 7.3 and 7.4). The study of [20], showed that, before a full-blown seizure may be seen, there is a gradual change in electrophysiological properties that becomes manifest as a build-up of the neuronal firing frequency variance index within the affected neuronal population (Figure 7.5). This implies that a proictal state with special

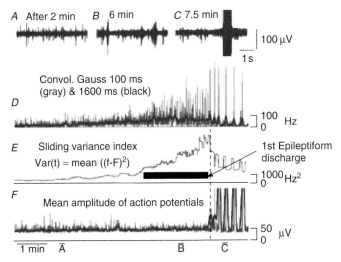

Fig. 7.5 Disinhibition-induced synchronization of CA3 population firing (perfusion with 10 μm bicuculline). Recordings obtained in hippocampal slices (CA3 sub-area) where an in vitro 'seizure' was induced by perfusion with 10 μm bicuculline. A–C, multi-unit activity recorded by an electrode in the CA3 stratum pyramidale at 2, 6 and 7.5 mins after perfusion. Note the disinhibition-induced synchronization of CA3 neuronal population firing. D–F, evolution of indices of excitability over the period of transition to the 'seizure.' D, frequency of all detected spikes convolved with Gaussian functions of 100 ms (black) and 1600 ms (gray) reveal fast and slow variations during the transition to synchrony. Slow variations reach a plateau frequency before the first epileptiform discharge (vertical dotted line). Fast oscillations increase in amplitude before the onset of fully synchronous firing. E, a sliding variance index reveals a progressive increase in coherence of fast fluctuations around the mean frequency. F, the mean amplitude of detected action potentials increased only after the onset of epileptiform activity. This implies that spike superpositions within 1–2 ms do not occur during the transition period. Black and gray traces correspond to smoothing Gaussians of 100 and 1600 ms, respectively. (Adapted from [20].)

properties can be identified. Thus these observations show that the evolution of the excitability state of the neuronal population in these hippocampal slices differs sharply from what was seen in the case of SWDs in absence epilepsy. Assuming that these *in vitro* observations are representative of the *in vivo* situation in patients with mesial temporal lobe epilepsy, and taking these findings together, we may draw the following conclusions.

Theoretically, we may consider that the transition to an epileptic seizure can occur basically according to two models:

- Bi- (or multi-stable) systems where jumps between two or more pre-existing attractors can take place, caused by stochastic fluctuations (noise) of any input – Case 1.

- Parametric alteration, or deformation, that may be caused by an internal change of conditions or an external stimulus (sensory in reflex epilepsies) – Case 2.

The main question in cases of the second type is how to detect the special properties of the proictal state. Many analytical methods have been proposed as presented and discussed in several chapters of this book and elsewhere (see [21,22]). Here I will consider only those methods that use a probe – i.e., a given stimulation protocol – in order to estimate changes in the excitability state of the neuronal networks that may be characteristic of this proictal state.

7.6
Active Observation: Stimulation with 'Carrier Frequency' – Changes in Phase Clustering Index (PCI)

We found previously that in cases of photosensitive epilepsy where epileptic patients were stimulated with intermittent light stimuli [23], the relative Phase Clustering Index (rPCI, Figures 7.6 and 7.7) in the gamma frequency band

$$Z_f^k$$

Phase clustering index (PCI)

$$|< Z_f^k >_k| = K_f <| Z_f^k |>_k \equiv K_f A_f$$

$$K_f = \frac{\left| < Z_f^k >_k \right|}{<| Z_f^k |>_k}$$

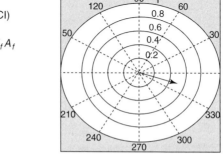

Fig. 7.6 Definition of phase clustering index (PCI) K_f. (For details see [24].)

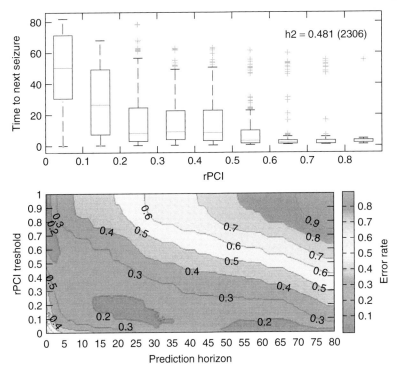

Fig. 7.7 rPCI as function of time preceding a seizure. Illustration (averages of six data from six patients shown in box plots) of the time-course of the rPCI (horizontal axis) in relation to the time to the next seizure (vertical axis, hours). The lower panel is a pseudo-color and contour plot of the prediction error rates, with respect to predicting a seizure within a certain time, as a function of the rPCI threshold (vertical axis) and the prediction horizon (in hours, horizontal axis). Note, for example, that for values of rPCI > 0.6 it may be predicted with 80 % accuracy that a seizure will occur in less than 2 h. (Adapted from [25].) (Please find a color version of this figure on the color plates.)

(see [24] for technical details) was significantly increased some seconds before the seizure occurred. This finding led us to investigate whether the rPCI of EEG signals in patients with mesial temporal lobe epilepsy could also have a predictive value using an active stimulation paradigm. With this objective we stimulated patients with indwelling electrodes in the hippocampal formation using short bursts of electrical pulses, five seconds long, at a frequency of 20 Hz, repeated at intervals of 20 seconds over long periods of hours, and even days. We may call this stimulation paradigm a carrier frequency modulation probe [25]. The relative phase clustering index (rPCI) computed for all signals and all stimulated epochs of six patients, was found to be larger for electrode sites near to the seizure onset site. Even more interesting was the finding that it was possible to forecast the probability of a seizure occurring within a certain time based on the values of rPCI estimated at a given moment in time. For

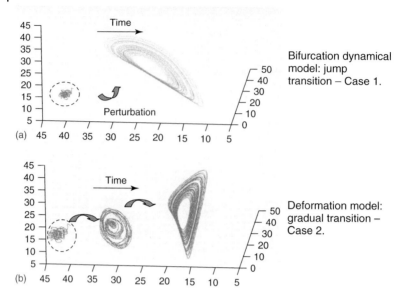

Fig. 7.8 Two main classes of model that may explain the transition to epilepsy. (a) The bifurcation dynamical model showing a jump transition – Case 1. (b) The deformation model showing a gradual transition – Case 2. Note that the attractors are virtual, they are designed only for illustration.

example, in the patient population analyzed in this study, a value of rPCI > 0.6 predicts that a seizure will occur in less than 2 h, with an accuracy >80 % (Figure 7.7).

7.7
Conclusion

We should consider there to be two main classes of models (Figure 7.8) which may explain the transition to epileptic seizures:

- The bifurcation dynamical model: jump transition – Case 1; in this case it does not appear that the transitions to epileptic seizures (absences) can be predicted.

- The deformation model: gradual transition preceding the epileptic seizure, i.e., a proictal state can be identified – Case 2; in this case a transition to a seizure may be predicted on condition that the proictal state can be appropriately detected. This detection may be based on using analytical techniques applied to the on-going EEG (passive approach) or it may be based on measurements made by way in an active approach, i.e., using a stimulation paradigm to probe changes in the excitability of the neuronal populations characteristic of the proictal state.

References

1 L. Glass, M. C. Mackey, *The Rhythms of Life*, (Princeton University Press, New Jersey, **1988**).

2 F. H. Lopes da Silva, J. P. M. Pijn, J. A. Gorter, E. van Vliet, E., E. W. Daalman and W. Blanes, *Rhythms of the Brain: Between Randomness and Determinism*, In *Chaos in Brain*, Eds: K. Lehnertz, J. Arhold, P. Grassberger, C. E. Elger (World Scientific, Singapore, **2000**).

3 F. H. Lopes da Silva, W. Blanes, S. N. Kalitzin, J. Parra, P. Suffczynski and D. N. Velis, *Epilepsia* **44**(s12), 72–83 (**2003**).

4 F. H. Lopes da Silva, W. Blanes, S. N. Kalitzin, J. Parra, P. Suffczynski and D. N. Velis, *IEEE Trans. Biomed. Eng.* **50**(5), 540–8 (**2003**).

5 R. G. Andrzejak, G. Widman and K. Lehnertz, *et al. Epilepsy Res.* **44**, 129–40 (**2001**).

6 A. M. Coenen, W. H. Drinkenburg, M. Inoue and E. L. van Luijtelaar, *Epilepsy Res.* **12**, 74–86 (**1992**).

7 E. L. van Luijtelaar and A. M. Coenen, *Neurosci. Lett.* **70**, 393–7 (**1986**).

8 H. K. Meeren, J. P. Pijn, E. L. van Luijtelaar, A. M. Coenen and F. H. Lopes da Silva, *J. Neurosci.* **2**(4), 1480–95 (**2002**).

9 H. K. Meeren, E. L. van Luijtelaar, F. H. Lopes da Silva and A. M. Coenen, *Arch. Neurol.* **62**(3), 371–6 (**2005**).

10 P. O. Polack, I. Guillemain, E. Hu, C. Deransart, A. Depaulis and S. Charpier, *J. Neurosci.* **27**(24), 6590–9 (**2007**).

11 D. Pinault and T. OÕBrien, *Thalamus & Related Systems* 1/23 (**2007**).

12 P. Suffczynski, S. Kalitzin and F. H. Lopes Da Silva, *Neuroscience* **126**(2), 467–84 (**2004**).

13 X.-J. Wang, D. Golomb and J. Rinzel, *PNAS* **92**, 5577–81 (**1995**).

14 W. W. Lytton, D. Contreras, A. Destexhe and M. Steriade, *J. Neurophysiol.* **77**, 1676–96 (**1997**).

15 A. Destexhe, *J. Neurosci.* **18**, 9099–111 (**1998**).

16 A. Destexhe, *Eur. J. Neurosci.* **11**, 2175–81 (**1999**).

17 P. Suffczynski, F. H. Lopes da Silva, J. Parra, D. N. Velis, B. M. Bouwman, C. M. van Rijn, P. van Hese, P. Boon, H. Khosravani, M. Derchansky, P. Carlen and S. Kalitzin, *IEEE Trans. Biomed. Eng.* **53**(3), 524–32 (**2006**).

18 B. M. Bouwman, P. Suffczynski, F. H. Lopes da Silva, E. Maris and C. M. van Rijn, *Eur. J. Neurosci.* **25**(9), 2783–90 (**2007**).

19 J. G. Milton, J. Gotman, G. M. Remillard and F. Andermann, *Epilepsia*, **28**, 471–8 (**1987**).

20 I. Cohen, G. Huberfeld and R. Miles, *J. Physiol.* **570**(3), 583–94 (**2006**).

21 M. Le Van Quyen, *C. R. Biol.* **328**(2), 187–98 (**2005**).

22 F. Mormann, R. G. Andrzejak, C. E. Elger and K. Lehnertz, *Brain* **130**(Pt 2), 314–33 (**2007**).

23 J. Parra, S. N. Kalitzin, J. Iriarte, W. Blanes, D. Velis and F. H. Lopes da Silva, *Brain* **126**(Pt 5), 1164–72 (**2003**).

24 S. N. Kalitzin, J. Parra, D. Velis and F. H. Lopes da Silva, *IEEE Trans. Biomedl Eng.* **49**(11), 1279–85 (**2002**).

25 S. N. Kalitzin, D. Velis, P. Suffczynski, J. Parra and F. H. Lopes da Silva, *Clin. Neurophysiol.* **116**(3), 718–28 (**2005**).

8

Neuronal Synchronization and the 'Ictio-centric' vs the Network Theory for Ictiogenesis: Mechanistic and Therapeutic Implications for Clinical Epileptology

Ivan Osorio, Mark G. Frei, Ying-Cheng Lai

8.1
Seizures and Neuronal Synchronization: Increased or Decreased Relative to Interictal Values?

Automated contingent ('closed-loop') seizure blockage using electrical stimulation (ES), an emerging and promising therapy for pharmaco-resistant epilepsies in humans [1,2], would benefit from development of mechanisms-based approaches. Its efficacy for pharmaco-resistant epilepsy is critically dependent upon knowledge of the mechanisms and the dynamics of seizure generation and termination. Currently, the choice of stimulation parameters, such as frequency, is largely based on a limited animal literature. Temporal evolution of spectral (decreases in frequency as the seizure progresses) and other signal features are ignored and possibly more importantly, neuronal phase information and, through it, the prevailing level of neuronal synchronization during seizures. Given that changes (increases or decreases) in neuronal synchronization levels appear to be causally related to spontaneous onset and termination of seizures [3–5], real-time quantification of their spatio-temporal evolution with reference to their interictal values may be valuable for selection, and if necessary adaptation of ES parameters, to optimize therapeutic efficacy. Investigation in humans of signal features correlated with seizure blockage and non-blockage, in response to local or remote ES of the epileptogenic zone(s) [1], would advance this therapeutic modality. This may be accomplished by performing statistical comparisons of certain signal features extracted from blocked and non-blocked seizures, using novel analysis techniques.

While recent findings [4–7] challenge the widely accepted notion that seizures are the expression of neuronal hypersynchrony, it is likely that departures in the level of neuronal synchronization from values underlying optimal brain function may manifest as epileptic seizures. Several additional lines of evidence do not support neuronal hypersynchrony as the only network mechanism underlying the generation of generalized, spike-slow wave, or partial ('focal') seizures:

1) Multi-site, simultaneous EEG, extracellular, and intracellular recordings from various neocortical and thalamic nuclei of penicillin-treated cats, revealed that what is recorded from the scalp as sudden, generalized and bilaterally synchronous spike-wave bursts, is the product of neuronal activity with time lags between their spike trains and changes in their temporal relations [8].

2) Dual-cell patch-clamp of CA1 pyramidal neurons in hippocampal slices revealed that, during seizure-like events, the degree of inter-neuronal synchronization decreased compared with interictal periods; moreover it was observed that synchronization increased as seizures neared their end [4]. Similar observations have been made in humans [5] and in a model of focal neocortical epilepsy (Figure 8.1).

3) Application by our research group of a novel measure of synchronization to ECoG from humans with intractable epilepsy showed increases during the ictus in certain cases and decreases in others [5–7] which underscores the importance of taking the synchronization level into account in the development and implementation of efficacious therapeutic strategies. For the case in hand, if the mechanism of spontaneous termination of certain seizures in certain regions is increased phase synchrony (Figure 8.1), interventions that mimic this putative mechanism are likely to reproduce that effect. Thus, in this case, 'synchronizing' influences [9] in the form of, let us say, low-frequency, low-intensity ES delivered to the site of origin, may be more likely to have a beneficial effect than those causing 'desynchronization' as they are likely to prolong the state of low neuronal phase synchrony that sustains the seizure [4, 10]. Parameters that induce activation ('desynchronization') [1, 11, 12] would be used if phase synchronization were increased, relative to its interictal value, as they have proved to be efficacious in blocking

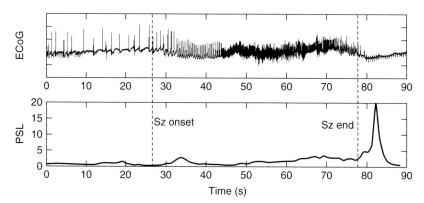

Fig. 8.1 Temporal evolution of PSL values (y-axis; only one channel shown) of a seizure induced with topical penicillin applied to rabbit cortex. PSL increases briefly and minimally shortly after onset (32–34 s), remaining at low levels until approximately 62 s, when it begins increasing slightly and gradually as the seizure nears its end; a marked and rapid increase occurs at the ictal–postictal transition. Similar changes were observed in human seizures.

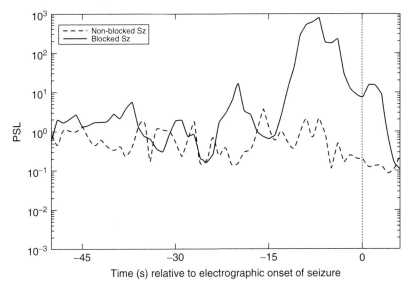

Fig. 8.2 PSL values (y-axis; logarithmic scale) of two seizures originating from the same subject/site (similar morphologies, amplitudes and frequencies), treated with electric stimulation shortly (≈ 4 s) after onset (vertical dotted line). One seizure (solid) responds and the other (dashed) does not respond to ES. Notice the large increase (almost three orders of magnitude) in PSL between 'early' interictal and 'late' (-15 s) interictal values in the blocked seizure, which were not observed in the non-blocked. These results reveal that PSL values of seizures originating from the same subject/site may or may not change relative to interictal values (ictal decreases in PSL also may occur but are not shown in this example) and that the response to ES may depend on signal features such as PSL which cannot be extracted from power spectral measures. These findings also point to the importance of tracking (in real-time) this measure and of tailoring ES parameters accordingly to optimize therapy. The lack of effect on one of the seizures may be related to the fact that delivery of 'desynchronizing' pulses (as was the case for both seizures) have no effect on seizures with low synchronization levels, unlike in those with high levels as they can be 'brought down' to interictal or postictal values.

epileptiform activity and seizures [1,13–15] (see Figure 8.2). Therapies based, among other factors, on level of neuronal synchronization would take into account that the probability of seizure occurrence depends on the state and degree of cortico-thalamic or intra-limbic interactions which manifest electrographically roughly as either synchronization or, activation ('desynchronization') [16–18]. The ability to track, in real time, the spatio-temporal behavior of measures such as Phase Synchronization Level (PSL) [7] during seizures and, based on the degree and direction (increased or decreased compared to interictal values) automatically select and if necessary, adapt ES parameters to optimize efficacy, may advance ES as a therapy. The fulfilment of this objective will lay the foundations for a 'mechanisms-based' approach for the treatment of seizures.

The recent availability of means for real-time automated quantitative seizure detection [2, 19, 20] and of implantable devices in humans [2] for contingent delivery of ES to epileptogenic tissue along with the preliminary evidence of safety, tolerability and efficacy in humans with pharmaco-resistant epilepsies, [1, 2] provides further impetus for investigating, how ES alters neuronal phase synchronization and in turn how these changes correlate with seizure blockage or with lack of a beneficial effect. This may be accomplished by investigating, in humans with pharmaco-resistant epilepsies, the existence of significant correlations between PSL, at the scale (macrocolumns) recorded using depth or subdural electrodes and the type of response (blockage vs non-blockage) to ES.

The Phase Synchronization Level [6] quantifies, from multi-channel time series, weak types of synchrony such as transient phase synchronization that are prevalent (over global or complete synchronization) in complex, non-stationary systems with high intrinsic (dynamical) noise levels, such as the human brain. PSL was selected over other existing measures of synchrony [21–23] because it is robust and sensitive, may be used in real-time (an important consideration for therapeutic applications with implantable devices) and having been developed by the proponents of this research, its strengths and limitations are well known to us.

Estimates of PSL in human seizures and of its correlation, if any, with the therapeutic response, will also increase understanding of seizure dynamics in the context of synchronization, shedding light into an area of great importance. Questions such as: When is the delivery of ES during a seizure most likely to block it? Is the probability of seizure blockage conditioned on the ictal PSL level relative to interictal values? Should 'desynchronizing' or 'synchronizing' pulses be used in a negative or positive feedback control loop? That is, should relatively 'desynchronized' seizures be treated with 'synchronizing' or 'desynchronizing' pulses?, may be properly and systematically investigated. This approach may: a) shed light on the relation between PSL and ictiogenesis on one hand and PSL and the probability of seizure blockage or of adverse effects with ES, on the other; and b) provide valuable information to improve timing, site of delivery of electrical currents and of the other parameters such as frequency and intensity of stimulation for increasing efficacy and decreasing adverse effects of ES. It should be mentioned that interpretation of measures of synchronization is confounded by volume conduction and the reference electrode [24].

8.2
The 'Focus' ('Ictio-centric') vs the Network Theory in Ictiogenesis

Historically in clinical epileptology, ictiogenesis has been anatomically and functionally restricted to the so-called 'focus' or epileptogenic zone, defined as that from where seizures originate and electrographic onset precedes or occurs simultaneously with clinical manifestations. The concept of an epileptogenic 'focus' (or 'foci') connotes that ictiogenesis is largely, if not exclusively, dependent on mechanisms or dynamics inherent to the neuronal assemblies that make up the 'focus,' as if independent of, or not susceptible to, intra- or inter-regional (global) dynamical influences. The 'seizure focus' theory that we label 'ictio-centric' ignores that the so-called 'focus' remains anatomically and functionally connected, albeit probably aberrantly or incompletely, to 'non-intrinsically epileptogenic' regions and ignores dynamical links and potential interactions with other regions, that have been identified in experimental animal models of seizures and epilepsy [25–30] and also probably in humans [1, 31, 32]. By way of example, the interactions between epileptogenic and non-epileptogenic sub-regions were investigated on 5 s windows of ECoG (Figure 8.3(b)) recorded with a 4 × 8 electrode grid (Figure 8.3(a)) placed on the dorso-lateral right frontal lobe region of a subject with pharmaco-resistant epilepsy undergoing invasive monitoring for surgical evaluation. High frequency low voltage seizure activity (waxing and waning) is seen on contacts 2, 10 and 18 and coexists with quasi-periodic epileptiform discharges (contacts 12 and 20) and slow activity (contacts 13, 21, 22, etc.). The phase synchronization level as depicted in Figure 8.3(b), represents the weighted level of synchronization of the signal recorded from each contact/sub-region within the grid and for purposes of visualization was assigned a color (red: highest; dark blue: lowest).

The phase synchronization level between each pair of electrodes was also assigned a color (red: highest; dark blue: lowest) and line width (thick: highest

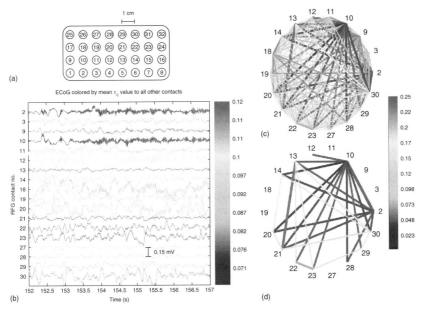

Fig. 8.3 (a) Grid used in recording ECoG signals. (b) Referential ECoG during a 5 s epoch colored according to mean phase synchronization level between the respective channel and all other intact contacts. (c) Graphical depiction of the bivariate phase synchronization measure between all pairs of channels during this epoch, with color and linewidths corresponding to phase synchronization level. (d) Illustration of the most significant pair-wise synchronization within the epoch. (Please find a color version of this figure on the color plates.)

and lowest levels; thin: moderate levels). The results (Figure 8.3(c)) show, at a glance, the temporo-spatial (only one frame is displayed) dynamical state of phase synchronization level of this region. The intricate structure of the graphic, mirrors the complex and varied interactions within a region where only certain sub-regions are epileptogenic: subregions with high and low phase synchronization relative to their surroundings coexist in time. To further facilitate visualization, intermediate synchronization values were removed leaving only the highest and lowest (Figure 8.3(d)). Two sets of contacts/regions adjacent (≈ 0.5 cm) to each other (2 & 10; 13 & 21), but separated (2 & 10 vs 13 & 21) by almost 3 cm (Figure 8.3) have, independently and simultaneously, high levels of phase synchronization; (i) the high phase synchrony between 2 & 10 is expected given the ECoG activity in those contacts, but 'unexpected' between 13 & 21; (ii) sub-regions 2 & 10 have, during the ictus, little 'functional connectivity' (blue lines or no lines) to all other parts in this region; (iii) sub-regions 13 & 21 have stronger 'functional connectivity' to other parts; and (iv) sub-region 18 which is adjacent to sub-region 10 and is also generating ictal activity, has only intermediate phase synchronization with 2 & 10. These graphics provides clues for systematically investigating and implementing in real-time therapeutic strategies/parameters. For example, one rational strategy in this case would be to deliver 'desynchronizing' ES to areas 2 & 10 and/or 13 & 21 and another to increase regional phase synchronization through delivery of appropriate ES parameters to parts of this region that are 'functionally disconnected' or 'weakly' connected to others.

The scientific and therapeutic implications of testing the ictio-centric vs the spatially distributed theory of ictiogenesis would have far-reaching implications. Prediction of seizures, an objective as valuable as it is elusive, may be feasible if signal monitoring and analyses are not limited in humans to the epileptogenic zone as conventionally defined, and for efficacy, contingent or non-contingent therapy is delivered not only to the epileptogenic zone but also to other regions in the network.

References

1 I. Osorio, M. G. Frei, S. Sunderam, J. Giftakis, N. C. Bhavaraju, S. F. Schaffner and S. B. Wilkinson, Ann. Neurol. 57(2), 258–68 (2005).

2 G. L. Barkley, B. Smith, G. Bergey, G. Worrell, J. Drazkowski, D. Labar, R. Duchrow, A. Murro, M. Smith, R. Gwinn, L. Fish, L. Hirsch and M. Morrell, Neurology (Suppl 2) A387 (2006).

3 E. R. Kandel, J. H. Schwartz and T. M. Jessell, Principles of Neural Science, 3rd Edition (Appleton and Lange, Norwalk, CT, 1991).

4 T. I. Netoff and S. J. Schiff, J. Neuroscience 22, 7297–307 (2002).

5 K. Schindler, C. E. Elger, K. Lehnertz, Clin. Neurophysiol. 118, 1955–68 (2007).

6 Y.-C. Lai, M. G. Frei and I. Osorio, Physical Review E 73, 026214 (2006).

7 Y.-C. Lai, M. G. Frei and I. Osorio, L. Huang, Phys. Rev. Lett. 98, 108102(1–4) (2007).

8 M. Steriade, D. Contreras and F. Amzica, Trend. Neurosci. 17, 199–208 (1994).

9 M. Monnier, L. Hosli and P. Krupp, *Electroencephalogr. Clin. Neurophysiol.* **24**, 97–112 (**1963**).

10 M. Steriade, *Arch. Ital. Biol.* **134**, 5–20 (**1995**).

11 P. A. Tass, *Biol. Cybern.* **87**, 102–15 (**2002**).

12 P. A. Tass and M. Majtanik, *Biol. Cybern.* **94**, 58–66 (**2006**).

13 A. Kreindler, E. Crighel and M. Steriade, *Prog. Brain Res.* **22**, 286–96 (**1968**).

14 R. Wagner, D. M. Feeny, F. P. Gullotta and I. L. Cote, *Electroencephalogr. Clin. Neurophysiol.* **39**, 499–506 (**1975**).

15 D. M. Durand and E. M. Warman, *J. Physiol.* **480** (Pt 3), 527–37 (**1994**).

16 M. Steriade and R. Llinas, *Physiol Rev.* **68**, 649–742 (**1988**).

17 D. Contreras and M. Steriade, *J. Neurosci.* **15**, 604–22 (**1995**).

18 S. Sunderam, I. Osorio, J. F. Watkins, *III*, S. B. Wilkinson, M. G. Frei and R. E. Davis, *Brain Res.* **918**, 60–66 (**2001**).

19 I. Osorio, M. G. Frei and S. B. Wilkinson, *Epilepsia* **39**(S16), 615–27 (**1998**).

20 I. Osorio, M. G. Frei, J. Giftakis, T. Peters, J. Ingram, M. Turnbull, M. Herzog, M. T. Rise, S. Schaffner, R. A. Wennberg, T. S. Walczak, M. W. Risinger and C. Ajmone-Marsan, *Epilepsia* **43**(12), 1522–35 (**2002**).

21 F. Mormann, K. Lehnertz, P. David and C. E. Elger, *Physica D* **144**, 358 (**2000**).

22 J. P. Lachaux, E. Rodrigues, J. Martiniere and F. J. Varela, *Human Brain Mapping* **8**, 194–208 (**1999**).

23 M. Winterhalder, B. Schelter, T. Maiwald, A. Brandt, A. Schad, A. Schulze-Bonhage and J. Timmer, *Clin. Neurophysiol.* **117**, 2399–413 (**2006**).

24 R. Guevara, J. L. Velazquez, V. Nenadovic, R. Wennberg, G. Senjanovic and L. G. Dominguez, *Neuroinformatics* **3**, 301–14 (**2005**).

25 M. Raisinghani and C. L. Faingold, *Brain Res.* **1032**, 131–40 (**2005**).

26 M. Raisinghani and C. L. Faingold, *Brain Res.* **1048**, 193–201 (**2005**).

27 A. H. Sheerin, K. Nylen, X. Zhang, D. M. Saucier and M. E. Corcoran, *Neuroscience* **125**, 57–62 (**2004**).

28 T. Kudo, K. Yagi and M. Seino, *Epilepsy Res.* **28**, 1–10 (**1997**).

29 M. Sitcoske OŌShea, J. B. Rosen, R. M. Post and S. R. Weiss, *Brain Res.* **873**, 1–17 (**2000**).

30 P. Mohapel and M. E. Corcoran, *Brain Res.* **733**, 211–18 (**1996**).

31 T. Wagner, H. Osterhage, C. E. Elger and K. Lehnertz, 3rd International Workshop on Seizure Prediction in Epilepsy, Abstract 3.9 (**2007**).

32 M. Staniek, A. Chernihovskyi, H. Osterhage, C. E. Elger and K. Lehnertz, 3rd International Workshop on Seizure Prediction in Epilepsy, Abstract 4.1 (**2007**).

9

Cellular Neural Networks and Seizure Prediction: An Overview

P. Fischer, F. Gollas, R. Kunz, C. Niederhöfer, H. Reichau, R. Tetzlaff

9.1
Introduction: Cellular Neural Networks

Cellular Neural Networks (CNN) were first introduced by Chua and Yang in 1988 [5]. They have been the subject of numerous investigations by a multidisciplinary scientific community and constitute a new paradigm of massive parallel computation [4, 20]. In principle a CNN is an array of locally coupled dynamical systems, so-called cells, usually described by a set of coupled nonlinear ordinary differential equations. Depending on cell coupling weights and initial and boundary conditions, a dynamical behavior which ranges from stable equilibrium states to chaotic dynamics or to emergent behavior, i.e., structure formation following self-organizing principles, can be observed. CNN behavior is also referred to as brain-like computing and can be used to model systems of high complexity. The hardware realizations combine high computational capacity with low power consumption and small size [2].

CNN algorithms take advantage of the spatio-temporal paradigm of cellular computation and yield the possibility of processing multiple signals, e.g., from various electrode probes together. By using miniaturized hardware realizations of CNN it might become possible to implement calculation intensive algorithms on an implantable device for detecting and treating impending seizures by either warning the patient or taking appropriate counteractive measures.

According to a general definition given by Chua, a CNN is a spatial arrangement of locally-coupled cells with an input, a state and an output which evolves according to prescribed dynamical laws. This covers a broad class of nonlinear systems. Previous and recent work is focused mainly on single-layer networks according to the standard model with translation invariant linear weight functions having $M \times N$ cells.

A standard CNN can be described by the state equation

$$\dot{x}_{ij}(t) = x_{ij} - (t) + \sum_{k,l \in N_{ij}(r)} a_{k,l} \cdot y_{kl}(t) + \sum_{k,l \in N_{ij}(r)} b_{k,l} \cdot u_{kl}(t) + z_{ij} \qquad (9.1)$$

Seizure Prediction in Epilepsy. Edited by Björn Schelter, Jens Timmer and Andreas Schulze-Bonhage
Copyright © 2008 WILEY-VCH Verlag GmbH & Co. KGaA, Weinheim
ISBN: 978-3-527-40756-9

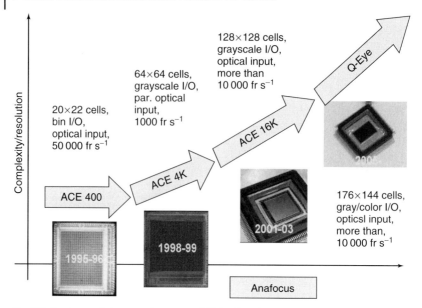

Fig. 9.1 Roadmap of the most important Cellular Visual Microprocessors (CVM).

Here $x_{ij}(t)$ denotes the state, $y_{ij}(t)$ the output and $u_{ij}(t)$ the input of a cell C_{ij}. The above summation has to be carried out by taking output and input values in the neighborhood according to $N_{ij}(r) = \{C_{kl} : \text{Max}(|k - i|, |l - j|) \leq r, 1 \leq k \leq M, 1 \leq l \leq N\}$. This means that, in the simplest case, $r = 1$, only direct neighboring cells to C_{ij} are considered. The feedback and feedforward coupling to neighboring cells is determined by a_{kl} and b_{kl}, which will usually be represented in a matrix form called a template of a CNN. In most cases the cell output $y_{ij}(t) = f(x_{ij}(t)) = \frac{1}{2}(|x_{ij}(t) + 1| - |x_{ij}(t) - 1|)$ is a piecewise linear function of the cell state $x_{ij}(t)$ and of the cell-dependent bias z_{ij}.

A CNN is used to solve a certain problem by applying a two-dimensional input; initial conditions and boundary conditions in many cases. These have to be specified for networks having different cell coupling structures.

CNN miniature devices are completely parallel computing systems with stored programmability having an approximate size of $1\,\text{cm}^2$. They have the computing capability of supercomputers (TeraOps) and can be used directly for real-time multidimensional signal processing. Compared to classical computers which are logic machines based on binary logic and arithmetic, CNN Universal Machine (UM) realizations are locally interconnected analog processor arrays acting on continuous signals in continuous time. Properties of the dynamical behavior of CNNs are used efficiently in recent work for the development of new methods in information processing and in practical applications [3, 6, 7, 24].

CNN can be endowed with optical sensor arrays, i.e., the cell inputs of, e.g., a full-range CNN [19a] are the outputs of optical sensors. There, sensing and processing is performed in one step. The roadmap of the most important Cellular Visual Microprocessors (CVM) is given in Figure 9.1.

9.2
Spatio-temporal Signal Prediction in Epilepsy by Delay-type Discrete-time Cellular Nonlinear Networks (DT-CNN)

In this approach a one-dimensional discrete time version of the Chua–Yang model

$$x_{\mathbf{r}}(t_{n+1}) = -x_{\mathbf{r}}(t_n) + \sum_{j \in \mathcal{N}(\mathbf{r})} \sum_{p=0}^{P} a_{j-\mathbf{r}}^{(p)} \, x_j(t_n)^p \tag{9.2}$$

is used. Considering further more a delay-type cell interaction the prediction equation

$$x_{\mathbf{r}}(t_{n+1}) = -x_{\mathbf{r}}(t_n) + \sum_{j \in \mathcal{N}(\mathbf{r})} \sum_{\tau=0}^{T} \sum_{p=0}^{P} a_{j-\mathbf{r}}^{(p)(\tau)} \, x_j(t_{n-\tau})^p \tag{9.3}$$

results, which is the state equation of a delay-type discrete-time CNN (DT-CNN). This polynomial DT-CNN can also be regarded as a spatio-temporal predictor of the form

$$\hat{x}_{\mathbf{r}}(t_{n+1}) = S[x_{\mathbf{r}-1}(t_n), \dots, \tag{9.4}$$

$$x_{\mathbf{r}-1}(t_{n-T}), x_{\mathbf{r}}(t_n), \dots,$$

$$x_{\mathbf{r}}(t_{n-T}), x_{\mathbf{r}+1}(t_n), \dots, x_{\mathbf{r}+1}(t_{n-T})],$$

with the prediction order T. In a signal prediction task the coefficients $a_i^{(p)(\tau)}$ have to be determined by minimizing the error function

$$e(m) = \sqrt{\frac{1}{N|\mathcal{M}|} \sum_{n=1}^{N} \sum_{(j) \in \mathcal{N}} \frac{(x_j(t_n) - \hat{x}_j(t_n))^2}{< x_j^2 >}}. \tag{9.5}$$

After each iteration the networks output value represents the estimated signal value $\hat{x}_i(t_{n+1})$.

In the following, the results for one exemplary case based on the delay-type DT-CNN (9.3) will be shown and discussed in detail. Taking EEG-signal segments of 2000 values length (corresponding to 10 s in time) leads to time series of predictor parameters and prediction errors. To find distinct changes previous to a seizure onset, which possibly reveal a pre-seizure state, these time series have been analyzed in detail.

The prediction performance obtained with different DT-CNN (9.3) has been studied in comprehensive investigations. By taking a varying polynomial order, different delays and neighborhood ranges into account, several learning algorithms have been applied.

In this contribution results obtained by a special class of predictory networks with linear weight functions $P = 1$, $T = 1$, and $r = 1$ have been studied in detail. The thus-defined network can be considered as a multivariate linear predictor.

In this case it should be noted that during the prediction procedure only the output of the center cell is taken to derive the prediction error in order to optimize the predictor networks parameters. This has been performed for all recordings of all patients treated so far during our investigations.

The underlying data consist of 16 consecutive sets with lengths between 45 minutes and 12 hours, representing a continuous long-term recording with a duration of nearly six days. Ten seizures occur from dataset 10 to dataset 16. During the first long-term examinations analyzing the prediction error an interesting shape behavior as function of time was revealed before the seizure onsets.

In Figure 9.2 the prediction error (9.5) is depicted for a typical dataset convering only interictal activity. The error retains a nearly constant moving average during the complete registration interval. This is assumed to be a typical interictal behavior. Different behavior can be observed for the dataset including three seizures at $(m = 70, 1700, 2379)$ (see Figure 9.3). Before each of these seizures an increase in the prediction error occurs. During this increase the value of $e(m)$ clearly exceeds the previously observed interictal mean. At the seizure onset a peak followed by a sharp drop can be noticed in many cases. Then the error mostly returns to its previous interictal mean value and again suddenly exceeds the mean prior to the next occurring seizure. This behavior of the prediction error in several data sets of the analyzed electrode points has been joined to have a different significance, possibly indicating the assumed pre-seizure states.

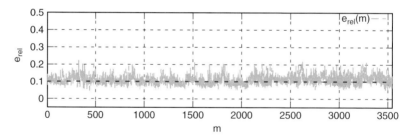

Fig. 9.2 Interictal data: $e(m)$.

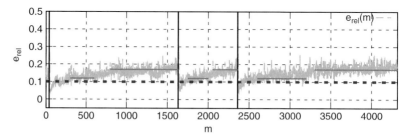

Fig. 9.3 Data with three seizures: $e(m)$ and two datasets for a male patient. In the interictal phase $e(m)$ is fluctuating around a so-called interictal mean. During the seizure phase $e(m)$ increases toward the seizure onset and exceeds the *interictal mean*.

9.3
Identification of EEG-signals by Reaction–Diffusion CNN

The approach described in the following is based on the assumption that non stationary EEG signals – divided into quasi-stationary, consecutive segments – may be represented by the output values of Reaction–Diffusion CNNs (RD-CNN). Particularly, for RD-CNN it has been shown that the existence of locally active cells is a necessary condition for emergent complex behavior. The signal models resulting for each segment then can be analyzed with respect to distinct changes before impending epileptic seizures, indicating a possible transient from an interictal to a preictal state [8] in the complex 'human brain' system.

Reaction–diffusion differential equations can be used to describe various complex phenomena, e.g., structure formation, particularly in biological systems [25]. One well-known example in the field of biology is the FitzHugh–Nagumo simplified nerve conduction model [9,10], a two-component partial differential equation (PDE) system. Reaction–diffusions are given by

$$\frac{\partial}{\partial t}\mathbf{x}(\mathbf{r}, t) = \mathbf{f}(\mathbf{x}(\mathbf{r}, t)) + \mathbf{D}\nabla^2\mathbf{x}(\mathbf{r}, t) \tag{9.6}$$

with the reaction part $\mathbf{f}(\cdot)$ and a diffusion term with the second spatial derivatives of $\mathbf{x}$, denoted by the Laplacian operator ∇^2, and the diffusion coefficient given by the diffusion matrix $\mathbf{D}$. By spatial discretization the PDE system (9.6) can be mapped to a ODE system [4,11]

$$\dot{x}_{1,r}(t) = f_1(x_{1,r}(t), x_{2,r}(t) \ldots x_{m,r}(t)) + D_1\tilde{\nabla}_r^2 x_1(t)$$
$$\dot{x}_{2,r}(t) = f_2(x_{1,r}(t), x_{2,r}(t) \ldots x_{m,r}(t)) + D_2\tilde{\nabla}_r^2 x_2(t)$$
$$\vdots \tag{9.7}$$
$$\dot{x}_{m,r}(t) = f_m(x_{1,r}(t), x_{2,r}(t) \ldots x_{m,r}(t)) + D_m\tilde{\nabla}_r^2 x_m(t)$$

representing an m-layer RD-CNN. Here the Laplacian has been replaced by its discretized version. Using first-order finite difference approximations the discretized Laplacian results in $\tilde{\nabla}^2 x(t) = x_{i+1} + x_{i-1} - 2x_i$ for the one-dimensional case.

In order to derive a CNN algorithm using certain properties of the network dynamics a concrete representation of the reaction function has to be chosen. With polynomial weight functions a nonlinear characteristic has been proposed [12,19] for CNN, capable of representing a wide range of nonlinear functions by a power series expansion. Thus polynomials are considered in $f(\cdot)$ and with this a RD-CNN state equation for the layer ℓ can be given by

$$\dot{x}_{\ell,i} = \mathcal{P}_{\ell,\ell}(x_{\ell,i}) + \mathcal{P}_{\ell,\ell-1}(x_{\ell-1,i}) + \mathcal{P}_{\ell,\ell+1}(x_{\ell+1,i}) + D_\ell\tilde{\nabla}^2 x(t)$$
$$\mathcal{P}_{\ell,\ell'}(x_i) = \sum_{p=0}^{P} a_{i,\ell\ell'}^{(p)}(x_{\ell',i})^p. \tag{9.8}$$

additionally with the assumption that only coupling between adjacent layers ℓ and $\ell \pm 1$ occurs.

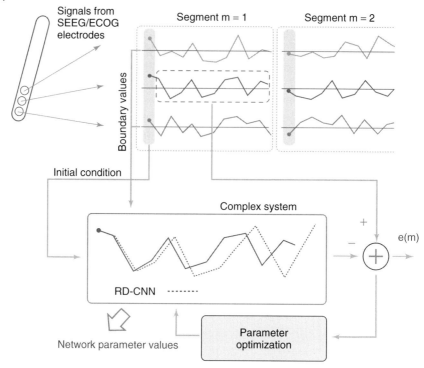

Fig. 9.4 Identification of RD-CNN using segmented EEG signals. The first value of the segment is taken as the initial condition. Signal values of neighboring cells are used as boundary conditions. The network weight function parameter values are determined by minimizing the error (9.9).

Networks derived from Reaction–diffusion systems with the state equations (9.8) will now be considered in order to identify the EEG-signal-generating neural network and further to analyze the parameters obtained by means of the Local Activity Theory [4] which gives a necessary condition for emergent behavior in the identifying network. We suppose that no detailed information regarding the exact dynamic law and the nonlinear weight functions can be derived easily from the underlying neural network. Thus the network weight function parameter values have to be determined in a supervised optimization process (Figure 9.4), i.e. by minimizing the relative mean square error

$$e(m) = \sqrt{\frac{1}{N} \sum_{t=1}^{N} \frac{(\chi(t) - x(t))^2}{\overline{\chi}^2}} \qquad (9.9)$$

for segment m with cell outputs x and EEG signal values χ.

In our investigations the numerical integration of the state equations has been performed by the fourth-order Runge–Kutta [1] integration method. Various

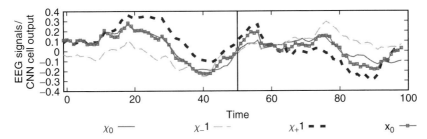

χ_0 ————— χ_{-1} — — χ_{+1} ■ ■ χ_0 —■—

Fig. 9.5 Cell outputs x_0 of RD-CNN determined according to (9.8) with polynomial order $P = 3$ and reference EEG electrode signal values χ_0 and neighboring electrode signal values χ_{-1} and χ_{+1} for two segments each of length 50. Only the signal values of the first segment have been used during parameter optimization and an error $e(m) = 0.17$ has been obtained. Then cell output values for the next segment $t = 51 \cdots 100$ have been determined resulting in an error of $e(m_1 + m_2) = 0.36$.

optimization methods have been studied [13] in which the Differential Evolution (DE) optimization method [22] showed the best performance.

In Figure (9.5) the cell output values of a one-layer RD-CNN according to (9.8) with polynomial order $P = 3$ and corresponding EEG signal values are depicted for two consecutive segments of length 50 while only the first segment has been used in order to determine the network in a supervised parameter optimization procedure, i.e., the time points $t = 1 \cdots 50$ have been taken into account only during error minimization. In the period $t = 51 \cdots 100$ EEG values have been approximated by output values of the previously obtained CNN. This result demonstrates that the shape of Brain Electrical Activity can be well represented, even for signal values that have not been considered during the supervised parameter optimization.

9.4
A CNN-based Pattern Detection Algorithm

Usually, the search of patterns in an EEG signal is associated with a direct detection of special signal behavior or significant signal shape, e.g., peak and drop occurrences, increasing peak frequency and so forth. In this contribution the proposed CNN pattern detection algorithm [14] is based on a statistical analysis of the level-crossing behavior of EEG signals.

The essential idea of pattern detection for predicting epileptic seizures supposes that involved areas of the brain are changing their usual way of acting and entering an other state of behavior. The second presumption is that these behavior alterations could be imaged by occurrence or non-occurence of specific patterns which represent different level-crossing behaviors of a signal. Regarding these two preconditions, one can derive different types of behavior changes of brain electrical activity as far as the occurrance of patterns is concerned. Two of these types are shown as an example in the following.

Type 1 behavior

Type 1 behavior is represented by patterns that occur in the preictal state but not in interictal states. Detection of this behavior provides a hint for an upcoming seizure.

Type 2 behavior

There are patterns that occur frequently in interictal periods. During this time the greatest distance between two pattern occurences in time will be taken as a so-called maximum distance. Detection will consider this non-occurence within the maximum distance as a hint for an upcoming seizure.

Although it would be sufficient for a successful detection of these two types of behavior at just one electrode, the exploration of several datasets has shown that these types occur at different electrodes for different patterns. Now, the question needs to be answered, as to how the location and exploitation of patterns in an EEG signal can be accomplished in order to detect the different types of behavior.

9.4.1
Preprocessing the Data

At first, there must be a preprocessing of the EEG data that allows us to adapt the data to the CNN to perform a pattern detection. Therefore the data first have to be sliced into segments with as many values as there are cells in the CNN. For a better understanding of how the pattern detection works on a CNN, a special example will be shown in the following. In this example a CNN with 72×72 cells will be used. Hence, one has to split the data of a chosen electrode into segments of 5184 timesteps.

The first step of preprocessing is to normalize the data to the interval $[-1,1]$. For performing the binarization of the normalized data in the next step a threshold has to be chosen. For this example a threshold of 0 will be used. Then, every value of the current segment will be switched to -1 if the value is below the threshold of 0 and to 1, if it is above. The preprocessing is shown in Figure 9.6(a).

9.4.2
Performing the Pattern Detection

The result of the preprocessing is a data segment which only contains the values -1 and 1. As convention -1 connotes white and $+1$ connotes black. Now the 5184 binarized values will be filled in the CNN row by row. This CNN then maps binary inputs to binary outputs. In the next step the patterns have to be created. Therefore a 3×3 CNN is used. The patterns are all possible permutations of black and white cells of this CNN. In the case of 3×3 patterns, there are $2^9 = 512$ possible patterns. Every pattern now will be used to shift it over the 72×72 CNN. In each place the black and white cells will be compared. If there is a match for this segment and electrode a pixel in a so-called pattern occurance image (POI) will be turend to black. The POI depicts the segment numbers against the patterns in a two-dimensional picture. After proving every pattern in each segment of one

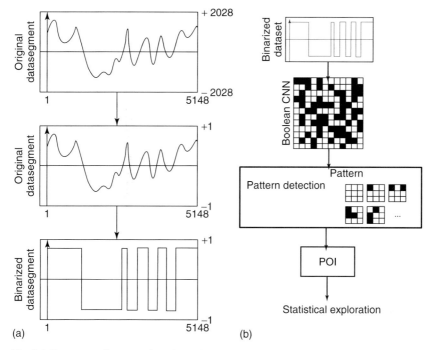

Fig. 9.6 Two steps of pattern detection. (a) Preprocessing.
(b) Creating the pattern occurence image (POI).

electrode, the POI visualizes the occurrence or non-occurence of all patterns within
the dataset. The creating of the POI is shown in Figure 9.6(b).

9.4.3
Detecting Seizures by POI Evaluation

With the POI, different methods and algorithms have been used for an analysis of
EEG data, e.g.:

1) **Typedetection.** This allows the detection of behavior changes, although there
   is the potential for the prediction of upcoming seizures by identifiying these
   changes in time to alert.

2) **Time-window analysis.** This takes a time-window with a fixed size, which is
   moved point by point over the POI. The information within this time-window
   will be statistically evaluated in order to improve the typedetection.

It has been found that short-term recordings show a strict type 1 and type
2 behavior. For long-term recordings a generalized pattern detection offers a
determination of preictal states. Actually, there are further investigations that are
concerned with the combination of patterns to obtain a more generalized feature
vector to refine the prediction quality of this method.

9.5

CNN for Approximation of the Effective Correlation Dimension in Epilepsy

In the following the approximation of the effective correlation dimension procedure [23] is briefly described in order to illustrate how a computationally complex feature extraction method has been represented effectively by a CNN algorithm. Lehnertz and Elger have shown [15, 16, 26] that a feature extraction, based on a so-called effective correlation dimension, allows characterization of an epileptogenic process for successive segments of multi-electrode EEG signals.

In order to enable an extraction of this feature in real-time by using implantable devices, new methods for the approximation of the dimension $D_2^*(k, m)$ by CNN output values have been developed. Therefore CNN with polynomial weight functions of order $p = [1, 9]$ and neighborhood radius of $r = [0, \ldots, 4]$ have been determined using a parameter optimization procedure [17]. In order to ensure the determination of fixed output values the error

$$
\text{EMSE} = \begin{cases} N^2 \sum_{i=1}^{N^2} \frac{(y_j(t_\tau) - \hat{y}_j)^2}{4} & : \quad \forall \dot{x} = 0 \\ 1 & : \quad \forall \dot{x} > 0 \end{cases}
\tag{9.10}
$$

was used in all simulations where $\hat{y}_j$ denotes the value of the reference taken for the optimization algorithm. Two different CNN analogic algorithms allowing the determination of the dimension approximation $D_{CNN}(k, m)$ have been studied. The schemes of the algorithms are given in Figure 9.7. An EEG data segment consisting of $N = 5184$ normalized values of brain electrical activity is taken as the initial condition for the first CNN calculating an intermediate result $D_{CNN}^*(k, m)^*$. The steady state output values of this network are taken as the inital condition to the second network, which performs the well known diffusion operation $\frac{dx_i}{dt} = \Delta x_i(t)$ by considering the Neuman boundary condition. In this way 5184 steady state cell outputs $y_j^{diff}(t_\tau) = D_{CNN}(k, m)$ at a certain gray scale level will be obtained as an approximation of $D_2^*(k, m)$. Thereby, different methods for calculating the nonlinear weight functions of the CNN have been investigated, and an intermediate optimization procedure (IRO) and a final result optimization (FRO) procedure have been studied.

In the first procedure the steady state output of the first network $D_{CNN}^*(k, m)$ will be taken for the determination of the EMSE by comparing it to the desired output $y_{opt}(t_\tau) = D_{CNN}(k, m)$. Thereby, the nonlinear weight functions of the first CNN have been calculated in an error minimization procedure. Then the second CNN is used only for a further improvement of the approximation accuracy. In the second procedure the cell outputs of the second CNN are taken in the error minimization process. While in the IRO procedure the weight functions of the first CNN have been optimized based on an intermediate result, in the FRO the optimization of the first CNN is performed as well but different from the IRO by taking the cell outputs of the second CNN.

The CNN approximation $D_{CNN}(k, m)$ of the $D_2^*(k, m)$ has been analyzed in detail leading to accurate results with an error MSE < 0.08 [14]. There the approximation

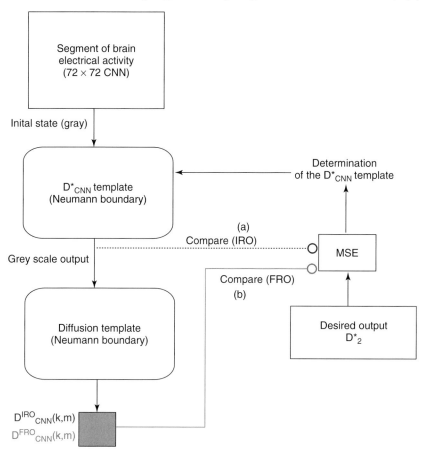

Fig. 9.7 (a) IRO procedure; $D_2^*(k, m)$ is used as the desired output for all cells of the first CNN calculation $D_{CNN}^*(k, m)$. A second CNN is used after obtaining the optimal result for the first CNN during the verification phase. (b) FRO procedure; the complete algorithm has already been applied in the determination procedure of the nonlinear weight functions. The desired value $D_2^*(k, m)$ is directly compared to all cell outputs $D_{CNN}(k, m)$ of the second CNN. Only the nonlinear weight functions of the first CNN will be altered.

accuracy has been verified by taking different recordings as in the parameter training. The investigations also have shown that, in most studied cases, higher polynomial orders in the coupling parameters ($p > 2$) do not lead to increasing accuracy, but increase the time of computation. It is remarkable that in all treated cases accurate approximations are obtained by taking only a few EEG segments of a certain recording in the approximation process.

These results and those obtained in the recent work of Lehnertz [21] clearly show that CNN can be efficiently used for the approximation of standard EEG features. Furthermore, by defining and analyzing new CNN procedures, we have proved that the application of these networks [18, 24] provide new methods for an improved analysis of the bio-electrical activity of the human brain. The results obtained in these investigations can lead to a better understanding of neural phenomena.

References

1 Peter Albrecht. The runge-kutta theory in a nutshell. *SIAM J. Numer. Anal.*, **33**(5), 1712–35 (**1996**).

2 Analogic, Neural Computing Laboratory Computer, and Automation Research Institute. Url: http://lab.analogic.sztaki.hu/. Technical report, Hungarian Academy of Sciences (**2007**).

3 M. Brendel and T. Roska. Adaptive image sensing and enhancement using the cellular neural network universal machine. *International Journal of Circuit Theory and Applications*, **30**, 287–312 (**2002**).

4 L. O. Chua. *A Paradigm of Complexity*, volume 31 of A. World scientific series on nonlinear science, University of California, Berkeley (**1998**).

5 L. O. Chua and L. Yang. Cellular neural networks: Theory and applications. *IEEE Transactions on Circuits and Systems*, **35**, 1257–72 (**1988**).

6 D. Feiden and R. Tetzlaff. Obstacle detection in planar worlds using cellular neural networks. In *IEEE International Workshop on Cellular Neural Networks and their Applications (CNNA 2002)*, pp 383–90, Frankfurt, Germany (**2002**).

7 D. Feiden and R. Tetzlaff. Coding of binary image data using cellular neural networks and iterative annealing. In *ECCTD'03 European Conference on Circuit Theory and Design*, number ISBN 83-88309-95-1, pp 229–32, Krakau, Polen (**2003**). ECS.

8 Fernando Lopes Da Silva, Wouter Blanes, S. N. Kalitzin, Jaime Parra, Piotr Suffczynski and D. N. Velis. Epilepsies as dynamical diseases of brain systems: Basic models of the transition between normal and epileptic activity. *Epilepsia*, **44**, 72–83 (**2003**).

9 R. FitzHugh. Impulses and physiological states in models of nerve membrane. *Biophysical Journal*, **1**, 445–66 (**1961**).

10 R. FitzHugh. *Mathematical Models of Excitation and Propagation in Nerve.*, volume Biological Engineering, chapter 1, pp 1–85. McGraw-Hill Book Co. (**1969**).

11 M. Gilli, T. Roska, L. O. Chua and P. P. Civalleri. On the relationship between cnns and pdes. In *Proceedings of IEEE Cellular Neural Networks and their Applications (CNNA)*, pp 16–24 (**2002**).

12 F. Gollas and R. Tetzlaff. Modeling complex systems by reaction-diffusion cellular nonlinear networks with polynomial weight functions. In *Cellular Neural Networks and Their Applications CNNA, 2005 9th International Workshop on*, Hsin-Chu, Taiwan (**2005**).

13 F. Gollas and R. Tetzlaff. Identification of eeg signals in epilepsy by cell outputs of reaction-diffusion networks. In *Proceedings of the IEEE World Congress on Computational Intelligence–WCCI 2006*, pages 10641–4, Vancouver, Canda (**2006**).

14 R. Kunz and R. Tetzlaff. Spatio-temporal dynamics of brain electrical activity in epilepsy: Analysis with cellular neural networks (cnns). *Journal of Circuits, Systems and Computation*, **12**(6), 825–44, December (**2003**). Dcnn training.

15 K. Lehnertz and C. E. Elger. Spatio-temporal dynamics of the primary epileptogenic area in

temporal lobe epilepsy character-ized by neural complexity loss. *Electroencephalography and clinical Neurophysiology*, **95**, 108–17 (**1995**).

16 K. Lehnertz and C. E. Elger. Can epileptic seizures be predicted? evidence from nonlinear time series analyses of brain electrical activity. *Phys. Rev. Lett*, **80**, 5019–22 (**1998**).

17 J. A. Nelder and R. Mead. A simplex method for function minimization. *Computer Jour-nal*, **7**, 308–13 (**1965**). Simplex optimization.

18 C. Niederhöfer and R. Tetzlaff. Recent results on the prediction of eeg signals in epilepsy by discrete-time cellular neural networks (dtcnn). In *Proceedings of the International Symposium on Circuits and Systems*, Kobe, Japan (**2005**).

19 F. Puffer, R. Tetzlaff and D. Wolf. Modeling nonlinear systems with cellular neural networks. In *Proceedings of the IEEE International Conference On Acoustics, Speech And Signal Processing (ICASSP 1996)*, pp 3513–6, Atlanta, USA (**1996**).

19a A. Rodriguez-Vazquez, S. Espejo, R. Dominguez-Castron, J. L. Huertas and E. Sanchez-Sinencio. Current-mode techniques for the implemen-tation of continous- and discrete-time Cellular Neural Networks. *Circuits and Systems II: Analog and Dig-ital Signal Processing, IEEE Trans-actions*, **40**(3), pp 132–46, March (**1993**).

20 T. Roska and L. O. Chua. The cnn universal machine: an ana-logic array computer. *IEEE Trans.*

on *Circuits and Systems II, Analog and Digital Signal Processing*, **40**, 163–73, (**1993**).

21 R. Sowa, F. Morman, A. Chernihovskyi, S. Florin, C. E. Elger and K. Lehnertz. Estimating synchronization in brain electrical activity from epilepsy patients with cellular neural networks. *Proc. IEEE International Workshop on Cellular Neural Networks and their Applica-tions (CNNA2004)*, pp 327–32 (**2004**).

22 R. Storn and K. Price. Differential evolution–a simple and efficient heuristic for global optimization over continuous spaces. *Journal of Global Optimization*, **11**, 341–59 (**1997**).

23 R. Tetzlaff, R. Kunz, C. Ames and D. Wolf. Analysis of brain electrical activity in epilepsy with cellular neural networks (cnn). In *Proc. European Conference on Circuit Theory and Design (ECCTD)*, volume Vol. II (**1999**). D2 Dimension.

24 R. Tetzlaff, C. Niederhöfer and P. Fischer. Automated detection of a preseizure state: Nonlinear eeg analysis in epilepsy by cellular nonlinear networks and volterra systems. *International Journal of Circuit Theory and its App-lications*, pp 89–108 (**2005**).

25 A.M. Turing. The chemical basis of morphogenesis. *Trans. Roy. Soc. London*, **237** (**1952**).

26 G. Widman, K. Lehnertz, P. Jansen, W. Meyer, W. Burr and C.E. Elger. A fast general purpose algorithm for the computation of auto- and crosscorrelation integrals from single chanel data. *Physica D*, **121**, 65–7 (**1998**).

10

Time Series Analysis with Cellular Neural Networks

Anton Chernihovskyi, Dieter Krug, Christian E. Elger, Klaus Lehnertz[1]

10.1
Introduction

Time series analysis is a general tool used to study the diversity of natural phenomena. Generally speaking, the main aim of time series analysis is to efficiently extract knowledge about the underlying dynamics of the investigated system and to predict its temporal evolution. Nowadays, numerous different linear and nonlinear techniques of time series analysis have been developed [1–6]. However, despite a rigorous theoretical background and reasonable success in various applications, a reliable analysis of noisy and non-stationary field data is still an unsolved problem. In the last years it also became obvious that an implementation of particularly nonlinear time series analysis techniques demands high computational resources, and is thus very time consuming. This limits the use of these techniques in cases where a large amount of information has to be processed in real time.

Artificial neural networks (ANN) are computational tools that have already found extensive utilization in solving many complex real world problems. In general, an ANN is an array of globally interconnected information processing units. In contrast to the conventional serial computer architecture ANN are characterized by intrinsic nonlinearity, high parallelism, fault and noise tolerance, and their ability to adapt, i.e., to learn a rule from a set of examples [7]. However, the practical use of ANN in real-world applications has revealed many obstacles. For instance, due to the global connectivity and an irregular alignment of the individual processing units, large-scale hardware implementations of ANN are still rather limited. The concept of Cellular Neural Networks (CNN) provides a possible solution to this problem. CNN are a subclass of ANN where individual processing units (called *cells*) are only *locally* coupled. This unique feature of CNN enables large-scale hardware implementations on a modern level of manufacturing technology, e.g., as very large scale integrated (VLSI) circuits. Despite the restriction to the local connectivity, the CNN paradigm still offers enough degrees of freedom to solve a great diversity

1) We would like to thank Ronald Tetzlaff and his group for their support. This work was supported by the Deutsche Forschungsgemeinschaft (Grant No. LE 660/2-3.)

of complex problems in various real-world applications [8]. Recent studies in the field of time series analysis have shown a potential use of CNN to process a large amount of field data in real time. It was shown that the CNN can be used to efficiently approximate various nonlinear statistical measures whose evaluation on a conventional serial computer demands a high number of computational resources [9–14].

We here present two complementary CNN-based methods for time series analysis. In Section 10.3 the phenomenon of signal-induced excitation waves in excitable media is introduced. This biologically motivated approach to time series analysis provides an alternative way to instantaneously characterize transient spectral patterns and synchronous activities in noisy and non-stationary time series. In cases where a direct and analytical CNN-based solution cannot be achieved, a CNN can be trained to approximate some statistical measure. In Section 10.4 we show that, after successful supervised learning on a set of examples, a CNN can be used to efficiently approximate various synchronization measures and, moreover, is capable of generalization. In the context of epileptic seizure prediction, previous studies have shown the existence of a *pre-seizure* state that is characterized by a significant deviation from the mean level of synchronization between different electroencephalographic recording sites observed during the interictal state [15, 16]. A robust detection of the degree of synchronization between noisy and non-stationary signals can improve the anticipation of epileptic seizures. Both methods provide a possibility of hardware implementations within the framework of CNN and thus can be used to design a miniaturized detector for transient spectral patterns and synchronous activities in the EEG. The following section provides a brief introduction to the concept of CNN and its applications. For further details we refer the reader to Chapter 9 in this book.

10.2
Cellular Neural Networks

In the late 1980s a new approach for simulating the dynamics of spatially extended systems was introduced [17–19]. It was shown that the diversity of dynamical phenomena observed in nonlinear media can be modeled by means of a spatial array of only locally coupled and rather simple dynamical units (cells). The interconnection of these units was called a *Cellular Neural Network* (CNN) that can either be simulated on a digital computer or can be realized as an array of locally coupled integrated electrical circuits. The original motivation to invent the CNN was to build a more realistic and practical neural network compared to the general theoretical architecture proposed just few years before by Hopfield [20, 21]. In a Hopfield network the number of interconnections between neurons is growing exponentially with the network size. The predefined global connectivity, however, forbids large-scale implementation. In contrast to Hopfield networks, CNN provide a more realistic approach to construct large-scale neural networks due to their locality. Unlike in the brain, where a massive array of globally interconnected excitable units (neurons) capture their energy from the oxidation of biochemical substrates, a CNN exhibits an active medium of only locally interconnected active elements (e.g., transistors) that are powered by external electrical energy suppliers. Each cell in a CNN is coupled only to its neighbors and represents a dynamical system evolving according to some dynamical law. In general, neither the cell's dynamics nor the local interaction template between cells are required to be spatially invariant. However, for the sake of simplicity, it is convenient to consider a translation-invariant architecture of CNN, which depicts spatially homogeneous nonlinear media. In this special case all cells evolve according to the same dynamical law and the resulting CNN dynamics is uniquely predefined by a set of real numbers, called a *connection template* or *CNN gene* that describes spatially invariant interconnection weights between each cell and its neighboring cells. These few numbers define a local connectivity rule, which governs the global asymptotic dynamics of the network. The CNN represents an array of analog processors or, in other words, continuous dynamical systems. In such systems information processing is regarded as a dynamical evolution of the initial states, which encode the input information, to some desired final states that are regarded as the result of computation. A CNN gene plays the role of a programming instruction, which is applied to a continuous information flow where the intrinsic dynamics of a network uniquely maps an input pattern to some desired output pattern.

Nowadays, the CNN architecture has found a broad range of applications in such fields as, e.g., image processing and pattern recognition (see [8, 22] for an overview). Nevertheless, the ultimate theory of the CNN gene design is still far from being completed. In addition, a straightforward or 'brute force' optimization of an arbitrary CNN dynamics to some desired spatial-temporal behavior, by means of various existing optimization strategies, cannot always be successfully performed within a reasonable amount of time. Thus, by analogy with digital computers (where algorithms are usually implemented in some form of a flow-chart of relatively simple instructions) it was quite natural to assume that the modeling of a complex spatial-temporal dynamics can in principle also be decomposed into a series of more simple intermediate steps. By definition, a sequence of CNN genes defines a *CNN chromosome* or a 'program' with each gene within this sequence representing an elementary instruction. A universe of all CNN genes is called a *CNN genome* that defines some sort of 'programming language' that allows one to program the spatial-temporal dynamics of CNN [8]. Having such a programming language, we can try to reduce a complex spatial-temporal dynamics to some CNN program or a CNN chromosome.

In 1993 a further development of the CNN paradigm, the so-called CNN Universal Machine (CNN-UM) was introduced [23]. In contrast to the conventional von Neumann computer architecture, the original CNN architecture lacks the ultimate property of stored programmability, which is a crucial feature of universal computation. Following the von Neumann definition of stored programmability, it was assumed that any universal computational device (based on a CNN architecture) must also possess additional local memory units, which may be used for both the storage and the successive retrieval of analyzed spatially-distributed information along with a set of applied instructions. A program or chromosome can thus be executed on the CNN, and the results obtained can then again be stored in memory units for further processing. In the last decade, a variety of successful applications of the CNN-UM paradigm eventually led to the development of a number of 'analogic' visual processors (see [22] and references therein). These platforms represent fully programmable arrays of analog processors along with local analog memory units that are used for the storage of analog information (such as connection templates, initial state distribution, etc.). A digital circuitry is used to control the global and local information flow and establishes, by means of analog-to-digital and digital-to-analog converters, an input-output process with an outside digital world. Alternative CNN-UM implementations also include purely digital approaches based on numerical integration of a cell's equation of motion by means of an array of digital signal processors, special-purpose processors, and different software packages (for a review see [22, 24]). Capturing computational universality of cellular automata ([25], for a review see [26]) and many important properties of recurrent neural networks (such as synaptic plasticity, massive parallelism, etc.), the CNN can be regarded as the next evolutionary step of the analog computer architecture. In contrast to conventional analog computers that may be used to simulate dynamics of some ordinary differential equations, the CNN may be used for the simulation of spatial-temporal phenomena in partial differential equations (PDE) which describe some nonlinear homogeneous medium. Thus, the CNN concept can, in principle, provide a unified framework for the study of the observed plethora of pattern formation and wave propagation phenomena in locally active nonlinear media.

The next two sections provide an introduction into the field of time series analysis with CNN. We present two complementary CNN-based methods for time series analysis. The first, analytical, method is based on the phenomenon of a frequency-selective induction of excitation waves in excitable media, simulated with CNN. The second, adaptive, method exploits the computational universality of CNN.

10.3
An Analytical CNN-based Method for Pattern Detection in Non-stationary and Noisy Time Series

Excitable media (EM) are spatially distributed excitable dynamical systems, which are capable of rapidly propagating impulses (excitations) over long distances without damping[2]. According to the theory of dynamical systems, excitable systems are *structurally unstable* and thus reside (in parameter space) near a bifurcation point [27]. Due to the intrinsic nonlinearity of such systems, even a tiny external perturbation may lead to a qualitative change in their dynamical behavior. However,

---

2) In contrast, in linear dispersive media any spatially localized moving wave-like object will rapidly lose its shape due to intrinsic linear dispersion.

in many cases an external perturbation has to exceed a certain *excitation threshold* to induce a transition from a steady (non-excitable) to an excitable state of the system. Generally, the shape of the generated excitation does not depend on the perturbation strength. Excitable systems are widespread in nature and more recently they have also found a variety of technical applications [28].

Traveling waves in excitable media have been observed in many contexts. It is now widely accepted that this phenomenon plays a vital role in the information processing in many biological systems. For example, neural tissue possesses the remarkable ability to generate and then to propagate electrical impulses (action potential) quite rapidly over long distances. The energy, which is consumed during the propagation of an action potential does not itself propagate along the direction of the action potential, but instead is locally supplied by some intracellular mechanism. Generally speaking, any excitable medium has to be *locally active*, i.e., has to possess local energy sources. As a direct consequence, the energy that is consumed during wave propagation must be regenerated before the next wave front can pass through the medium. This results in a *refractory behavior*, which indeed can be observed in excitable media such as neural tissue [29–31]. The refractory period (i.e., the time during which a medium is not capable of wave-propagation) defines an upper limit of the frequency of an induced wave train. In contrast to waves produced in linear passive media, which can easily propagate through each other, the collision of two waves generated in locally active excitable media will lead to their mutual annihilation. Traveling regenerative waves are also observed in a family of oscillating chemical reactions that are known as *Belousov–Zhabotinskii* reactions. Other examples include such diverse phenomena as propagation of forest fires, or currents of sodium and potassium ions in the cardiac muscle [29]. Despite the different nature of the presented examples, the propagation of traveling excitation waves in all EM has many common characteristics. The above discussed features are thus common to all EM systems.

EM are often considered as a collection of locally coupled individual elements. Neighboring elements of an EM interact with each other by a diffusion-like transport process. In many cases, diffusion leads to a spatially homogeneous steady state of the medium. However, recent studies have shown that the notion of excitability might be further extended to the case of a spatially extended excitable system whose unperturbed steady state resembles spatio-temporal chaos [32, 33]. In contrast to EM with a spatially homogeneous steady state, these systems show chaotic behavior that tends to actively destroy long-range spatial and temporal correlations. This prevents the propagation of a single supra-threshold excitation induced by a short-lasting local perturbation. In order to facilitate a wave propagation phenomenon one therefore has to repeatedly perturb the medium with perturbations of appropriate amplitudes and certain (resonance) frequencies. Only in this case will the medium actively support the propagation of localized excitation waves (cf. Figure 10.1). Thus, the observed phenomenon of *frequency-selective propagation of excitation waves* in EM provides a possible way to extend the conventional (i.e., amplitude-selective) notion of excitability in dynamical systems [34, 35].

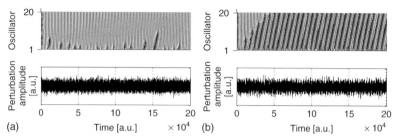

Fig. 10.1 Examples for the induction of spatial-temporal patterns in an excitable medium (20 *FitzHugh–Nagumo* oscillators) due to a perturbation of the first oscillator. The range of amplitude values of this medium is encoded with colors ranging from blue (minimum amplitudes) to red (maximum amplitudes). (b) A noisy signal containing a profound rhythmic component leads to coherent periodic patterns. (a) Only few excitation waves are induced if the medium is perturbed with a non-periodic or, in general, a non-correlated signal (here white noise). (Please find a color version of this figure on the color plates.)

The rapid and robust recognition of transient and broadband patterns in a temporally varying and noisy acoustic environment is one of the major tasks performed by the auditory system of mammals. Sensory neurons within the auditory pathway exhibit frequency-selective firing of action potentials with respect to the spectral content of perceived acoustic stimuli. Already in 1863 Helmholtz had pointed out that our hearing organ – the mammalian cochlea – performs some sort of a spatially distributed Fourier transform of acoustic stimuli [36]. In his model, he assumed that the cochlea comprises a spatial array of resonators showing maximum excitations only for given stimulus frequencies. It is now widely accepted that the mammalian cochlea operates as an active sensor, amplifying some frequencies, and suppressing others. Formally, this biological system provides an important example of a nonlinear excitable medium that actively supports the frequency-selective propagation of excitation waves in response to local perturbations. Probably the most tractable mathematical model of a nonlinear excitable medium showing qualitative properties of excitable neural tissue is the *FitzHugh–Nagumo* (FHN) reaction – diffusion PDE [37, 38]. The discretized version of this equation – comprising a chain of diffusively coupled FHN oscillators (neurons) – provides a generic mathematical model for excitable media. Recent numerical studies show that this type of simulated excitable media exhibits the phenomenon of frequency-selective induction of excitation waves with respect to the frequency of a locally applied periodic perturbation [34, 35]. Thus the medium acts as a narrow-band frequency filter. By perturbing several such excitable media, each tuned to a different characteristic frequency, a broad frequency band of the applied perturbation can be scanned for the presence of a rhythmic component. The resulting system resembles the functioning of the mammalian cochlea, and thus represents some sort of a filter bank instantaneously converting analog waveforms into excitation patterns [39]. Such systems provide a novel tool for the spectral analysis of non-stationary and noisy signals. This provides the possibility to efficiently detect complex (broadband) spectral patterns (e.g., stereotyped waveforms observed during epileptic seizures)

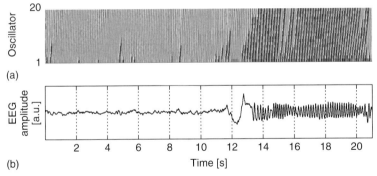

(a)

(b)

Fig. 10.2 (a) Spatial-temporal pattern induced in an excitable medium (20 *FitzHugh–Nagumo* oscillators due a perturbation of the first oscillator with an EEG signal (b). The range of amplitude values of this medium is encoded with colors ranging from blue (minimum amplitudes) to red (maximum amplitudes). The EEG signal was recorded intracranially from a patient suffering from medial temporal lobe epilepsy. Data were sampled at 200 Hz within the frequency band of 0.5–85 Hz using a 16 bit analog-to-digital converter. (Please find a color version of this figure on the color plates.)

in non-stationary signals (cf. Figure 10.2). Conventional approaches to time series analysis require stationarity or at least approximate stationarity of time series and have frequently been designed for a retrospective evaluation and usually performed in a moving-window fashion (i.e., the time series is decomposed into a sequence of overlapping or non-overlapping segments). This segmentation should represent a reasonable trade-off between approximate stationarity of the time series and a sufficient number of data points that is needed to obtain statistically significant results for the calculation of some measure. The price we pay for this is the need to presume stationarity over some time interval that defines a temporal resolution of the applied methods. The method of frequency-selective excitation waves allows almost instantaneous detection of short-lasting transient patterns in non-stationary signals (with a high noise level) and can be regarded as a good complement to already existing techniques based on statistical data evaluation.

After spectral decomposition the cochlea transforms incoming acoustic stimuli into neural excitation patterns, which are projected via a variety of tonotopically ordered parallel nerve fibers (i.e., each nerve fiber responds optimally to a particular frequency) into a collection of specialized nuclei in the brain stem, where the process of spatial localization of sound sources is taking place. Early anatomical studies revealed that these nuclei are the first sites in the ascending auditory pathway receiving massive converging binaural inputs from both ears [40]. One of the acoustical cues about spatial localization of sound sources along the azimuth is given by interaural time differences (ITD), i.e., differences in the times of arrival of stimuli at both ears. It is now known that some neurons in the brain stem are highly sensitive even to microsecond ITD. Our abilities to localize sound sources along the azimuth axis thus partly stems from the capability of these neurons to precisely discriminate phase-relationships in acoustic stimuli between both ears. Following this idea our recent numerical studies [41] have shown that a combination of two EM with a coincidence detector (that is realized as a nonlinear oscillator, e.g., an integrator neuron) provides a possible approach to detect synchronization phenomena even in cases where it is difficult to directly apply phase-based measures of synchronization due to a relatively low signal-to-noise ratio (cf, Figure 10.3). Such an analog approach to time series analysis – in which correlated changes in the dynamics of the system under investigation are characterized by their influence on the global dynamics of a perturbed system – is conceptually different from a variety of statistical approaches currently in use. In this context, recent studies have shown that a simple leaky integrate and fire model of neurons allows the on-line identification of epileptic seizure onsets and the detection of pre-seizure changes in EEG recordings [42, 43].

10.4
An Adaptive CNN-based Method to Measure Synchronization

Research over recent years has shown that the analysis of synchronization phenomena in the epileptic brain can contribute significantly to the field of seizure prediction (see [44] for an overview). As opposed to previously used – mostly linear and nonlinear – univariate time series analysis approaches, different measures for synchronization offer a significant predictive performance above chance level [16, 45–48]. Despite these promising features there are also disadvantages that render an application of synchronization measures in a clinical setting problematic. In current clinical or neuroscientific investigations the number of sensoring electrodes typically ranges between 100 and 200, and algorithms' runtime grows quadratically with the number of channels. This leads to strong restrictions when using currently available personal computers, particularly for real-time applications. CNN can be regarded as an alternative computational tool for measuring synchronization due to their attractive information-processing characteristics already discussed above. However, in contrast to the analytical CNN-based method presented in the previous chapter, no CNN templates are currently available that would allow one to estimate the various types of synchronization in an analytical way. In this section we present an adaptive CNN-based method for the estimation of phase and generalized synchronization in EEG time series via supervised learning.

Synchronization phenomena can be observed in nearly all fields of science [4, 48]. Despite the very intuitive use of the term *synchronization* in everyday life situations, several obstacles in finding a universal definition of the term occur since various types of synchronization can be observed. In the literature, synchronization was mentioned for the first time at the end of the 17th century by Christiaan Huygens. He noticed that two pendulum clocks, suspended from a bar, can adapt their rhythms, and their phase difference vanishes, which obviously depicts some kind of *phase synchronization*. Mathematically, Huygens defined this phenomenon as a locking of phases ϕ [49]:

$$\phi_j^a - \phi_j^b = \text{const.} \tag{10.1}$$

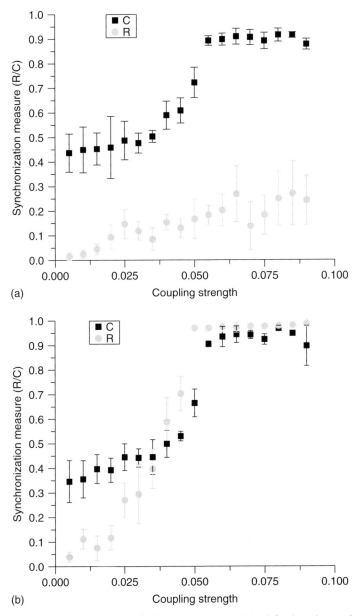

(a)

(b)

Fig. 10.3 Means and standard deviations (10 realizations) of the mean phase coherence R [15] and the coincidence rate C measured between the x-components of two coupled noise-free (b) and noisy ((a) SNR $= -6$ dB) Rössler systems for increasing coupling strengths. Coincidence rate C is defined as the number of coincidences between the last oscillators of the excitable media (approximated as a diffusively coupled chain of the FitzHugh–Nagumo oscillators). The first oscillators were perturbed by the two time series.

Almost 300 years later, it was observed that nonlinear and particularly chaotic dynamical systems show synchronization behavior similar to periodic oscillators. Since then it turned out that the classical definition of phase synchronization is not sufficient in certain cases, and it became clear that extended definitions are required. A straightforward approach was the definition of a phase for non-periodic oscillators using the Hilbert or the wavelet transform. These more generally defined phases can be exploited to construct measures that determine the degree of phase synchronization between dynamical systems. A widely used concept is the *mean phase coherence R* [15], based on a circular statistics of the phases:

$$R = \left| \frac{1}{K} \sum_{j=1}^{K} e^{i\left[\phi_j^a - \phi_j^b\right]} \right|,$$
(10.2)

where K denotes the number of data points in the time series.

A more general definition for synchronization is based on state space properties of the observed systems. The concept of *generalized synchronization* [50] claims the existence of some functional relationship between the states of the systems. Properties of this functional (e.g., smoothness, differentiability, etc.) can then be exploited for a characterization of the systems' synchronization state. Though being conceptually very simple the application to real world systems is difficult since the functional is *a priori* not known in most cases and, in general, cannot be determined analytically. In order to avoid these shortcomings, Arnhold and colleagues developed the concept of *nonlinear interdependences*, which is based on the measurement of distances in state space [51]. Given some knowledge about the systems, nonlinear interdependence measures can be regarded as estimators for the strength and the direction of coupling, thus allowing one to detect driver-responder-relationships.

Consider two dynamical systems V and W and their reconstructed state space vectors $\vec{v}_n = (v_n, \ldots, v_{n-(m-1)d})$ and $\vec{w}_n = (w_n, \ldots, w_{n-(m-1)d})$, where m denotes the embedding dimension and d the time delay [52]. Let $r_{n,j}$ and $s_{n,j}$, $j = 1 \cdots k$ denote the time indices of the k nearest neighbors of $\vec{v}_n$ and $\vec{w}_n$, respectively. The W-conditioned mean squared Euclidean distance is defined as

$$Q_n^{(k)}(V|W) = \frac{1}{k} \sum_{j=1}^{k} \left(\vec{v}_n - \vec{v}_{s_{n,j}} \right)^2 .$$
(10.3)

Following [53] the nonlinear interdependence measure N is then defined as

$$N^{(k)}(V|W) = \frac{1}{K} \sum_{n=1}^{K} \frac{Q_n(V) - Q_n^{(k)}(V|W)}{Q_n(V)},$$
(10.4)

where $Q_n(V)$ is the mean distance of one vector to all other state space vectors:

$$Q_n(V) = \frac{1}{K-1} \sum_{j \neq n} \left(\vec{v}_n - \vec{v}_j \right)^2 .$$
(10.5)

Low values of $N^{(k)}(V|W)$ indicate the independence of systems V and W (also slightly negative values are possible), while $N^{(k)}(V|W) \to 1$ for identical systems. The fact that the opposite interdependence $N^{(k)}(W|V)$ is, in general, not equal to $N^{(k)}(V|W)$ can be exploited to obtain information about driver – responder relationships. Quantitatively this can be realized by a symmetric and an asymmetric composition of $N^{(k)}$

$$N_s = \frac{N^{(k)}(V|W) + N^{(k)}(W|V)}{2}, \quad N_a = \frac{N^{(k)}(V|W) - N^{(k)}(W|V)}{2}, \tag{10.6}$$

where N_s can be regarded as a measure for the strength of synchronization and N_a identifies the *more active* and the *more passive* system [51]. For further details we refer the reader to Chapter 5 in this book.

10.4.1
Learning Synchronization in EEG Time Series with CNN

Due to the lack of specific CNN templates for measuring synchronization in time series, we considered the concept of supervised learning to find a CNN structure that allowed us to approximate the different measures for phase and generalized synchronization [13, 14]. In the following, let M denote a measure for synchronization (mean phase coherence R or nonlinear interdependence N). For an in-sample optimization of the network we compiled a training set that consists of L elements where half of the chosen pairs of EEG segments exhibit high synchronization values $M_{\max}$ and the other half low synchronization values $M_{\min}$. We set the desired outputs of each CNN cell to $y^{\text{ref}} = +1$ and $y^{\text{ref}} = -1$, respectively, in order to exploit the whole range of values of the CNN. Optimization of the network was performed by minimizing the cost function Γ for all elements of the training set simultaneously:

$$\Gamma = \frac{1}{L} \sum_{l=1}^{L} \left(\frac{1}{4K} \sum_{k=0}^{K-1} \left[y_{l,k}(\tau_{\text{trans}}) - y_l^{\text{ref}} \right]^2 \right) \tag{10.7}$$

Although this optimization procedure is time consuming and computationally demanding, we were able to identify CNN templates that allowed us to approximate the different synchronization values in the training set with an acceptable accuracy. In order to assess the performance of our CNN in an out-of-sample validation study we used a linear rescaling of the approximated measure M^{CNN}:

$$M^{\text{CNN}} = \left(\frac{1}{K} \sum_{k=0}^{K-1} \frac{y_k(\tau_{\text{trans}}) + 1}{2} \right) (M_{\max} - M_{\min}) + M_{\min}. \tag{10.8}$$

In [13] we applied the adaptive CNN-based concept to approximate the mean phase coherence R in long-term (more than five days), intracranial EEG recordings from an epilepsy patient. Using approximately five minutes of EEG recording for optimizing the CNN only, we achieved an average deviation of 5.5 % between analytically calculated R values and estimated R^{CNN} values. Evaluating the temporal evolution of R^{CNN} we observed a long-lasting drop in synchronization (cf. [54]) prior to seven of the ten seizures recorded in this patient. These findings indicate that a differentiation between interictal and preictal states is in principle possible using R^{CNN}, although the achieved performance might not be sufficient for all cases. Nevertheless, we note that our CNN allowed a sufficient approximation of R even during and after the seizures, although we did not use data from these states for the training.

In [14] we recently applied the adaptive CNN-based concept to approximate both the symmetric N_s and the asymmetric N_a nonlinear interdependence measures in the EEG recording already studied in [13]. Again, only approximately five minutes of in-sample training data sufficed to reproduce – in an out-of-sample study – the long-term fluctuations of N_s and N_a in the long-term EEG recording with an acceptable accuracy (cf. Figure 10.4). For the symmetric nonlinear interdependence measure we assessed the suitability of our adaptive CNN-based concept for a possible application in seizure prediction studies. Using receiver-operating characteristic statistics (ROC) and assuming that for each of the ten seizures a preictal state with a duration of 4 h exists (cf. [16]), we compared the frequency distributions of N_s and of N_s^{CNN} values from the preictal and interictal interval, excluding data from the seizure states and from the 30 min following the seizures. For N_s^{CNN} we achieved a prediction performance of ROC = 0.74, which was lower than for the calculated nonlinear interdependence N_s (ROC = 0.83). Given the observed higher variance of the CNN-based estimates, this deviation was to be expected to some extent. The achieved performance, however, can still be regarded as promising and indicates that a differentiation of preictal states from the interictal interval is in principle possible using our CNN-based approach.

We recently investigated within-subject and across-subject generalization properties of the adaptive CNN-based concept to measure symmetric nonlinear interdependences. For the within-subject generalization study we used the CNN settings derived in [14] and estimated – for the data of the same patient investigated there – N_s^{CNN} values for all (i.e., 210) channel combinations. A comparison with the analytically derived N_s values revealed an average deviation of about 7 % (mean over the measure profile lasting more than five days) for the channel combination that was used for optimizing the CNN. For the remaining 209 channel combinations the mean deviation (mean over all channel combinations and over the measure profile) amounted to about 11 %. This degradation was to be expected (at least to some extent), given the fact that the optimization was not performed for these channel combinations separately. When investigating the detectability of preictal states for each channel combination using ROC and seizure time surrogates [55] we observed that the number of channel combinations for which the statistical tests indicate a significant detection performance decreased (47 channel combinations using N_s vs 25 channel combinations using N_s^{CNN}). We note that the remaining 25 channel combinations captured EEG activity from the seizure onset zone and from homologous contralateral structures.

We observed similar effects when investigating across-subject generalization properties of our approach. Here we used CNN settings optimized for a single

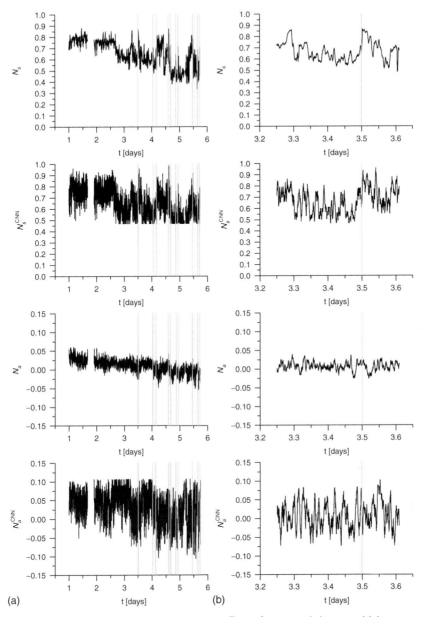

Fig. 10.4 From top to bottom. (a) Time courses of the calculated symmetric and asymmetric nonlinear interdependence measures N_s and N_a and their estimated counterparts (N_s^{CNN} and N_a^{CNN}) using our CNN-based approach for a long-term intracranial EEG recording from a patient suffering from a medial temporal lobe epilepsy. (b) Zoom on the period prior to, during, and after the first seizure. All profiles are smoothed using a five-minute moving-average filter for better visualization. Gray vertical lines mark seizure onsets.

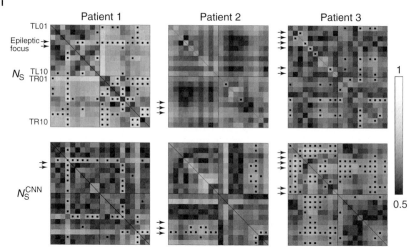

Fig. 10.5 Across-subject generalization properties of the adaptive CNN-based concept of measuring the strength of generalized synchronization. Color-coded ROC-areas obtained from numerically derived (upper row) and approximated (lower row) profiles of the symmetric nonlinear interdependence N_s for all electrode combinations. Black dots indicate a significant ROC-area value (using 19 seizure time surrogates) for a given electrode combination. Data from a single channel combination from patient #1 was used for optimizing the CNN. This optimized CNN was used to estimate N_s^{CNN} values for the remaining channel combinations in this patient, and for the data from patient #2 and #3. Arrows indicate seizure onset area. (Please find a color version of this figure on the color plates.)

channel combination from one patient (patient #1), estimated N_s^{CNN} values for long-lasting (on average 8.8 days) multi-channel EEG data recorded via 20 intrahippocampal depth electrodes from another four patients, and proceeded as described above. Here we observed that the number of channel combinations for which the statistical tests indicated a significant performance in detecting a preictal state even increased in three patients. In two patients the, thus defined, channel combinations were confined to the seizure onset zone (cf. Figure 10.5). Although it is too early to draw final conclusions about the generalization properties of our approach, our preliminary findings indicate that the adaptive CNN-based concept of measuring synchronization appears to be capable of focusing on spatial and temporal information in the EEG that might be of relevance for a further improvement of seizure prediction algorithms.

10.5
Conclusions and Outlook

We have presented approaches for the analysis of interacting complex systems with the nonlinear dynamics of interacting nonlinear elements. We here considered two complementary approaches, an analytical and an adaptive one, that are based on the concept of Cellular Neural Networks (CNN). The analytical approach exploits the phenomenon of frequency-selective induction of excitation waves in excitable media. We showed that this can be a novel tool for the spectral analysis of non-stationary and noisy time-series. By combining two excitable media and feeding their output to a coincidence detector, we implemented a system that allows one to detect synchronization phenomena even in cases where it is difficult to directly apply phase-based measures for synchronization due to a relatively low signal-to-noise ratio. The adaptive approaches exploit the computational universality of CNN. We showed that a CNN can be trained (via supervised learning) to *learn* synchronization phenomena in complex time series such as phase and generalized synchronization.

In contrast to other artificial neural networks, CNN are characterized by a local coupling of processing units. This unique feature enables large-scale hardware implementations, e.g., as very large scale integrated (VLSI) circuits, and may lead to the development of miniaturized analysis systems. We thus expect that our approaches can complement the various existing methods used for the prediction of epileptic seizures, e.g., by providing a means for real-time analyses of long-lasting, multichannel EEG data. Together with currently available seizure-prevention techniques this may eventually lead to the development of a miniaturized – and possibly implantable-seizure prediction and prevention device.

References

1 D. Brillinger. *Time Series: Data Analysis and Theory.* Holden-Day, San Francisco (**1981**).

2 M. B. Priestley. *Nonlinear and Non-Stationary Time Series Analysis.* Academic Press, London (**1988**).

3 J. S. Bendat and A. G. Piersol. *Random Data Analysis and Measurement Procedure.* Wiley, New York (**2000**).

4 A. S. Pikovsky, M. G. Rosenblum and J. Kurths. *Synchronization – A universal concept in nonlinear sciences.* Cambridge University Press, Cambridge, UK (**2001**).

5 H. Kantz and Th. Schreiber. *Nonlinear Time Series Analysis.* Cambridge Univ. Press, Cambridge, UK, second edition (**2003**).

6 G. V. Osipov, J. Kurths and C. Zhou. *Synchronization In Oscillatory Networks.* Springer, Berlin, New York (**2007**).

7 T. L. H. Watkin, A. Rau and M. Biehl. The statistical mechanics of learning a rule. *Rev. Mod. Phys.*, **65**, 499–556 (**1993**).

8 L.O. Chua. *CNN: A paradigm for complexity.* Singapore: World Scientific (**1998**).

9 R. Tetzlaff, R. Kunz, C. Ames and D. Wolf. Analysis of brain electrical activity in epilepsy with Cellular Neural Networks (CNN). In C. Beccari, M. Biey, P.P. Civalleri and M. Gilli, editors, *Proc. IEEE European Conference on Circuit Theory and Design*, pages 1007–10, Turin, Italy (**1999**). Levrotto & Bella.

10 A. Potapov and M. K. Ali. Neural networks for estimating intrinsic dimension. *Phys. Rev. E*, **65**, 046212 (**2002**).

11 R. Kunz and R. Tetzlaff. Spatio-temporal dynamics of

brain electrical activity in epilepsy: analysis with cellular neural networks CNNs. *J. Circuit Syst. Comp.*, **12**, 825–44 (**2003**).

12 R. Tetzlaff, T. Niederhofer and P. Fischer. Automated detection of a preseizure state: Non-linear EEG analysis in epilepsy by Cellular Nonlinear Networks and Volterra systems. *Int. J. Circ. Theor. Appl.*, **34**, 89–108 (**2006**).

13 R. Sowa, A. Chernihovskyi, F. Mormann and K. Lehnertz. Estimating phase synchronization in dynamical systems using cellular nonlinear networks. *Phys. Rev. E*, **71**, 061926 (**2005**).

14 D. Krug, H. Osterhage, C. E. Elger and K. Lehnertz. Estimating nonlinear interdependences in dynamical systems using cellular nonlinear networks. *Phys. Rev. E*, **76**, 041916 (**2007**).

15 F. Mormann, K. Lehnertz, P. David and C. E. Elger. Mean phase coherence as a measure for phase synchronization and its application to the EEG of epilepsy patients. *Physica D*, **144**, 358–69 (**2000**).

16 F. Mormann, T. Kreuz, C. Rieke, R. G. Andrzejak, A. Kraskov, P. David, C. E. Elger and K. Lehnertz. On the predictability of epileptic seizures. *Clin. Neurophysiol.*, **116**, 569–87 (**2005**).

17 L. O. Chua and L. Yang. Cellular neural networks: theory. *IEEE Trans. Circ. Syst.*, **35**, 1257–72 (**1988**).

18 L. O. Chua and L. Yang. Cellular neural networks: applications. *IEEE Trans. Circ. Syst.*, **35**, 1273–90 (**1988**).

19 L. O. Chua, M. Hasler, S. Moschytz and J. Neirynck. Autonomous cellular neural networks: A unified paradigm for pattern formation and active wave propagation. *IEEE Trans. Circ. Syst.*, **42**, 559–77 (**1995**).

20 J. J. Hopfield. Neural networks and physical systems with emergent computational abilities. *Proc. Natl. Acad. Sci. USA*, **79**, 2554–8 (**1982**).

21 J. J. Hopfield. Neurons with graded response have computational

properties like those of two-state neurons. *Proc. Natl. Acad. Sci. USA*, **81**, 3088–92 (**1984**).

22 L. O. Chua and T. Roska. *Cellular Neural Networks and Visual Computing*. Cambridge UK: Cambridge University Press (**2002**).

23 T. Roska and L. O. Chua. The CNN universal machine: An analogic array computer. *IEEE Trans. Circ. Syst. II: Analog and Digital Signal Processing*, **40**, 163–73 (**1993**).

24 T. Roska and A. Rodriguez-Vazquez. *Towards the Visual Microprocessor: VLSI Design and the use of Cellular Neural Network Universal Machines*. Chichester: J. Wiley (**2000**).

25 S. Ulam. Random processes and transformations. *Proc. Int. Congress Mathem.*, **2**, 264–75 (**1952**).

26 S. Wolfram. Statistical mechanics of cellular automata. *Rev. Mod. Phys.*, **55**, 601–44 (**1983**).

27 E. M. Izhikevich. *Dynamical Systems in Neuroscience: The Geometry of Excitability and Bursting*. The MIT Press, Cambridge, Massachusetts, London, England (**2007**).

28 A. Adamatzky. *Computing in Nonlinear Media and Automata Collectives*. IoP Publishing, Bristol (**2001**).

29 J. D. Murray. *Mathematical Biology*. New York: Springer (**1989**).

30 E. Meron. Pattern formation in excitable media. *Phys. Rep.*, **218**, 1–66 (**1992**).

31 M. C. Cross and P. C. Hohenberg. Pattern formation outside from equilibrium. *Rev. Mod. Phys.*, **65**, 854–1112 (**1993**).

32 G. Baier, R. S. Leder and P. Parmananda. Human electroencephalogram induces transient coherence in excitable spatiotemporal chaos. *Phys. Rev. Lett.*, **84**, 4501–4 (**2000**).

33 G. Baier and M. Müller. Nonlinear dynamic conversion of analog signals into excitation patterns. *Phys. Rev. E.*, **70**, 037201 (**2004**).

34 G. Baier and M. Müller. Frequency-selective induction of excitation

waves near sub- and super-critical Hopf bifurcation. *Phys. Lett. A.*, **330**, 350–7 (**2004**).

35 A. Chernihovskyi, F. Mormann, M. Müller, C.E. Elger, G. Baier and K. Lehnertz. EEG analysis with nonlinear excitable media. *J. Clin. Neurophysiol.*, **22**, 314–29 (**2005**).

36 H. L. F. Helmholtz. *On the Sensations of Tone as a Physiological Basis for the Theory of Music.* New York: Dover Publications (**1954**).

37 R. FitzHugh. Impulses and physiological states in theoretical models of nerve membrane. *Biophys. J.*, **1**, 445–66 (**1961**).

38 J. S. Nagumo, S. Arimoto and S. Yoshizawa. An active pulse transmission line simulating nerve axon. *Proc. IRE*, **50**, 2061–70 (**1962**).

39 A. Chernihovskyi, C. E. Elger and K. Lehnertz. Effect of inhibitory diffusive coupling on frequency-selectivity of excitable media simulated with cellular neural networks. In V. Tavsanoglu and S. Arik, editors, *Proc. of the 2006 10th IEEE International Workshop on Cellular Neural Networks and their Applications*, pp 292–6. IEEE-Press (**2006**). Catalog No: 06TH8915, ISBN 1-4244-0639-0.

40 E. R. Kandel, J. H. Schwartz and T. M. Jessell. *Principles of Neural Science.* McGraw-Hill Publishing Co. (**2000**).

41 A. Chernihovskyi and K. Lehnertz. Measuring synchronization with nonlinear excitable media. *Int. J. Bifurcation Chaos*, **17**, 3425–9 (**2007**).

42 K. Schindler, R. Wiest, M. Kollar and F. Donati. Using simulated neuronal cell models for detection of epileptic seizures in foramen ovale and scalp EEG. *Clin. Neurophysiol.*, **112**, 1006–17 (**2001**).

43 K. Schindler, R. Wiest, M. Kollar and F. Donati. EEG analysis with simulated neuronal cell models helps to detect pre-seizure changes. *Clin. Neurophysiol.*, **113**, 604–14 (**2002**).

44 F. Mormann, R. G. Andrzejak, C. E. Elger and K. Lehnertz. Seizure prediction: the long and winding road. *Brain*, **130**, 314–33 (**2007**).

45 M. Le Van Quyen, J. Soss, V. Navarro, R. Robertson, M. Chavez, M. Baulac and J. Martinerie. Pre-ictal state identification by synchronization changes in long-term intracranial EEG recordings. *Clin. Neurophysiol.*, **116**, 559–68 (**2005**).

46 B. Schelter, M. Winterhalder, T. Maiwald, A. Brandt, A. Schad, A. Schulze-Bonhage and J. Timmer. Testing statistical significance of multivariate time series analysis techniques for epileptic seizure prediction. *Chaos*, **16**, 013108 (**2006**).

47 H. Osterhage, F. Mormann, M. Staniek and K. Lehnertz. Measuring synchronization in the epileptic brain: a comparison of different approaches. *Int. J. Bifurcation Chaos*, **17**, 3425–9 (**2007**).

48 S. Boccaletti, J. Kurths, G. Osipov, D. L. Valladare and C. S. Zhou. The synchronization of chaotic systems. *Phys. Rep.*, **366**, 1–101 (**2002**).

49 C. (Hugenii) Huygens. *Horologium Oscillatorium.* Apud F. Muguet, Parisiis (**1673**).

50 V. S. Afraimovich, N. N. Verichev and M. I. Rabinovich. General synchronization. *Izv. VUZ. Radiophiz.*, **29**, 795–803 (**1986**).

51 J. Arnhold, P. Grassberger, K. Lehnertz and C. E. Elger. A robust method for detecting interdependences: application to intracranially recorded EEG. *Physica D*, **134**, 419–30 (**1999**).

52 F. Takens. Detecting strange attractors in turbulence. In D. A. Rand and L. S. Young, editors, *Dynamical Systems and Turbulence*, volume 898 of *Lecture Notes in Mathematics*, pp 366–81. Springer-Verlag, Berlin (**1980**).

53 R. Quian Quiroga, A. Kraskov T. Kreuz and P. Grassberger. Performance of different synchronization measures in real data: A case study on electroencephalographic

signals. *Phys. Rev. E*, **65**, 041903 (2002).

54 F. Mormann, R. Andrzejak, T. Kreuz, C. Rieke, P. David, C. Elger and K. Lehnertz. Automated detection of a preseizure state based on a decrease in synchronization in intracranial electroencephalogram

recordings from epilepsy patients. *Phys. Rev. E*, **67**, 021912 (**2003**).

55 R. G. Andrzejak, F. Mormann, T. Kreuz, C. Rieke, A. Kraskov, C. E. Elger and K. Lehnertz. Testing the null hypothesis of the nonexistence of a preseizure state. *Phys. Rev. E*, **67**, 010901(R) (**2003**).

11
Intrinsic Cortical Mechanisms which Oppose Epileptiform Activity: Implications for Seizure Prediction

Andrew J. Trevelyan

11.1
Introduction

Epilepsy is amongst the most common serious neurological conditions. Yet any consideration of what we know about cortical anatomy or function begs the question 'why is epilepsy not more common still?' Rhythmic synchronized firing appears to be a fundamental feature of cortical function: there are multiple recognized synchronizing mechanisms (for instance, intrinsic bursting of principal neurons, sensory inputs, glutamate release from glia, amongst others), and many reports of synchronized activity patterns dating back to the first EEG recordings by Berger in 1929. Furthermore, when one examines the firing patterns of individual neurons in various *in vivo* and *in vitro* preparations, one notable and common pattern of activity is for the neuron to enter a sustained depolarized state for hundreds of milliseconds, and fire repeatedly (this pattern of activity is often referred to as an 'UP state'). While the function of these various physiological states remains unclear, if one considers these activity patterns in the context of the known recurrent pattern of connectivity of cortical networks (reciprocal connections not only between individual neurons but also local populations and also cortical areas), then the likelihood of the network becoming locked into a cycle of rhythmic, re-entry excitation appears great.

Why then, are we not seizing continuously? One proposal suggested 40 years ago by Prince and Wilder [1], is that the spread of epileptiform activity is prevented by a surround inhibition. The inhibitory surround theory, if correct, deserves a great deal of attention, since it immediately suggests a likely cause of epilepsy arising from deficits in this protective mechanism. The mechanism might also be a target for pharmacotherapy, seeking to bolster the inhibitory mechanism in patients who suffer focal seizures. Yet in spite of the obvious importance of this mechanism, we are little nearer understanding the network basis for the inhibitory surround than when Prince first suggested his hypothesis 40 years ago. A significant factor slowing our progress has been the difficulty of studying this phenomenon *in vivo*. Recently, however, we have been able to demonstrate a means of examining the inhibitory surround mechanism in an *in vitro* preparation, with all the benefits that ensue from such studies (enhanced control over the

Seizure Prediction in Epilepsy. Edited by Björn Schelter, Jens Timmer and Andreas Schulze-Bonhage
Copyright © 2008 WILEY-VCH Verlag GmbH & Co. KGaA, Weinheim
ISBN: 978-3-527-40756-9

experimental regime, improved access both for imaging and for electrophysiological recordings, etc.).

In this chapter, I will review the evidence for such a mechanism from *in vivo* experimental work, and follow this with a brief discussion of the relative merits of *in vitro* and *in vivo* studies and the interpretation of these different data sets. I will then describe my own studies, which stemmed initially from a theoretical consideration, but which then led to my developing a means of examining this important protective cortical mechanism *in vitro*. Finally, I will discuss what implications the inhibitory surround hypothesis has for seizure prediction.

11.2
The Inhibitory Surround in Cortex: *In Vivo* Studies

The concept of the inhibitory surround dates back to the first intracellular recordings made in cortical tissue by Powell and Mountcastle [2] who showed that when a focal stimulation is applied to the cortex, the surrounding territories have their activity suppressed (as an aside, Tom Powell, the first author of that study, taught this author neuroanatomy in his first year at Oxford in 1985, thus providing a personal link over this 50 year research project!). Prince and Wilder [1] showed that the same suppression of surround cortex also occurs with epileptic activity. They induced epileptiform activity with a focal application of penicillin, and recorded interictal events using field electrodes. At the same time, they also recorded single neurons and divided these into two groups according to the pattern of synaptic barrage during these events. In one group, the cells experienced strong paroxysmal depolarizing events synchronous with the interictal event. These cells were mostly located close to the ictal focus. The second group, which formed the predominant cell type in the surrounding territory, received inhibitory barrages during the events, and had their firing suppressed. Dichter and Spencer [3] then showed an almost identical phenomenon in hippocampus, inducing their foci with local strychnine injections. More recently, similar differences between the focus and the surrounding territories have been demonstrated using using intrinsic optical imaging [4] and voltage-sensitive dyes [5].

Further evidence for the inhibitory surround mechanism comes from immuno-histochemical studies subsequent to focal tetanus injections in rats. In contrast to the models discussed in the previous paragraph, focal tetanus toxin injection generally (although not always) triggers seizures only after a latent period of several days [6]. The seizures in those first few days after the initial event are the most intense and tend to generalize, but subsequently, seizure activity appears reduced and much more localized [6]. This temporal evolution is reflected in the expression patterns of a number of proteins. At the stage when seizures are most severe, BDNF is upregulated throughout the cortex, but after activity settles to a more focal pattern, BDNF upregulation is also restricted to the area surrounding the initial tetanus injection [7]. At this late stage, at the injection site, there is upregulation of the the immediate early genes, zif268 and c-fos, as well as GAD67 (GABA synthesis enzyme whose expression is known to be modulated by activity) and NR1 (NMDA receptor subunit 1), and down regulation of CaMKII (a pyramidal specific marker) and reduction also in GluR2 mRNA (expressed in both pyramids and interneurons) [8, 9]. In contrast, all these markers showed the exact opposite expression pattern in the surround area. Undoubtedly there will be changes in expression of other molecular species too, and much more work is required before we can ascribe a precise cause and effect to these changes relative to the changes in seizure pattern. Suffice to say that, as far as our current state of knowledge allows us to speculate, these annular expression patterns could be interpreted as providing a restraint which, after the initial seizure activity, subsequently keeps the pathological activity localized.

11.3
In Vitro Studies: Strengths and Weaknesses

In vitro preparations provide a greatly simplified network to study, and therein lies both their strength and their weakness. One has almost complete control over the extracellular solution, and can rapidly switch the bathing media to apply or remove pharmacological agents. The control does not extend to the small localized, transient fluxes in the extracellular environs, most obviously involving the release of neurotransmitters and ionic species, in particular K^+ when activity levels are especially high. But then, such local fluxes are desirable since they represent the network behavior that we want to understand. The important fact is that the general environment can be maintained very stably because the large volume of the bathing solution overwhelms any homeostatic capabilities of the reduced preparation.

The second big advantage of *in vitro* preparations is the accessibility. One can visualize the network, especially in submerged slices, and identify certain neuron classes with relative ease using DIC (differential interference contrast) imaging, or fluorescence microscopy if these neurons have been prelabeled with a fluorescent marker such as GFP. One can thus target particular neurons to patch clamp, and furthermore, patch on to multiple cells locally. In recent years, Henry Markram's group have pursued this to an extreme level, demonstrating an ability to patch clamp up to 12 closely apposed neurons simultaneously. These kinds of experiments are difficult, if not impossible, *in vivo*.

The prime drawback of *in vitro* experiments is that all long-range axonal tracts are severed during the preparation. The most relevant of these tracts for epilepsy research are the pathways to and from thalamic nuclei, those that connect the two hemispheres through the corpus callosum and the commissural connections between the two hippocampi, and the long range intrahemispheric connections. This is a significant limitation, and to address it, various researchers have developed a number of *in vitro* preparations that preserve particular long-range tracts. The simple fact though, is that if one wants to have all long-range tracts intact, then

Mouse: *in vitro*

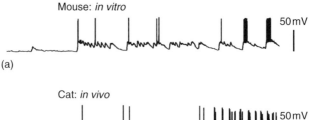

50 mV

(a)

Cat: *in vivo*

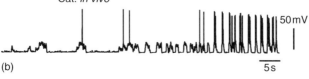

50 mV

(b)   5 s

Fig. 11.1 Comparable pattern of recruitment of cells to ictal events as recorded *in vitro* and *in vivo*. Typical pattern of rhythmic depolarizations without action potentials recorded in pyramidal cells prior to their recruitment to an ictal event. Note the extreme similarity between the recording from a brain slice of a juvenile mouse (a) and that from an adult cat *in vivo* (b) recording provided by Igor Timofeev.

one must work with the whole brain. The absence of these pathways *in vitro* means that one will have a compromised view of the pattern of generalization of ictal activity. It also means that we have to validate these models using whatever means of comparison is available – generally utilizing activity patterns recorded either intracellularly or extracellularly. And it is important to emphasize here that when we do just this, many *in vitro* models, while obviously being 'reduced preparations,' do appear to reflect particular aspects of *in vivo* activity really rather well (Figure 11.1).

For instance, bathing brain slices in artificial cerebrospinal fluid (CSF) lacking in Mg^{2+} ions provokes a wide range of epileptiform activity including transient interictal events with a low level of neuronal participation [10, 11], slow and rapid patterns of generalization [12–14], and full ictal events with a clear tonic-clonic pattern apparently involving all neurons in the network [10,11]. This model has also been used to explore pharmaco-resistant status epilepticus [15,21,22]. Furthermore, the pattern of membrane potential (E_m) fluctuations during the critical time period when a neuron is being recruited to an ictal event is demonstrably the same in focal neocortical seizures *in vivo* as it is *in vitro* (Figure 11.1), as is the pattern of high-frequency activity in extracellular recordings. Finally, the wide range of propagation speeds recorded *in vivo* is well matched by the thousand-fold range of speeds seen in $0\,Mg^{2+}$ *in vitro* [13]. Thus, in so far as there are *in vivo* recordings to compare with, the activity seen in the *in vitro* $0\,Mg^{2+}$ model appears well matched to certain *in vivo* epileptic events. One implication is that what regulates the recruitment of a neuron to an ictal event is preserved within the *in vitro* preparation: the critical factors are likely the local synaptic network interactions and the intrinsic cellular properties of the neurons. This then is where we benefit, because our *in vitro* experiments have allowed us to go far beyond what has been observed *in vivo*, and we can start to unveil the network basis of the inhibitory surround mechanism.

11.4
Inhibitory Surround in an *In Vitro* Preparation

We recorded epileptiform activity using two extremely powerful research tools: fast Ca^{2+} imaging of network activity in tandem with patch clamp recordings. The usefulness of Ca^{2+} imaging comes from the fact that every time neurons fire action potentials, they experience a sharp, transient rise in intracellular Ca^{2+}, primarily due to opening of voltage-dependent Ca^{2+} channels in the cell membrane. Ca^{2+} dyes can indicate these fluxes because their fluorescent properties change with the local Ca^{2+} concentration. There are, of course, technical and interpretative problems as there are with any experimental protocol, and I would encourage the interested reader to read up about these issues in more specialist texts [23]. The fact remains though that Ca^{2+} dyes provide an unprecedented view of neuronal networks in action, allowing hundreds of neurons to be monitored simultaneously with single-cell resolution. Because of the kinetics of the dyes (fast onset, slow decay), they most readily lend themselves to identifying when a neuron's activity increases, which is of course exactly what is desired when studying the recruitment of neurons to epileptiform events.

Simultaneous with our Ca^{2+} imaging of the network, we made patch clamp recordings of neurons.Initially we targeted for patching just one specific class, the large layer-5 pyramidal cells. It was important to have a set of 'reference' recordings which would allow us to relate different experiments, and these cells seemed a good choice for this purpose: they are readily identifiable under DIC optics, and generally easy to patch; their dendritic trees span virtually the entire cortical axis from their apical tuft lying in layer 1 down to their basal dendrites sampling layer 6, so they provide a good 'microphone' for listening to synaptic activity within the slice; and these cells had previously been implicated in ictogenesis. To separate out inhibitory and excitatory drives onto these cells, we held them roughly halfway between the reversal potentials for glutamate (E_{glut}) and GABA (E_{GABA}). Our reference recordings then, were voltage clamp (V_{clamp}) recordings of these layer-5 pyramidal cells held at $-30\,mV$.

We were further helped in our analysis of ictal activity patterns by finding that adjacent layer-5 pyramidal cells experienced almost identical synaptic barrages at these times (Figure 11.2). This was a statistical effect stemming from the sheer intensity of the local activity, because at other, more quiescent times, the synaptic currents in the paired recordings were not particularly well correlated. During ictal events though, this correlation of synaptic drive presented us with an extremely useful research tool: one cell could be recorded in 'reference mode,' that is in V_{clamp} mode at $-30\,mV$, and the other cell could be recorded in a different mode to examine other facets of the ictal activity pattern. These paired recordings of adjacent pyramidal cells then allowed us to tease apart the synaptic drives onto these cells and their firing patterns. For instance, the second cell could be recorded in current clamp or cell attached modes, to derive the firing

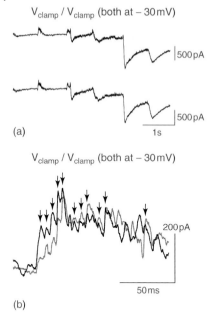

V_{clamp} / V_{clamp} (both at – 30 mV)

| 500 pA

| 500 pA

(a)   1s

V_{clamp} / V_{clamp} (both at – 30 mV)

| 200 pA

50 ms

(b)

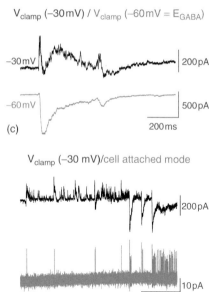

V_{clamp} (–30 mV) / V_{clamp} (–60 mV = E_{GABA})

–30 mV | 200 pA

–60 mV | 500 pA

200 ms

(c)

V_{clamp} (–30 mV)/cell attached mode

| 200 pA

| 10 pA

5 s

(d)   ↑
                    Recruitment

Fig. 11.2 Example of paired recordings from adjacent pyramidal cells. (a) Paired recording with both cells held in the 'reference mode,' voltage clamped at −30 mV, approximately midway between the reversal potential for glutamate (approx. 0 mV), and GABA (approx. −60 mV). The convention is for inward currents (depolarizing, excitatory currents) to be plotted as downward deflections. The period shows the transition period from predominantly inhibitory to predominantly excitatory barrages, which is coincident with the ictal wavefront invading the local territory. (b) Paired recordings of two adjacent pyramidal cells in 'reference mode' at an expanded timescale. The recording shows an interictal event, and synchronous IPSCs are readily seen in the two traces (arrowed), suggesting that the currents in both cells are caused by synaptic events from the same presynaptic interneuron. (c) One cell is held at the reference holding potential (−30 mV) and the other is held at the GABAergic reversal potential (E_{GABA}). These recordings show that, what appear to be predominantly inhibitory events from the −30 mV recordings, also have a huge excitatory component. Indeed the excitatory drive at these times would induce a typical ictal discharge with many action potentials in the absence of any inhibitory drive. (d) The restraining period is best visualized by holding one cell at the reference holding potential (−30 mV), whilst recording the output of the adjacent cell. Here we see a protracted period of almost 15 s when the second cell is held in check, firing just four action potentials in this time. We know from recordings like that shown in (c) that, during this whole time, the cell is being bombarded with intense excitatory barrages, yet the vetoing inhibition is the dominant force.

pattern corresponding to the synaptic barrages. Alternatively, the second cell might be recorded at the GABA reversal potential, thereby allowing the excitatory drive to be assessed in relative isolation. These were key experiments, and it is important to understand that such paired recordings of adjacent cells are completely impossible *in vivo*. These experiments could only have been done in an *in vitro* preparation.

Further details of these experiments are to be found in our recent publications [12, 13, 24, 25]. I will restrict myself here to a summary of the main findings. The model we chose to examine is one of the most widely studied *in vitro* seizure models, induced by removing extracellular Mg^{2+} ions ($0\,Mg^{2+}$ model) [10, 16–20, 38]. Mg^{2+} ions normally provide the voltage-dependent blockade of NMDA receptors. Removing Mg^{2+} thus increases the level of excitatory drive and it is this that causes the epileptiform activity. The inhibitory circuitry on the other hand, is relatively preserved, at least initially.

The earliest full ictal events, following the removal of extracellular Mg^{2+}, progress very slowly across the cortical network, and we were able to demonstrate that the speed of propagation was the direct result of an inhibitory restraint ahead of the ictal wavefront. If we take the perspective of an individual layer-5 pyramidal cell, such as that shown in Figure 11.1, then the sequence of events as the ictal wavefront approaches is the following. When the wavefront is still several hundred microns away, the cell starts to experience rhythmic synaptic barrages at approximately δ frequency (0.5–2 Hz), that is temporally matched to the rhythmic network activity of the cells upstream, which have already been recruited to the event (Figure 11.3). This temporal correspondence tells us that the synaptic barrage arises from the ictal event, feeding forward onto the adjacent territories. The feedforward synaptic drive includes a very powerful excitatory drive that ordinarily would exceed threshold many times over, yet this drive does not activate the cells because it is vetoed by an even more powerful feedforward inhibition.

The basic δ frequency rhythm of the ictal event means that the network experiences repeated crises at this rhythm. These are the times when new neurons are recruited to the event, and if recruitment is resisted at these crisis points, then the ictal wave does not progress. The eventual recruitment of cells can be delayed in this way for tens of seconds after the initial barrages. We showed that the speed at which an ictal event traverses a region of cortex is inversely proportional to how many crises were resisted [13]. This simple relationship explained propagation speeds over a thousand-fold range of speeds. On occasions, the restraint event outlasts the ictal activity, and the event is aborted altogether. These experiments showed us, therefore, that the recruitment of pyramidal cells to an ictal event occurs not when the initial excitation occurs, but rather when the inhibitory restraint fails.

The task ahead of us then, is to identify which cells provide the restraint, and why it should fail. The fact that we can now study this important phenomenon in an *in vitro* preparation will be a huge help in this endeavour. It is clear from other work, however, that inhibition need only be slightly compromised to nullify the restraint altogether. Thus, in brain slices bathed in a normal, non-epileptogenic medium, focal electrical stimulation only activates the local territory, and this activity is rapidly extinguished in the surrounding territory, much as *in vivo*. Barry Connors' group explored what then happens if one perfuses increasing levels of $GABA_A$ antagonist [26]. Initially, at levels which suppress evoked IPSCs only by about 10–20%, the focal activation is still contained, albeit not as well, activating roughly double the area of cortex. When disinhibition was increased just a fraction more, however, the focal activation spread with great speed, and without diminution,

Occipital neocortex:
coronal slice

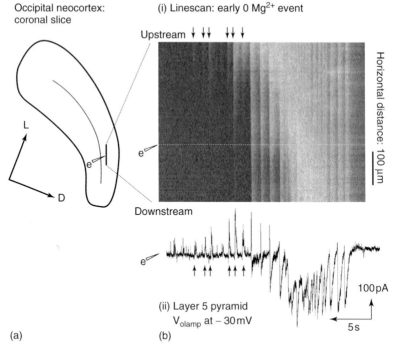

(i) Linescan: early 0 Mg²⁺ event

Fig. 11.3 (a) Schematic of a coronal slice showing the recording electrode ('e') and the location of a 'line scan' through layer 5. Imaging was done at 30 Hz using a 10 × objective. (bi) Line scan showing the neuropil Ca²⁺ fluorescence as an epileptiform event progresses from top to bottom – distance is plotted on the ordinate, and time on the abscissa. Each row of pixels represents the neuropil Ca²⁺ fluorescence from a bin 40 μm high by 4 μm wide (165 bins = 660 μm total length). The location of the recording electrode ('e') is shown. (bii) The V_{clamp} recording from the layer-5 pyramidal cell plotted below on the same time scale. The cell was held at −30 mV, approximately half-way between the reversal potentials for GABA and glutamate, to distinguish inhibitory drives (upward deflections) and excitatory drives (downward deflections). Note the prominent inhibitory volleys (arrowed) corresponding with upstream activity in the line scan. Figure used with permission from [12].

throughout the slice. Notably, further increasing the level of GABA$_A$ antagonist had little effect on the propagation pattern or speed. In complete agreement with Connors' experiments, we also found that these same levels of disinhibition negated the inhibitory restraint, allowing all spontaneously generated epileptiform events to spread with great rapidity and with no delays across the entire tissue.

To summarize our *in vitro* work, in an epilepsy model with intact inhibition, there is a powerful feedforward inhibition ahead of the ictal wavefront which can veto the episodic excitatory barrages also emanating from the wavefront. The inhibitory restraint though is relatively fragile, being compromised by relatively low levels of GABA$_A$ antagonists. When present, however, the inhibitory restraint can explain a very wide range of propagation speeds across cortical networks.

11.5
Models of the Inhibitory Surround: The Importance of how Synaptic Inputs are Distributed

The key feature of the inhibitory surround is that the inhibitory effect of focal activation extends further than the excitatory effect. How is this achieved? A good place to start is the excellent monograph by Traub and Miles [27]. Their insights into epileptic activity derived from a gradual elaboration of a network model incorporating thousands of connected multicompartment cellular models. Their initial aim was to model activity patterns in the *in vitro*, disinhibition, experimental model of epilepsy, identifying this as perhaps the most tractable of the experimental models since it removed the requirement for modeling fast inhibition. They then developed the model further to explore the effects of inhibition on epileptiform activity, including a suggested basis for the inhibitory surround. Many of these insights stand the test of time extremely well, although I believe that there is one additional facet of the network which plays an important role in the inhibitory surround: namely, the arrangement of synaptic terminals on the soma and proximal dendrites of the pyramidal cells.

One way in which the inhibitory surround might extend beyond the focal excitation is for inhibitory neurons to have a larger radius of axonal arborizations than excitatory ones. While long-range inhibitory axons clearly do exist [28], they appear to be at an extremely low density and rather patchily distributed. The vast majority of inhibitory interactions are very local, and so it seems unlikely that distant monosynaptic inhibitory interactions are the basis of the inhibitory surround. A further argument against a long range-inhibitory axonal explanation is that our experiments using brain slices consistently show evidence of the inhibitory restraint in every recording: if the inhibitory restraint were based on long-range connections, one might expect it to be very sensitive to the sectioning process. The evidence then is that the inhibitory surround arises not through extended monosynaptic pathways, but rather through disynaptic inhibitory pathways.

The key feature of Traub and Miles' model is that epileptiform events only propagate if excitatory neurons fire bursts of action potentials. The reason is that excitatory neurotransmission between cortical neurons is very weak. Consequently, recruitment of the next population of neurons requires protracted barrages of excitatory synaptic events, and is thus the rate-limiting step in the propagation process. This insight explains why propagation in the completely disinhibited slice is about a factor of ten slower than axonal propagation. When inhibition is gradually increased in their model, it brought about a precipitous drop in the likelihood of epileptiform events arising from a single cell. This precipitous cut-off also mirrors very well our [12] and other's [26] results that increasing levels of GABAergic antagonists have an all-or-nothing effect suppressing the inhibitory restraint. The explanation is that pyramidal cell bursting in the model is peculiarly sensitive to inhibition, an insight that received experimental support from dendritic patch clamp experiments which showed that inhibitory drives onto the apical axonal trunk disconnect the excitable apical tuft from the soma [29].

In contrast to pyramidal cells, many interneuron cell classes are activated rather more easily, and thus one sees a preferential recruitment of interneurons in the surrounding territories. Furthermore, the differential recruitment of interneurons over pyramidal cells would be expected to increase at greater distances from the focus of activity. This then appears to explain the more distant inhibitory surround, but our experiments suggest that there exists a vetoing inhibition that suppresses monosynaptic excitatory drive directly. Thus the inhibitory surround can be manifest within a narrower confine than suggested by Traub and Miles' proposed mechanism.

This more adjacent inhibitory restraint, I believe, arises from the peculiar arrangement of synaptic drives on to pyramidal cells and interneurons, as noted from electron-microscopy (EM) studies of cortical networks [30–32]. These studies showed that, while interneurons receive excitatory drive across their entire somatodendritic axis, pyramidal cells only receive excitatory drive onto their more distal dendritic branches. The somata and proximal dendrites (up to about $50\,\mu$m from the soma) of pyramidal cells receive only inhibitory synapses arising from a particular subpopulation of interneurons, termed basket cells. (For this discussion, I will ignore a second class of interneuron, the chandelier cell, which targets the pyramidal axon initial segment, since their long presumed role vetoing action potential generation in pyramidal cells has recently been confused by the suggestion that they may be excitatory due to anomalies in local intracellular Cl^- concentration.)

There is now a large body of evidence that basket cells can coordinate firing in pyramidal cell populations [33], and particularly during γ rhythms (30–80 Hz) [39–42]. This entrainment appears to occur by permitting pyramidal firing only between inhibitory postsynaptic currents (IPSCs); because IPSCs are synchronized on large numbers of pyramidal cells, the windows of opportunity for pyramidal firing are also synchronized. Basket cells can also fire at much higher rates than γ frequencies, up to ≈ 300 Hz, and a simple consideration of the arrangement of their output onto pyramidal cells suggests that this firing intensity can veto any amount of excitatory synaptic drive on to the pyramids [43]. Notably, an almost identical arrangement of proximal inhibitory drive exists onto neurons which control the escape reflex in crayfish [44]. In the same way that pyramidal neurons can be restrained by an inhibitory veto [12], the proximal inhibition in crayfish can veto the escape mechanism.

Thus there appears to be a nested set of inhibitory effects that underlie the inhibitory surround: a very powerful inhibitory veto exerted directly onto the somata and proximal dendrites of pyramidal cells, and a second less intense effect arising from the preferential activation of interneurons over pyramidal cells at a distance from the ictogenic focus.

11.6
Surround Inhibition: Implications for Seizure Prediction

If the inhibitory restraint is mediated in part through high-frequency firing of basket cells, does this have any bearing on the association of fast EEG ripples with seizure initiation? Several studies have noted that the critical stage of recruitment of neurons to an ictal event is accompanied by the appearance of high-frequency oscillations in the local field recordings, leading to the proposal that these high-frequency ripples are in some way causing the seizure. Indeed, high-frequency oscillations are considered to be one of the most useful features of EEG recordings in our efforts to develop seizure prediction algorithms.

Careful analysis of the source of high-frequency ripples associated with certain physiological events in the hippocampus, indicated that the peak density of current is in the pyramidal cell layer [34]. These authors proposed that the predominant current was through $GABA_A$ channels opened by high-frequency basket cell discharges. My recent work suggests that the inhibitory restraint may be mediated through this same pathway, giving a rather different view of fast ripples: that these high-frequency ripples in the EEG are in fact an epiphenomenon of the inhibitory restraint.

At the outset, it is important to realise that the proposed mechanism of Ylinen et al. cannot explain high-frequency ripples in the non-synaptic *in vitro* model of epilepsy, induced by bathing brain slices in a Ca^{2+}-free medium [35]. The fast ripples in the $0\,Ca^{2+}$ model, however, have a rather different pattern to those recorded in models with intact synaptic transmission: in the non-synaptic model, the ripples shows a gradual build up over many minutes, with slight modulations of the amplitude. In this model, the high-frequency events appear to be action potentials, recorded extracellularly from many different neurons, whose activity gradually becomes synchronized, either by ephaptic means [36] or though gap-junction coupling of their axonal plexus [37]. In contrast, in synaptic models, the ripples come in relatively short bursts lasting a few hundred milliseconds. A significant component of the fast ripples in these synaptic models, I suggest, is from perisomatic high-frequency inhibitory barrages.

How does this new interpretation of fast ripples change our view of them as a predictive measure of seizures? It certainly necessitates a more nuanced view of fast ripples. Their presence in the EEG trace should still be indicative of a high risk of imminent seizures, since the implication is that the cortical network is under threat. Their presence though is also indicative of attempts by the network to restrain that threat. If, on the other hand, the inhibitory restraint was severely compromised for whatever reason, then one might anticipate seizures to evolve very suddenly with no warning in the form of high-frequency EEG ripples. Such a pattern is seen in myoclonic epilepsy, an epilepsy phenotype that has proved notoriously resistant to our attempts to predict seizures.

References

1 D. A. Prince and B. J. Wilder, *Arch. Neurol.* **16**, 194–202 (1967).

2 T. P. S. Powell and V. Mountcastle, *Bulletin of the John Hopkins Hospital* **105**, 133–62 (1959).

3 M. Dichter and W. A. Spencer, *J Neurophysiol* **32**, 649–62 (1969).

4 T. H. Schwartz and T. Bonhoeffer, *Nat Med* **7**, 1063–7 (2001).

5 D. Derdikman, R. Hildesheim,
E. Ahissar, A. Arieli and A. Grinvald,
J. Neurosci. **23**, 3100–5 (**2003**).

6 J. Mellanby, G. George, A. Robinson
and P. Thompson, *J. Neurol. Neuro-surg. Psychiatry* **40**, 404–14 (**1977**).

7 F. Liang, L. D. Le and E. G. Jones,
Cereb. Cortex. **8**, 481–91 (**1998**).

8 F. Liang and E. G. Jones, *J. Neurosci.*
17, 2168–80 (**1997**).

9 F. Liang and E. G. Jones, *Brain Res*
778, 281–92 (**1997**).

10 W. W. Anderson, D. V. Lewis, H. S.
Swartzwelder and W. A. Wilson,
Brain Res. **398**, 215–9 (**1986**).

11 A. C. Flint, U. S. Maisch and A. R.
Kriegstein, *J. Neurophysiol.* **78**,
1990–6 (**1997**).

12 A. J. Trevelyan, D. Sussillo,
B. O. Watson and R. M. Yuste,
J. Neurosci. **26**, 12447–55 (**2006**).

13 A. J. Trevelyan, D. Sussillo and
R. M. Yuste, *J. Neurosci.* **27**, 3383–87
(**2007**).

14 B. Y. Wong and D. A. Prince,
Epilepsy Res. **7**, 29–39 (**1990**).

15 J. P. Dreier and U. Heinemann,
Neurosci. Lett. **119**, 68–70 (**1990**).

16 J. P. Dreier and U. Heinemann, *Exp.
Brain Res.* **87**, 581–96
(**1990**).

17 I. Mody, J. D. Lambert and
U. Heinemann, *J. Neurophysiol.* **57**,
869–88 (**1987**).

18 A. M. Thomson and D. C. West,
Neuroscience **19**, 1161–77 (**1986**).

19 H. Walther, J. D. Lambert, R. S.
Jones, U. Heinemann and
B. Hamon, *Neurosci. Lett.* **69**, 156–61
(**1986**).

20 F. Weissinger, K. Buchheim,
H. Siegmund, U. Heinemann and
H. Meierkord, *Neurobiol. Dis.* **7**,
286–98 (**2000**).

21 J. P. Dreier, C. L. Zhang and
U. Heinemann, *Acta. Neurol. Scand.*
98, 154–60 (**1998**).

22 M. Pfeiffer, A. Draguhn,
H. Meierkord and U. Heinemann,
J. Pharmacol. **119**, 569–77 (**1996**).

23 R. M. Yuste and A. Konnerth,
*Imaging in Neuroscience and
Development: A Laboratory Man-ual*, (Cold Spring Harbor Labo-ratory Press, New York, **2005**).

24 A. J. Trevelyan, T. Baldeweg,
W. van Drongelen, R. Yuste and
M. A. Whittington, *J. Neurosci.* **27**,
13513–9 (**2007**).

25 A. J. Trevelyan, O. Watkinson,
I. Timofeev, D. Sussillo and
R. Yuste, *Soc. Neurosci. Abstr.* (**2007**).

26 D. J. Pinto, S. L. Patrick, W. C.
Huang and B. W. Connors,
J. Neurosci. **25**, 8131–40 (**2005**).

27 R. D. Traub and R. Miles, *Neuronal
Networks of the Hippocampus*, (Cam-bridge University Press, Cambridge,
1991).

28 R. Tomioka, K. Okamoto, T. Furuta,
F. Fujiyama, T. Iwasato, Y. Yanagawa,
K. Obata, T. Kaneko and
N. Tamamaki, *Eur. J. Neurosci.* **21**,
1587–600 (**2005**).

29 M. E. Larkum, J. J. Zhu and
B. Sakmann, *Nature* **398**, 338–41
(**1999**).

30 A. I. Gulyas, M. Megias, Z. Emri
and T. F. Freund, *J. Neurosci.* **19**,
10082–97 (**1999**).

31 M. Megias, Z. Emri, T. F. Freund
and A. I. Gulyas, *Neuroscience* **102**,
527–40 (**2001**).

32 A. Peters and E. G. Jones, *In
Cerebral Cortex*, A. Peters and
E. G. Jones, (eds), pp. 107–21,
(Plenum Press, London, **1984**).

33 S. R. Cobb, E. H. Buhl, K. Halasy,
O. Paulsen and P. Somogyi, *Nature*
310, 685–7 (**1995**).

34 A. Ylinen, A. Bragin, Z. Nadasdy,
G. Jando, I. Szabo, A. Sik and
G. Buzsaki, *J. Neurosci.* **15**, 30–46
(**1995**).

35 A. Draguhn, R. D. Traub, D. Schmitz
and J. G. Jefferys, *Nature* **394**,
189–92 (**1998**).

36 J. G. Jefferys, *Physiol. Rev.* **75**,
689–723 (**1995**).

37 R. D. Traub and A. Bibbig,
J. Neurosci. **20**, 2086–93 (**2000**).

38 B. W. Connors, *Nature* **310**, 685–7
(**1984**).

39 M. O. Cunningham, M. A.
Whittington, A. Bibbig, A. Roopun,
F. E. LeBeau, A. Vogt, H. Monyer,

E. H. Buhl and R. D. Traub, *Proc. Natl. Acad. Sci. USA* **101**, 7152–7 (2004).

40 T. Klausberger, P. J. Magill, L. F. Marton, J. D. Roberts, P. M. Cobden, G. Buzsaki and P. Somogyi, *Nature* **421**, 844–8 (2003).

41 E. O. Mann, J. M. Suckling, N. Hajos, S. A. Greenfield and O. Paulsen, *Neuron* **45**, 105–17 (2005).

42 M. A. Whittington, R. D. Traub and J. G. Jefferys, *Nature* **373**, 612–5 (1995).

43 A. J. Trevelyan and O. Watkinson, *Prog. Biophys. Mol. Biol.* **87**, 109–43 (2005).

44 E. T. Vu and F. B. Krasne, *Science* **255**, 1710–12 (1992).

12
Is Prediction of the Time of a Seizure Onset the Only Value of Seizure-prediction Studies?[1]

Anatol Bragin, Jerome Engel Jr

12.1
Purpose of Seizure-prediction Research

The principal goal of research on seizure prediction is to determine when a seizure is going to occur. The narrower the latency of the prediction, and the lower the number of false detections, the greater the value of the prediction software to permit seizure prevention, or protection against injury. In addition to this important clinical objective, it is worthwhile to consider whether the results of seizure prediction research might also provide new insights into the glial neuronal mechanisms responsible for ictal initiation.

In most publications dealing with seizure prediction, the seizure itself is regarded as an amorphous phenomenon characterized by a singular pathological activity and subsequent ictal clinical manifestations. There has been little concern regarding the fact that available seizure prediction data all appear to demonstrate that the electrical activity of the brain during the preictal state changes in only one direction. Depending on terminology, this has been defined as a decrease in dynamical similarity of electrical signal recorded from different electrodes, reduction in correlation dimension, or reduction of complexity of the signal [1–6]. All of these changes in nonlinear characteristics reflect a decrease in the degree of synchronicity of brain electrical activity. In contrast to published results on non-lineal analysis of preictal EEG, which implies that the transition periods of all seizures share a similar 'route' toward a clinical ictal manifestation, there are, in fact, many different types of epileptic seizures, which are believed to result from a variety of different underlying glial neuronal ictogenic mechanisms.

1) This work was supported by NIH grants NS-33310, NS-02808.

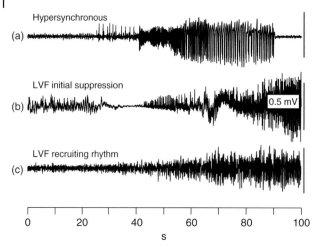

Fig. 12.1 Examples of different types of seizure onsets in patients with mesial temporal lobe epilepsy. (a) Hypersynchronous onset. (b) Low-voltage fast onset with an initial suppression. (c) Low-voltage fast onset with recruiting rhythm.

12.2
Seizure Onsets in Patients with MTLE

According to ILAEs Classification Task Force there are over 25 different seizure types [7]. In spite of immense efforts clinicians have not reached a consensus on a uniform classification of seizures on the basis of their electrographic patterns. Patients with mesial temporal lobe epilepsy (MTLE) show a variety of different patterns of seizure onsets. However, two main types have been described [8–10]. The most common is a hypersynchronous (HYP) onset (Figure 12.1a), which is characterized by the occurrence, or increases in the rate of regularly repetitive interictal spikes (IIS) before seizure spread. The second is a low-voltage fast (LVF) onset (Figure 12.1b), which is characterized by an initial suppression of the EEG amplitude and an increase in frequency of EEG activity, and can involve a monotonous increase in the amplitude of the wide-band frequencies of brain electrical activity referred to as a recruiting rhythm (Figure 12.1c) [9].

An important question with respect to the different types of electrographic seizure onset patterns is whether underlying glial-neuronal mechanisms are the same or different. Another question is whether the networks in which the disturbances occur are the same or different.

12.3
Factors Triggering Seizure Activity

Multiple precipitating factors that trigger seizure activity have been described in many publications (see [11–13] for review). Among them are: a) failure of effective

GABA inhibition while the strength of excitatory connections between principal cells remains normal; b) increase in glutamatergic excitation, as might occur with failure of glial cells to clear glutamate from the extracellular space; c) impairment of the K^+ pump leading to the accumulation of potassium in the extracellular space; and d) abnormal influx of Ca^{2+} in soma and dendrites. All of these factors could initiate seizures by disrupting the networks that prevent abnormal synchrony of neuronal discharges. It is reasonable to assume that at least some of these factors, or groups of factors, will evoke different electrographic patterns of seizure onset.

12.4
Simulating Human Electrographic Patterns of Seizure Onsets in Acute *In Vivo* Animal Experiments

It is generally accepted that the imbalance between inhibitory and excitatory networks leads to seizure activity. We asked if different types of seizure onset patterns could be produced depending on whether they were triggered by weakening the GABA-ergic system or enhancing glutamatergic systems. We evoked the imbalance between excitation and inhibition using two methods: 1) by injecting bicuculline (1 μl of 100 μM) into area CA3 of hippocampus, which causes blockade of the GABAA receptors; and 2) by injecting kainic acid (KA, $\sim$2 nanoM/0.2 μl) into area CA3 as a glutamatergic excitatory agent. Recording electrodes were implanted into the dentate gyrus (DG) adjacent to the point of injection. Experiments were carried out under freely moving conditions. We compared the pattern of EEG epileptiform activity of both drugs in each rat at an interval of two days or greater.

In all rats treated with bicuculline, the initial changes in baseline activity started with the occurrence of high-amplitude epileptiform spikes (Figure 12.2a). The appearance of the behavioral seizure component coincided with an interruption

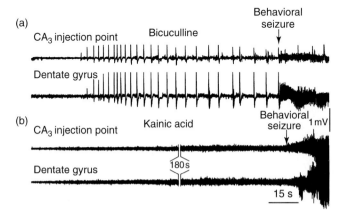

Fig. 12.2 Seizure onsets induced by intrahippocampal injection of bicuculline (a) and kainic acid (b) in the CA3 area of hippocampus and dentate gyrus in the awake rat.

of epileptiform spike activity and its replacement by high-amplitude sinusoidal 5–20 Hz activity. In all rats treated with KA the pattern of epileptiform events prior to the appearance of the behavioral component of the seizure was different from the pattern that occurred after bicuculline injection (Figure 12.2b). It consisted of a gradual increase in the amplitude of the EEG signal with steep increase in amplitude near the appearance of the behavioral seizure. This gradual increase in amplitude occurred primarily because of an increase in the power of beta and gamma activity and an increase in the amplitude of population spikes. EEG activity in the frequency band between 0.1 Hz and 10 Hz decreased during the transition period, and reappeared during the appearance of the behavioral seizure (not shown).

12.5
Conclusions

Our results demonstrated that manipulation of GABAergic inhibition in the rat hippocampus produced a very different EEG pattern during transition to ictus than manipulation of the glutamatergic system. Bicuculline injection led to a pattern resembling the HYP ictal onsets in patients with MTLE, while KA produced a recruiting rhythm similar to LVF ictal onsets in patients with MTLE. It is unclear at this point why a blockade of inhibition would produce hypersynchronization, which is believe to require enhanced inhibition [14–17], although undoubtedly bicuculline blocks some inhibitory mechanisms and not others. Similarly, it is unclear why KA should produce a recruiting rhythm, which is believed to represent a breakdown in inhibitory mechanisms [18–24]. Although future research is needed to elucidate the neuronal networks responsible for these different ictal onset patterns, they strongly suggest that the transition to ictus does not utilize a common final pathway for all types of epileptic seizures. It is reasonable to assume, therefore, that preictal changes leading up to the transition would also vary depending on the underlying triggering mechanisms and ultimate ictal manifestations. Why, then, have clinical studies of seizure prediction resulted in similar EEG changes regardless of the types of seizure under study?

Using microelectrode recordings, we have previously shown that high-frequency oscillations, termed 'Fast Ripples,' could represent the basic pathological disturbance underlying epileptogenicity in human MTLE and animal models of this disorder [25–29]. We have further shown that Fast Ripples are generated in widely dispersed, small, discrete clusters of neurons that can be difficult to locate, and that the size of these clusters may be increased by altering excitatory/inhibitory influences [25]. Based on this, we suggested that a possible mechanism of ictogenesis might involve a gradual consolidation and synchronization of areas generating Fast Ripples until a critical mass is reached that permits propagation to distant areas [30]. Although Fast Ripples can be recorded by standard electrodes at seizure onset [31, 32], if the glial neuronal mechanisms underlying ictogenesis actually begin in very small discrete areas, it is unlikely that they would be easily detectable by macroelectrodes during the preictal period. Rather, an alternative explanation for the non-lineal EEG alterations revealed by seizure prediction studies might be that these do not reflect ictal mechanisms at all, but, rather, homeostatic protective mechanisms that the brain naturally generates in order to prevent the changes that would eventually result in an epileptic seizure. This might also explain why the preictal electrical activity appears to be a decrease in synchronicity, when ictal activity is known to involve pathologically increased synchronization. Studies of seizure prediction in animal models of different types of epileptic seizure, generated by different precipitating mechanisms, would be useful to determine exactly what the preictal alterations in EEG activity identified by seizure prediction research actually represent; specifically whether they reflect ictogenic, or seizure-suppressing mechanisms. Such research will not only aid in the principal goal of seizure prediction research, but might also help to devise novel approaches to prevent seizure occurrence once this has been predicted.

References

1 C. E. Elger and K. Lehnertz, *European Journal of Neuroscience* **10**, 786 (**1998**).

2 M. Le van Quyen, J. Martinerie, M. Baulac and F. Varela, *Neuroreport* **10**, 2149 (**1999**).

3 I. Osorio, M. G. Frei, B. F. Manly, S. Sunderam, N. C. Bhavaraju and S. B. Wilkinson, *J. Clin. Neurophysiol.* **18**, 533 (**2001**).

4 B. Litt and K. Lehnertz, *Curr. Opin. Neurol.* **15**, 173 (**2002**).

5 L. D. Iasemidis, *IEEE Trans. Biomed. Eng.* **50**, 549 (**2003**).

6 F. Mormann, R. G. Andrzejak, C. E. Elger and K. Lehnertz, *Brain* **130**, 314 (**2007**).

7 J. Engel, *Epilepsia* **47**, 1558 (**2006**).

8 J. Engel, *Electroencephalogr. Clin. Neurophysiol.* **76**, 296 (**1990**).

9 J. Engel, *Can. J. Neurol Sci.* **23**, 167 (**1996**).

10 A. L. Velasco, C. L. Wilson, T. L. Babb and J. Engel, *Neural Plast.* **7**, 49 (**2000**).

11 D. A. Prince, B. W. Connors and L. S. Benardo, *Adv. Neurol.* **34**, 177 (**1983**).

12 D. A. McCormick and D. Contreras, *Ann. Rev. Physiol.* **63**, 815 (**2001**).

13 I. Timofeev and M. Steriade, *Neuroscience* **123**, 299 (**2004**).

14 P. A. Schwartzkroin and M. M. Haglund, *Epilepsia* **27**, 523 (**1986**).

15 S. R. Cobb, E. H. Buhl, K. Halasy, O. Paulsen and P. Somogyi, *Nature* **378**, 75 (**1995**).

16 C. J. McBain, T. F. Freund and I. Mody, *Trends Neurosci.* **22**, 228 (**1999**).

17 J. L. Velazquez and P. L. Carlen, *Eur. J. Neurosci.* **11**, 4110 (**1999**).

18 B. Lancaster and H. V. Wheal, *Brain Res.* **295**, 317 (**1984**).

19 N. W. Milgram, T. Yearwood, M. Khurgel, G. O. Ivy and R. Racine, *Brain Res.* **551**, 236 (**1991**).

20 C. Bernard and H. V. Wheal, *J. Physiol.* **495**, 127 (**1996**).

21 A. Rodriguez-Moreno, O. Herreras and J. Lerma, *Neuron* **19**, 893 (**1997**).

22 J. Lerma, *FEBS Lett.* **430**, 100 (**1998**).

23 M. Frerking, C. C. Petersen and R. A. Nicoll, *Proc. Natl Acad. Sci. U S A* **96**, 12917 (**1999**).

24 A. Rodriguez-Moreno, J. C. Lopez-García and J. Lerma, *Proc. Natl Acad. Sci. U S A* **97**, 1293 (**2000**).

25 A. Bragin, I. Mody, C. L. Wilson and J. Engel, *J. Neurosci.* **22**, 2012 (**2002**).

26 A. Bragin, C. L. Wilson, R. J. Staba, M. Reddick, I. Fried and J. Engel, *Ann. Neurol.* **52**, 407 (**2002**).

27 R. J. Staba, C. L. Wilson, A. Bragin, I. Fried and J. Engel, *J. Neurophysiol.* **88**, 1743 (**2002**).

28 J. Engel, C. Wilson and A. Bragin, *Epilepsia* **44** Suppl. 12, 60 (**2003**).

29 A. Bragin, C. L. Wilson, J. Almajano, I. Mody and J. Engel, *Epilepsia* **45**, 1017 (**2004**).

30 J. J. Engel, *In* M. R. T (Ed), *Seizure Freedom: Clinical, Research and Quality Life Perspectives* (Clarius Press, Ltd., 29–54, **2006**).

31 J. Jirsch, F. Dubeau, E. Urrestarazu, P. LeVan and J. Gotman, *Epilepsia* **46**, 267 (**2005**).

32 J. D. Jirsch, E. Urrestarazu, P. LeVan, A. Olivier, F. Dubeau and J. Gotman, *Brain* **129**, 1593 (**2006**).

13

High-frequency Pre-seizure Activity and Seizure Prediction

Premysl Jiruska, John G.R. Jefferys

Identifying unambiguous changes in cortical activity during the tens of seconds to minutes before epileptic seizures has the potential radically to improve treatment for people with epilepsy. For instance, it can be used to trigger adaptive treatments, such as therapeutic electrical stimulation (discussed elsewhere in this monograph) or focal drug delivery, which could interfere with mechanisms involved in transition to seizure. Methods for seizure prediction are described extensively in this monograph. They mainly use EEG data from patients with analyses ranging from the relatively simple to the exceedingly complex. However, the ideal method with a high sensitivity and low false positive rate has yet to be found. There is an argument that understanding the neuronal and/or population behavior preceding seizures will greatly help to identify better approaches to prediction [1], and may also lead to improved therapeutic targets. Here we will focus on high-frequency pre-seizure activity (HFA) as a candidate population behavior for this role. Fast activity has been linked with epileptic foci since the early 1990s [2, 3], and has been implicated in the processes of ictogenesis [4] and epileptogenesis [5]. In this article we will focus on its role in transition to seizure, briefly summarize current knowledge about HFA and suggest possible future developments. The run-up to focal seizures has been associated with the build-up of HFA >80 Hz [2, 3, 6, 7]. Several human studies demonstrated the presence of the fast activity at the beginning of, or early in, seizures from invasive recordings before any clinical symptoms occurred [2, 3, 6, 7]. Experimental observations of HFA at seizure onset have been made in both *in vivo* [8] and *in vitro* [4] models. However, HFA at transition to seizure usually lasts only several seconds. HFA lasting tens of seconds before seizure onset are exceptional [7]. Periods of abnormal activity lasting just a few seconds before the seizure make reliable seizure prediction exceedingly difficult, if not impossible: if the detection occurs after the seizure has already started then interfering with ictogenic mechanisms is likely to be more difficult than if reliable prediction can be achieved earlier (although in this case identifying false positives becomes more of a challenge).

Seizure prediction research depends on the assumption that focal seizures are not sudden (abrupt) events, but that they are preceded by gradual changes in

Seizure Prediction in Epilepsy. Edited by Björn Schelter, Jens Timmer and Andreas Schulze-Bonhage
Copyright © 2008 WILEY-VCH Verlag GmbH & Co. KGaA, Weinheim
ISBN: 978-3-527-40756-9

brain and neuronal circuit dynamics. The terms preictal, or pre-seizure, period are useful to describe a gradual transition from the purely interictal state to the seizure state. However, much work remains to be performed to demonstrate clearly whether the preictal period really exists, and to identify its biological markers and its biological mechanisms [9]. Recent work *in vitro* demonstrated that HFA progressively increased in advance of the onset of electrographic seizures and served as a preictal marker [10, 11]. Similar observations have been made in human patients [7]. This study showed that the energy of HFA progressively increased more than 20 minutes preceding the onset of seizure and giving it the potential to be used as a marker for the preictal state in humans. It has been shown that fast activity is highly specific for epileptic foci: removal of the areas generating HFA is associated with increased success of epilepsy surgery. However, many questions remain. Do ictal HFA and preictal HFA share similar cellular and subcellular mechanisms? What are their spatio-temporal dynamics? Perhaps the most important is whether fast activity really is part of the pathophysiology or whether it is just an epiphenomenon? If the latter, it still could be used as a marker of upcoming seizure and marker defining the preictal state. If the fast activity also had a pathogenic role, then understanding mechanisms involved in its genesis and its spatiotemporal dynamics would facilitate development of therapeutic approaches which directly target ictogenesis. Several theories on the origins of fast activity have been suggested by experimental observation.

One argues that HFA results from complex synaptic interactions between interneurons and pyramidal cells which then results in synchronized IPSPs generated at the level of pyramidal cells' somata. This mechanism is involved mainly in physiological high-frequency activity which is probably involved in memory consolidation [12, 13]. Draguhn et al. describe fast activity which survives a block of chemical synaptic transmission and must be generated by non-synaptic mechanisms [14]. Experiments and computer modeling suggested that this fast activity is due to synchronization of pyramidal cells action potential firing via axonal gap junctions [6, 14], although the poor specificity of agents that block gap junctions and the low incidence of coupled pyramidal cells suggests we cannot exclude other mechanisms such as ephaptic interactions [15]. Gap-junction blockers have suggested similar mechanisms may apply in neocortical high-frequency activity [8]. The fast activity at the onset of seizure activity in the low-calcium model *in vitro* reported by Bikson et al. has a non-synaptic origin [4] (Figure 13.1), and our recent work with the high-potassium model (Figure 13.2) argues that similar mechanisms apply even if chemical synapses are not blocked.

A further complication for the search for mechanisms (and functional consequences) is the classification of different frequency bands of fast activity, notably: around 80 Hz, around 200 Hz (ripples) or very fast activities between 250 and 500 Hz (fast ripples). Whether these classifications are functionally significant, or even whether they should be subdivided further remain open questions. For example, Bragin et al. stress that fast ripple activity is much more specific for epileptic foci and that spatial and depth profiles of fast ripple and ripple activity differ [16] suggesting different underlying mechanisms. For practical purposes of

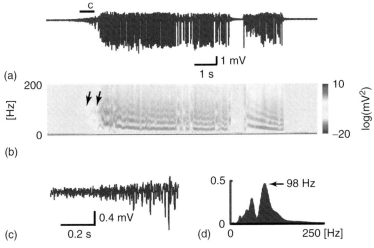

Fig. 13.1 Hippocampal slice perfused in low-calcium artificial cerebrospinal fluid leads to development of spontaneous recurrent electrographic seizures. (a) Recording from CA1 region (DC removed). (b) Seizure onset is characterized by occurrence of low-amplitude high-frequency activity, which is well demonstrated in corresponding wavelet spectrogram (arrows). (c) Detail of seizure onset. (d) Corresponding wavelet power spectrum demonstrating peak frequency at 98 Hz. (Please find a color version of this figure on the color plates.)

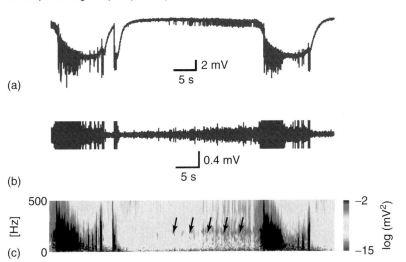

Fig. 13.2 Hippocampal slice perfused with high-potassium ACSF. (a) Recording from CA1 region shows presence of repeated electrographic seizures. Seizures are superimposed on large DC shifts. (b) Bandpass filtered data (80–250 Hz) show that seizures are preceded by gradual build-up of high-frequency activity. (c) Corresponding wavelet spectrogram demonstrates that the build-up of high-frequency activity has a peak frequency of 200 Hz (arrows). (Please find a color version of this figure on the color plates.)

seizure anticipation it is necessary to determine which frequencies are relevant for the epileptic focus and seizure genesis, and whether they differ between different classes of epilepsy and/or different regions.

While the association between HFA and epileptic foci seems to be strengthening, much remains to be resolved on both its roles and its mechanisms. Critical comparisons of the HFA found in experimental observations *in vitro*, *in vivo* and human studies will provide a better understanding of both its mechanisms and role in ictogenesis. We believe that a key element in these investigations will be improved signal analysis methods including the analysis of interactions between multiple recording channels [17, 18]. A crucial practical issue remains. Both experimental and clinical data show that HFA is often highly restricted spatially. This can be an advantage for localization. However, it complicates answering the question of whether all focal seizures are preceded by a build-up of fast activity. Negative results do not necessarily mean that HFA is absent: the electrode placement could miss the small regions of the brain that generate HFA, or, on a more mundane note, many clinical recording systems lack the technical specifications needed to record and/or detect HFA. Therefore, the implementation of methods of seizure prediction based on detecting of HFA will require the development or modification of many recording protocols currently used in the clinic. Wide-band recording setups with electrodes small enough to avoid spatial averaging of the small HFA signal are necessary requirements. Smaller electrodes will lead to the need for larger numbers of electrodes to cover large brain areas with a good enough spatial resolution to find regions of preictal HFA. One approach to this problem that has proved successful is the use of hybrid electrodes which combine conventional invasive electrodes with microwire electrodes [19].

References

1 F. Mormann, R. G. Andrzejak and C. E. Elger, *Brain* **130**, 314–33 (**2007**).

2 P. J. Allen, D. R. Fish and S. J. Smith, *Electroencephalogr. Clin. Neurophysiol.* **82**, 155–9 (**1992**).

3 R. S. Fisher, W. R. Webber, R. P. Lesser, S. Arroyo and S. Uematsu, *J. Clin. Neurophysiol.* **9**, 441–8 (**1992**).

4 M. Bikson, J. E. Fox and J. G. Jefferys, *J. Neurophysiol.* **89**, 2330–3 (**2003**).

5 A. Bragin, C. L. Wilson and J. Engel, Jr, *Epilepsia* **41**, S144–S152 (**2000**).

6 R. D. Traub, M. A. Whittington, E. H. Buhl, *et al.*, *Epilepsia* **42**, 153–70 (**2001**).

7 G. A. Worrell, L. Parish, S. D. Cranstoun, R. Jonas, G. Baltuch and B. Litt, *Brain* **127**, 1496–506 (**2004**).

8 F. Grenier, I. Timofeev and M. Steriade, *J. Neurophysiol.* **89**, 841–52 (**2003**).

9 K. Lehnertz, F. Mormann, H. Osterhage, *et al.*, *J. Clin. Neurophysiol.* **24**, 147–53 (**2007**).

10 V. I. Dzhala and K. J. Staley, *J. Neurosci.* **23**, 7873–80 (**2003**).

11 H. Khosravani, C. R. Pinnegar, J. R. Mitchell, B. L. Bardakjian, P. Federico and P. L. Carlen, *Epilepsia* **46**, 1188–97 (**2005**).

12 A. Ylinen, A. Bragin, Z. Nadasdy, *et al.*, *J. Neurosci.* **15**, 30–46 (**1995**).

13 G. Buzsaki, Z. Horvath, R. Urioste, J. Hetke and K. Wise, *Science* **256**, 1025–7 (**1992**).

14 A. Draguhn, R. D. Traub, D. Schmitz and J. G. R. Jefferys, *Nature* **394**, 189–92 (**1998**).

15 J. G. R. Jefferys, *Physiol Rev.* **75**, 689–723 (**1995**).

16 A. Bragin, I. Mody, C. L. Wilson and J. Engel, Jr., *J. Neurosci.* **22**, 2012–21 (**2002**).

17 M. Le Van Quyen and A. Bragin, *Trends Neurosci.* **30**, 365–73 (**2007**).

18 X. Li, D. Cui, P. Jiruska, J. E. Fox, X. Yao and J. G. R. Jefferys, *J. Neurophysiol.* (**2007**).

19 G. A. Worrell, M. Stead, R. Marsh, F. Meyer, G. Cascino E. So and B. Litt, *Epilepsia* **47**[Suppl. 4], 4 (**2006**).

14

Characterizing the Epileptic Process with Stochastic Qualifiers of Brain Dynamics

Jens Prusseit, Christian E. Elger, Klaus Lehnertz[1]

14.1
Introduction

Over recent years, linear and nonlinear analyses of electroencephalographic (EEG) time series have provided valuable insights into the complex spatial-temporal dynamics of physiological and pathophysiological brain functions (see [1, 2] for an overview). These processes, however, are far from being fully understood. In epilepsy, particularly nonlinear time series analysis techniques provided information relevant for diagnostic purposes by allowing an improved characterization of intermittent dysfunctioning of the brain between epileptic seizures [3–6]. Moreover, there are indications that these approaches might be able to extract characteristic features from the continuous EEG that are predictive of an impending seizure. This could be of great value for the development of seizure warning systems and for a further improvement of seizure prevention techniques (see [7, 8] for an overview). Nevertheless, there are a number of problems that can be related to the fact that the aforementioned analysis techniques preferentially focus on the low-dimensional deterministic part of the dynamics. Thus they might not be able to capture crucial aspects of the EEG that, in many cases, have to be regarded as stochastic (high-dimensional). In [9, 10] a time series analysis approach was introduced that explicitly takes into account the stochastic nature of signals. Using this technique, we recently analyzed multi-day, multi-channel EEG data recorded intracranially from patients suffering from refractory focal epilepsies and showed that this approach allows an improved characterization of pathological brain dynamics [11].

This time series analysis approach relies on the well known fact that dissipative dynamical systems under the influence of noise can often be successfully modeled by Fokker–Planck equations [12–15]. For the analysis of empirical data (such as the EEG) the mathematics of diffusion processes is used to estimate drift and diffusion coefficients at a number of points in the state space of the dynamical system

1) We are grateful to Joachim Peinke and Rudolf Friedrich for useful discussions and valuable comments. This work was supported by the Deutsche Forschungsgemeinschaft (SFB-TR3 sub-project A2 and LE660/4-1).

(e.g., the brain). As a result, a general Langevin equation or a Fokker–Planck equation can be extracted from the measured time series data. However, since the reconstruction of Fokker–Planck equations from observed time series data suffers strongly from finite sampling rates, correction terms [16–18] as well as extensions [19] of the original approach have been devised, which allow a more robust estimation of the diffusion terms. Addressing the issue of non-stationarity, an averaging procedure has been presented recently in [20] that allows one to reliably estimate the dynamics of diffusive Markov processes in certain situations. Moreover, in [21, 22] the analysis technique was extended further to handle both dynamical and measurement noise.

Besides having been tested successfully on artificially generated time series from well known dynamical systems (including nonlinear oscillators and chaotic systems [10, 23–25]), over the last decade the approach has already found applications in a variety of disciplines. Apart from physics [9, 26–32] we mention engineering [25, 33], economics [34, 35], sociology [36], and meteorology [37]. In addition, the method has been introduced in the biomedical domain by investigating different types of physiological and pathophysiological tremor [24], heart-rate fluctuations under normal [38] and pathological (congestive heart failure) conditions [39], and, more recently, rhythmic movement in humans [40]. These studies indicate the potential of the approach to gain deeper insights into the underlying dynamical processes, and also for diagnostic purposes.

We here present an overview of the method with a special emphasis on analyses of EEG recordings from epilepsy patients and discuss its relevance for a spatial and temporal characterization of the epileptic process. After a brief introduction of the basics of the time series analysis approach in Section 14.2, we show, in Section 14.3, results from an analysis of intracranial EEG recordings. We first discuss, in Section 14.3.1, results that are typical for intracranial EEG recordings from patients suffering from focal epilepsies. In Section 14.3.2 we present a framework to characterize epileptic brain dynamics using data-driven Fokker–Planck models. Specifically we present our results obtained from analyzing long-term, multi-channel EEG recordings in the context of localizing the epileptic focus during the seizure-free interval and in the context of seizure prediction. In Section 14.4 we draw our conclusion.

14.2
Data-driven Fokker–Planck Models

A stochastic dynamical system that can be described by a one-dimensional Langevin equation

$$\dot{x}(t) = h(x(t), t) + g(x(t), t)\Gamma(t) \tag{14.1}$$

is completely determined by the functions h and g. The state of the system is denoted by $x(t)$, and $\Gamma(t)$ is a delta-correlated Gaussian noise process with vanishing mean:

$$\langle\Gamma(t)\rangle = 0 \quad \text{and} \quad \langle\Gamma(t)\Gamma(t')\rangle = \delta(t - t'). \tag{14.2}$$

In (14.1) the function h describes the deterministic part of the dynamics and g is the amplitude of the driving noise force. If g depends on the state x, the stochastic part is referred to as multiplicative dynamical noise, otherwise as additive dynamical noise. Because of the delta-correlation of the noise equation (14.1) generates realizations of a Markovian stochastic process $X(t)$ whose conditional probability density function (PDF) obeys a Fokker–Planck equation [14]:

$$\frac{\partial}{\partial t} p(x, t|x', t') = \left(-\frac{\partial}{\partial x} D^{(1)}(x, t) + \frac{\partial^2}{\partial x^2} D^{(2)}(x, t) \right) p(x, t|x', t').$$ (14.3)

The functions $D^{(1)}$ and $D^{(2)}$ are called drift and diffusion coefficients of the process. The Fokker–Planck equation is a special case of a more general evolution equation for continuous Markov processes, namely the Kramers–Moyal expansion, which reads

$$\frac{\partial}{\partial t} p(x, t|x', t') = \left[\sum_{n=1}^{\infty} \frac{(-\partial^n)}{\partial x^n} D^{(n)}(x, t) \right] p(x, t|x', t').$$ (14.4)

The coefficients $D^{(n)}$ can be defined in a statistical sense using the conditional moments of the stochastic variable $X(t)$ [14]

$$D^{(n)}(x, t) = \frac{1}{n!} \lim_{\tau \to 0} \frac{1}{\tau} \langle [X(t + \tau) - X(t)]^n \rangle_{X(t)=x}$$ (14.5)

where $\langle \rangle_{X(t)=x}$ denotes the ensemble average over all realizations of X for which $X(t) = x$ at time t.

For processes which can be described by a Langevin equation, i.e., processes that are driven by delta-correlated Gaussian noise, the functions $D^{(n)}$ are related to the functions h and g of the corresponding Langevin equation by [14]:

$$D^{(1)}(x, t) \overset{(I)}{=} h(x, t)$$ (14.6)

$$D^{(1)}(x, t) \overset{(S)}{=} h(x, t) + g(x, t) \frac{\partial}{\partial x} g(x, t)$$ (14.7)

$$D^{(2)}(x, t) = (g(x, t))^2$$ (14.8)

$$D^{(i)}(x, t) = 0, \quad \forall i > 2.$$ (14.9)

The relation for the first coefficient depends on whether Ito's (I) [41] or Stratonovich's (S) [42] definition of stochastic integrals is used for the interpretation of the Langevin equation. Relation (14.9) is a consequence of the so-called Pawula theorem [43], which states that the Kramers–Moyal expansion either has an infinite number of terms or simplifies to a Fokker–Planck equation if the first and/or the second term are retained. For any other finite number of non-vanishing coefficients the PDF that solves (14.4) becomes negative. It is possible to show that, if one coefficient with $n > 2$ (where n is an even integer number) is zero, then all coefficients with $n > 2$ vanish. This is the case if the driving noise force $\Gamma(t)$ in the Langevin equation (14.1) has a Gaussian distribution.

For a stationary (and ergodic) process $X(t)$ the ensemble averages in (14.5) can be replaced by time averages over one realization of the process. It is therefore possible to estimate the coefficients of the Kramers–Moyal expansion from time series by evaluating the conditional moments in (14.5) for finite time steps τ, and then extrapolate to $\tau = 0$ [10, 24]. For this purpose, the condition $X(t) = x$ in (14.5) has to be replaced by $X(t) \in \mathcal{U}(x)$, where $\mathcal{U}(x)$ is an interval containing x that is usually fixed by some partitioning of the state space. Because of this discretization of the state space the values of the coefficients are estimated at fixed discrete values of x and can later be fitted by analytical functions, if necessary.

For this analysis method to be applicable, the time series under consideration has to be Markovian, i.e., a process without memory, which requires the condition

$$p(x_k, t_k | x_{k-1}, t_{k-1}; \ldots, x_1, t_1) = p(x_k, t_k | x_{k-1}, t_{k-1}) \tag{14.10}$$

to be fulfilled for arbitrary k and all values of $t_1 \leq t_2 \leq \ldots \leq t_k$. For experimentally derived time series with a limited number of data points the Chapman–Kolmogorov equation can be checked more reliably

$$p(x_3, t_3 | x_1, t_1) = \int dx_2 p(x_3, t_3 | x_2, t_2) p(x_2, t_2 | x_1, t_1) \tag{14.11}$$

where $t_1 < t_2 < t_3$. This equation is a necessary (but, in general, not sufficient) condition for the process to be Markovian.

14.3
EEG Analysis

In this section we present findings obtained from applying the time series analysis method described above to multi-channel EEG time series that were recorded from patients suffering from refractory temporal lobe epilepsy. For these patients, complete seizure control can be achieved by surgically removing the part of the brain responsible for seizure generation (epileptic focus). Since a clear cut localization of the epileptic focus could not be accomplished by means of scalp EEG recordings and other evaluation techniques, intracranial electrodes (cf. Figure 14.1) were implanted during presurgical evaluation. All patients achieved complete seizure control after surgery so the epileptic focus can be assumed to be contained within the resected area. The EEG time series analyzed here were sampled continuously over a longer period (5–12 days) with bandpass filter setting of 0.5–85 Hz (12 dB/octave) using a common average reference. The sampling interval Δt was 5 ms, and analog–digital conversion was performed at 16-bit resolution.

14.3.1
Markov Property and Characteristics of Estimated Kramers–Moyal Coefficients

We first present exemplary results obtained from analyzing EEG time series that were recorded during the seizure-free interval from within the epileptic focus

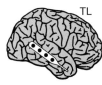 

Fig. 14.1 Scheme of intracranial electrodes used for the presurgical evaluation of epilepsy patients. Depth electrodes (10 contacts each, D) were implanted symmetrically into the hippocampal formations. Strip electrodes were implanted onto the lateral (4–16 contacts, TL) and basal regions (4 contacts each, TB) of the neocortex.

(denoted as *focal* time series) and from distant brain regions (denoted as *non-focal* time series). An EEG recording is an important example of a non-stationary time series [44] where especially robust changes occur during epileptic seizures. In the literature the period of time within which an EEG recorded during the seizure-free interval can be considered as approximately stationary varies from seconds to minutes [45, 46]. For the analyses presented here, we used EEG time series with $N = 100\,000$ and $N = 50\,000$ data points (corresponding to recording durations of 8.3 min and 4.2 min respectively). For these window sizes we obtained qualitatively similar results, and the observed effects became less pronounced for smaller window sizes.

As already mentioned in Section 14.2, for the analysis method to be applicable, the time series under consideration has to be Markovian. This property, however, cannot be proven in a strict mathematical sense when dealing with time series of finite size. Nevertheless, in order to get an indication for the property (14.10) to be approximately fulfilled, one might consider the Chapman–Kolmogorov equation (14.11) to find the smallest time shift τ (in units of the sampling interval Δt), for which the time series appears to be Markovian. For this purpose the left- and right-hand sides of the Chapman–Kolmogorov equation with a fixed time shift $t_3 - t_2 = t_2 - t_1 = \tau$

$$p(x_3, t + 2\tau | x_1, t) = \int dx_2\, p(x_3, t + 2\tau | x_2, t + \tau) p(x_2, t + \tau | x_1 t) \qquad (14.12)$$

have to be estimated and the resulting conditional PDFs can then be compared. In Figure 14.2 we show results for exemplary EEG time series recorded from a distant brain region and from within the epileptic focus (Figure 14.2(a)). For both EEG time series it can be observed that (14.12) appears to be fulfilled, at least approximately, for the smallest available time shift $\tau = 1$ sampling interval. Except for small deviations, which could be caused by the limited amount of data, the two distributions coincide in each case (cf. Figure 14.2(b) and (c)), and we take this as an indication for Markovian properties on that time scale.

Having fixed the time shift to $\tau = 1$ sampling interval as a possible Markov time scale, the next step is to estimate the first, second, and fourth Kramers–Moyal

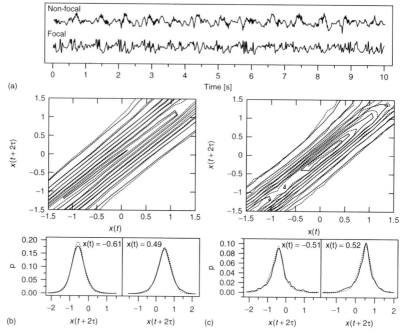

Fig. 14.2 Sections of exemplary time series recorded from a distant brain region (denoted as *non-focal*) and from within the epileptic focus (denoted as *focal*) (a). Comparison of the left- (dashed lines) and right-hand side (solid lines) of the Chapman–Kolmogorov equation for the two exemplary EEG time series of size $N = 100\,000$ data points: non-focal time series (b), focal time series (c). Contour lines (upper parts of (b) and (c)) and cuts through the resulting conditional PDFs (lower parts of (b) and (c)) for a time shift of $\tau = 1$. Contour plots were generated using an increment between contour lines of 0.02 for (a) and 0.012 for (b).

coefficient according to (14.5). For an automated and computationally inexpensive estimation of the coefficients (e.g., when analyzing long-term multi-channel EEG-data; cf. Section 14.3.2) a fixed time shift τ can be used, i.e., the limit $\tau \rightarrow 0$ is not performed. When neglecting the limit one has to keep in mind though that this can result in erroneous contributions of order $\mathcal{O}(\tau^2)$ in the estimation of the conditional moments according to (14.5), which can alter the functional characteristics of the estimated coefficients (see [16–18]).

In Figure 14.3 we show estimated coefficients that are characteristic for intracranially recorded focal and non-focal EEG time series using the one-dimensional approach described in Section 14.2. All coefficients usually assume larger absolute values for focal EEG time series than for non-focal ones. In particular, the magnitude of the values of the fourth coefficient $D^{(4)}$ is often larger by a factor of about twenty for the focal EEG, which – assuming Markov properties – indicates a strong deviation from a Gaussian behavior of the driving noise force (cf. Section 14.2). As shown in [11] this corresponds to the finding that a description of EEG recordings by one-dimensional Fokker–Planck models might be less appropriate

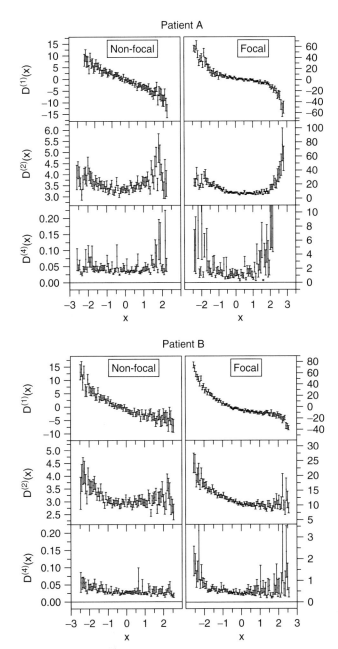

Fig. 14.3 Estimated Kramers–Moyal coefficients $D^{(1)}$, $D^{(2)}$, and $D^{(4)}$ for exemplary EEG time series from two patients (for each patient left column: from a distant brain region (denoted as *non-focal*); right column: from within the epileptic focus (denoted as *focal*)). Shown are estimates for time series consisting of 100 000 data points (squares), and error bars indicate the statistical errors of the estimation of the averages according to (14.5) for each value of x.

for pathophysiological activity recorded from within the epileptic focus than for activities recorded at distant sites. For the non-focal EEG time series the first coefficient $D^{(1)}$ typically exhibits an overall linear damping behavior, whereas for focal time series $D^{(1)}$ often shows pronounced nonlinearities toward higher amplitude values, which is in line with findings obtained from applying nonlinear time series analysis methods (e.g. [6]). The influence of the noise generally appears to be multiplicative as indicated by non-constant coefficients $D^{(2)}$. In principle, deviations from an additive behavior (i.e., $D^{(2)} =$ const.) of estimated second Kramers–Moyal coefficients can originate from higher order correction terms in the time shift τ if a finite τ is used for the estimation instead of performing the limit $\tau \to 0$. Whether this is indeed the case can be checked by considering a Taylor expansion of the second conditional moment [17, 18]. For the examples shown in Figure 14.3 it can be shown – by calculating the higher order terms in τ using the estimated first coefficients along with different constant second coefficients of varying magnitude – that the influence of these higher order terms cannot explain the observed multiplicativity of the estimated second coefficients. In addition, it can often be observed that the second coefficient $D^{(2)}$ appears to be more asymmetric for focal than for non-focal EEG time series.

Summarizing this section, we conclude that there are indications that intracranially recorded EEG time series can be regarded as Markovian. Furthermore, there seem to be specific characteristics of the estimators of the Kramers–Moyal coefficients that allow one to differentiate between physiological and pathophysiological activities, and may thus be useful for a characterization of epileptic brain dynamics.

14.3.2
Relevance for a Spatial and Temporal Characterization of the Epileptic Process

Based on the findings presented in the previous section we now define measures that allow one to extract information from the estimated Kramers–Moyal coefficients, which appears to be characteristic for pathological processes in the epileptic brain. In order to allow an automated processing of EEG data using a moving-window technique we applied the following preprocessing steps. For each window we generated an amplitude histogram, which was, in general, unimodal for EEG recordings from the seizure-free interval. Then the amplitude values to the left and to the right of the mode were determined for which the relative frequency dropped to 5% of the relative frequency of the mode of the distribution. Only amplitude values confined to the thus-defined interval were considered for further analyses and were normalized to zero mean and unit variance. This procedure excludes occasional high-amplitude artifacts and allows one to use a fixed number of bins when estimating the probability densities, thereby ensuring a comparable coverage of the EEG amplitude range for each window.

As *stochastic qualifiers* of brain dynamics we here consider the range covered by the values of the estimated first and second coefficients $D^{(1)}(x)$ and $D^{(2)}(x)$:

$$R_{1,2} := \left| \max\left(D^{(1,2)}(x)\right) - \min\left(D^{(1,2)}(x)\right) \right|. \tag{14.13}$$

To account for statistical fluctuations, only $D^{(1,2)}(x)$ values entered the calculation that were determined by more than 100 amplitude values in the bin $\mathcal{U}(x)$ of the partitioning of the state space that contains x.

In Figure 14.4 we show exemplary sections of the spatial-temporal distributions of R_1 and R_2 values. The EEG data were recorded during the seizure-free interval from a patient with an epileptic focus located in the right medial temporal lobe. Highest values of R_2 can be observed for recording sites confined to regions close to or within the epileptic focus (the sixth and seventh contact of the right depth electrode D). Although R_1 also assumes its highest values for contacts from this brain region, there are other – ipsilateral and contralateral – regions, which exhibit R_1 values of comparable magnitude. Since these effects were stable for long recording periods, we calculated, for each channel, temporal averages $\langle R_1 \rangle$ and $\langle R_2 \rangle$ presented in Figure 14.4(b) and Figure 14.4(d), respectively. With R_2 a localization of the epileptic focus was possible for this patient using our approach without observing actual seizure activity. However, when analyzing EEG data from a group of eight patients, we observed that such a clearcut identification of the epileptic focus was not always possible. Nevertheless, we could show that R_2 allowed to correctly lateralize the focal hemisphere using data from the seizure-free interval only [11].

We next address the question whether our approach allows one to identify characteristic temporal changes of EEG activity predictive of an impending seizure. We here followed [47] and performed a retrospective evaluation assuming the existence of a preictal state of fixed duration (240 minutes), and applied the concept of seizure time surrogates [48] to assess the statistical significance of our findings. Using our approach, we analyzed intracranial EEG data that were recorded with 48 electrode contacts for approximately five days (patient B in [49]). During this time period, the patient had 10 typical seizures that originated from the left medial temporal lobe. We calculated R_1 and R_2 using a moving-window technique with data windows of size $N = 50\,000$ data points, and the windows overlapped by 50%. We discarded any seizure activity and the postictal periods (30 minutes duration) and evaluated the amplitude distributions of R_1 and R_2 for preictal and interictal periods using receiver operating characteristics (ROC). In Figure 14.5 we show the amplitude distributions of R_1 and R_2 for the preictal and interictal periods from an EEG channel that exhibited the highest values of the area under the ROC curve (0.77 for R_1 and 0.73 for R_2). This channel was located close to the epileptic focus. Both measures attained higher values during the preictal periods. However, when testing for statistical significance with 19 seizure times surrogates only the result for R_1 turned out to be significant.

Summarizing this section, our findings indicate that the approach presented here can be regarded as an alternative to characterize complex brain dynamics. We addressed the issue of a spatial-temporal characterization of the epileptic process and focused on an interictal focus localization and on possibilities to identify a preictal state. We observed that particularly measures derived from the second Kramers–Moyal coefficient $D^{(2)}$ appear to capture more relevant information for a spatial characterization than measures derived from the first coefficient $D^{(1)}$. Interpreting $D^{(2)}$ as the diffusion coefficient of the Fokker–Planck equation one may conclude that additional and relevant information about the epileptic process

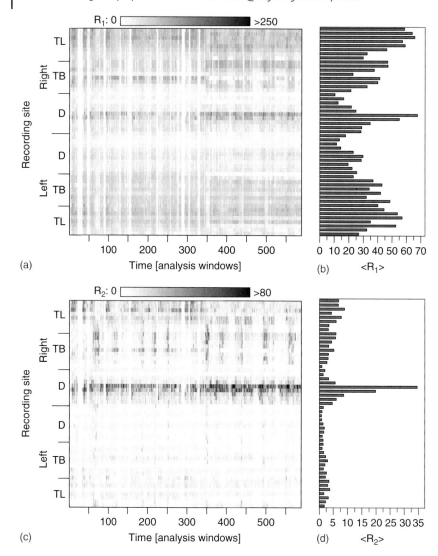

Fig. 14.4 Values of R_1 (a) and R_2 (c) calculated from a multi-channel EEG recording (approx. 21 hours) from a patient suffering from seizures of right medial temporal lobe origin. For abbreviations of recording sites see Figure 14.1. Temporal averages of R_1 (b) and R_2 (d) for all recording sites (same ordinates as in (a) and (c)). R_1 and R_2 were calculated using a moving-window technique with data windows of size $N = 50\,000$ data points, and the windows overlapped by 50%.

can be achieved when taking into account stochastic influences on brain dynamics. The better performance of R_1 for characterizing the seizure generating process probably indicates that also measures derived from the first Kramers–Moyal coefficient can yield valuable information. This preliminary finding, however, requires validation with data from a larger patient group.

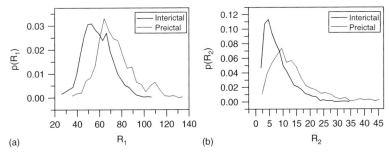

Fig. 14.5 Amplitude distributions of R_1 (a) and R_2 (b) for preictal and interictal periods. The duration of the assumed preictal period was 240 minutes. Data from an EEG channel close to the epileptic focus.

14.4
Conclusions

We have presented a time series analysis method that is based on the reconstruction of a Fokker–Planck equation from EEG recordings for an improved characterization of the epileptic process. We have derived stochastic qualifiers of epileptic brain dynamics that allowed a more comprehensive spatial characterization of the epileptic process during the interictal state, particularly when focusing on the stochastic part of the dynamics. Preliminary findings also indicate that the method can be regarded as helpful in identifying dynamical aspects of seizure precursors that are not fully captured by techniques preferentially focusing on deterministic structures. We expect that our approach, along with further improvements, can yield valuable information for diagnostic purposes and can advance our understanding of the complicated dynamical system epileptic brain.

References

1 C. J. Stam. Nonlinear dynamical analysis of EEG and MEG: Review of an emerging field. *Clin. Neurophysiol.*, **116**, 2266–301 (**2005**).

2 E. Pereda, R. Quian Quiroga and J. Bhattacharya. Nonlinear multivariate analysis of neurophysiological signals. *Prog. Neurobiol.*, **77**, 1–37 (**2005**).

3 K. Lehnertz and C. E. Elger. Spatio-temporal dynamics of the primary epileptogenic area in temporal lobe epilepsy characterized by neuronal complexity loss. *Electroencephalogr. Clin. Neurophysiol.*, **95**, 108–17 (**1995**).

4 R. G. Andrzejak, G. Widman, K. Lehnertz, C. Rieke, P. David and C. E. Elger. The epileptic process as nonlinear dynamics in a stochastic environment: an evaluation on mesial temporal lobe epilepsy. *Epilepsy Res.*, **44**, 129–40 (**2001**).

5 K. Lehnertz, R. G. Andrzejak, J. Arnhold, T. Kreuz, F. Mormann, C. Rieke, G. Widman and C. E. Elger. Nonlinear EEG analysis in epilepsy: Its possible use for interictal focus localization, seizure anticipation, and prevention. *J. Clin. Neurophysiol.*, **18**, 209–22 (**2001**).

6   R. G. Andrzejak, F. Mormann, G. Widman, T. Kreuz, C. E. Elger and K. Lehnertz. Improved spatial characterization of the epileptic brain by focusing on nonlinearity. *Epilepsy Res.*, **69**, 30–44 (2006).

7   M. Morrell. Brain stimulation for epilepsy: can scheduled or responsive neurostimulation stop seizures? *Curr. Opin. Neurol.*, **19**, 164–8 (2006).

8   F. Mormann, R. G. Andrzejak, C. E. Elger and K. Lehnertz. Seizure prediction: the long and winding road. *Brain*, **130**, 314–33 (2007).

9   R. Friedrich and J. Peinke. Description of the turbulent cascade by a Fokker–Planck equation. *Phys. Rev. Lett.*, **78**, 863–6 (1997).

10 S. Siegert, R. Friedrich and J. Peinke. Analysis of data sets of stochastic systems. *Phys. Lett. A*, **243**, 275–80 (1998).

11 J. Prusseit and K. Lehnertz. Stochastic qualifiers of epileptic brain dynamics. *Phys. Rev. Lett.*, **98**, 138103 (2007).

12 P. Hänggi and H. Thomas. Stochastic processes: time evolution, symmetries and linear response. *Phys. Rep.*, **88**, 207–319 (1982).

13 N. G. van Kampen. *Stochastic Processes in Physics and Chemistry*. North-Holland Publishing Company, Amsterdam (1981).

14 H. Risken. *The Fokker–Planck equation*. Springer, Berlin, 2nd edition (1989).

15 H. Haken. *Information and Self-Organization: A Macroscopic Approach to Complex Systems*. Springer, Berlin (2004).

16 M. Ragwitz and H. Kantz. Indispensable finite time corrections for Fokker–Planck equations from time series data. *Phys. Rev. Lett.*, **87**, 254501 (2001).

17 R. Friedrich, C. Renner, M. Siefert and J. Peinke. Comment on: Indispensable finite time corrections for Fokker–Planck equations from time series data. *Phys. Rev. Lett.*, **89**, 217–22 (2002).

18 P. Sura and J. Barsugli. A note on estimating drift and diffusion parameters from timeseries. *Phys. Lett. A*, **305**, 304–11 (2002).

19 D. Kleinhans, R. Friedrich, A. Nawroth and J. Peinke. An iterative procedure for the estimation of drift and diffusion coefficients of Langevin processes. *Phys. Lett. A*, **346**, 42–6 (2005).

20 A. M. van Mourik, A. Daffertshofer and P. J. Beek. Estimating Kramers–Moyal coefficients in short and nonstationary data sets. *Phys. Lett. A*, **351**, 13–7 (2006).

21 M. Siefert, A. Kittel, R. Friedrich and J. Peinke. On a quantitative method to analyze dynamical and measurement noise. *Europhys. Lett.*, **61**, 466–72 (2003).

22 F. Boettcher, M. Siefert and J. Peinke. A generalized method to distinguish between dynamical and measurement noise in complex dynamical systems. *Physics in Image and Signal Processing (Proceedings), Toulouse* (2005).

23 S. Siegert, R. Friedrich and P. Hänggi. Analysis of fluctuating data sets of diffusion processes. In K. Lehnertz, J. Arnhold, P. Grassberger and C. E. Elger, editors, *Chaos in brain?*, pp 226–9, Singapore (2000). World Scientific.

24 R. Friedrich, S. Siegert, J. Peinke, S. Lück, M. Siefert, M. Lindemann, J. Raethjen, G. Deuschl and G. Pfister. Extracting model equations from experimental data. *Phys. Lett. A*, **271**, 217–22 (2000).

25 J. Gradisek, S. Siegert, R. Friedrich and I. Grabec. Analysis of time series from stochastic processes. *Phys. Rev. E*, **62**, 3146–55 (2000).

26 S. Lück, J. Peinke and R. Friedrich. Uniform statistical description of the transition between near and far field turbulence in a wake flow. *Phys. Rev. Lett.*, **83**, 5495–8 (1999).

27 C. Renner, J. Peinke and R. Friedrich. Experimental indications for Markov properties of

small-scale turbulence. *J. Fluid Mech.*, **433**, 383–409 (**2001**).

28 H. U. Boedeker, M. C. Roettger, A. W. Liehr, T. D. Frank, R. Friedrich and H.-G. Purwins. Noise-covered drift bifurcation of dissipative solitons in a planar gas-discharge system. *Phys. Rev. E*, **67**, 056220 (**2003**).

29 M. Waechter, F. Riess, H. Kantz and J. Peinke. Stochastic analysis of different rough surfaces. *Europhys. Lett.*, **64**, 579–85 (**2003**).

30 G. R. Jafari, S. M. Fazeli, F. Ghasemi, S. M. Vaez Allaei, M. R. Rahimi Tabar, A. Irajizad and G. Kavei. Stochastic analysis and regeneration of rough surfaces. *Phys. Rev. Lett.*, **91**, 226101 (**2002**).

31 T. D. Frank, P. J. Beek and R. Friedrich. Identifying noise sources of time-delayed feedback systems. *Phys. Lett. A*, **328**, 219–24 (**2004**).

32 T. D. Frank, M. Sondermann, T. Ackemann and R. Friedrich. Parametric data analysis of bistable stochastic systems. *Nonlinear Phenomena in Complex Systems*, **8**, 193–9 (**2005**).

33 J. Gradisek, E. Govekar and I. Grabec. Qualitative and quantitative analysis of stochastic processes based on measured data, II: Applications to experimental data. *J. Sound Vibr.*, **252**, 563–72 (**2002**).

34 R. Friedrich, J. Peinke and C. Renner. How to quantify deterministic and random influences on the statistics of the foreign exchange market. *Phys. Rev. Lett.*, **84**, 5224–7 (**2000**).

35 C. Renner, J. Peinke and R. Friedrich. Evidence of Markov properties of high frequency exchange rate data. *Physica A*, **298**, 499–520 (**2001**).

36 S. Kriso, J. Peinke, R. Friedrich and P. Wagner. Reconstruction of dynamical equations for traffic flow. *Phys. Lett. A*, **299**, 287–91 (**2002**).

37 P. Sura and S. T. Gille. Interpreting wind-driven southern ocean variability in a stochastic framework. *J. Mar. Res.*, **61**, 313–34 (**2003**).

38 T. Kuusela, T. Shepherd and J. Hietarinta. Stochastic model for heart-rate fluctuations. *Phys. Rev. E*, **67**, 061904 (**2003**).

39 T. Kuusela. Stochastic heart-rate model can reveal pathologic cardiac dynamics. *Phys. Rev. E*, **69**, 031916 (**2004**).

40 A. M. van Mourik, A. Daffertshofer and P. J. Beek. Deterministic and stochastic features of rhythmic human movement. *Biol. Cybern.*, **94**, 233–44 (**2006**).

41 K. Ito. On stochastic differential equations. *Mem. Amer. Math. Soc.*, **4**, 1–51 (**1951**).

42 R. L. Stratonovich. A new representation for stochastic integrals and equations. *J. SIAM Control*, **4**, 362–71 (**1966**).

43 R. F. Pawula. Approximation of the linear Boltzmann equation by the Fokker–Planck equation. *Phys. Rev.*, **162**, 186–8 (**1967**).

44 M. Palus. Nonlinearity in normal human EEG: cycles, temporal asymmetry, nonstationarity and randomness, not chaos. *Biol. Cybern.*, **75**, 389–96 (**1996**).

45 S. Blanco, H. Garcia, R. Quian Quiroga, L. Romanelli and O. A. Rosso. Stationarity of the EEG series. *IEEE Eng. Med. Biol.*, **4**, 395–9 (**1995**).

46 C. Rieke, F. Mormann, R. G. Andrzejak, T. Kreuz, P. David, C. E. Elger and K. Lehnertz. Discerning nonstationarity from nonlinearity in seizurefree and preseizure EEG recordings from epilepsy patients. *IEEE Tans. Biomed. Eng.*, **50**, 634–9 (**2003**).

47 F. Mormann, T. Kreuz, C. Rieke, R. G. Andrzejak, A. Kraskov, P. David, C. E. Elger and K. Lehnertz. On the predictability of epileptic seizures. *Clin. Neurophysiol.*, **116**, 569–87 (**2005**).

48 R. G. Andrzejak, F. Mormann, T. Kreuz, C. Rieke, A. Kraskov, C. E. Elger and K. Lehnertz. Testing the null hypothesis of the non-existence of the preseizure state. *Phys. Rev. E*, **67**, 010901 (**2003**).

49 K. Lehnertz and B. Litt. The first international collaborative workshop on seizure prediction: summary and data description. *Clin. Neurophysiol*, **116**, 493–505 (**2005**).

15
Bivariate and Multivariate Time Series Analysis Techniques and their Potential Impact for Seizure Prediction

Hannes Osterhage, Stephan Bialonski, Matthäus Staniek, Kaspar Schindler, Tobias Wagner, Christian E. Elger, Klaus Lehnertz[1]

15.1
Introduction

The ambition to predict epileptic seizures has been one of the main driving forces in the analysis of electroencephalographic (EEG) time series since the 1970s (see [1–7] for comprehensive overviews). In the following years, univariate methods from linear time series analysis were predominantly applied in this field of research, particularly analysis techniques that exploit the spectral properties [8, 9] or occurrence rates of epileptic spikes in the EEG [10, 11]. With the beginning of the 1990s, methods from nonlinear time series analysis have increasingly been used. Initially, univariate measures such as the largest Lyapunov exponent [12, 13], the correlation dimension [14–16], the correlation density [17], and the dynamical similarity [18–21] were used to identify preictal states in EEG time series. The ongoing developments in the field, together with the increasing computational power, soon resulted in the introduction of bivariate measures, which aim at the investigation of relationships between pairs of recording sites. Prominent examples are the difference between the largest Lyapunov exponents calculated from two time series [22] and different measures for synchronization [23–25].

With N recording sites, the number of non-redundant combinations is equal to $N(N - 1)/2$, which may result in the comprehensive analysis of the available data being computationally demanding. Moreover, this often high number of possible combinations renders a statistical validation of any retrospectively optimized prediction algorithm indispensable [26–28]. Despite the fact that such a validation is very exigent for those techniques, recent studies [29–33] have revealed that bivariate measures exhibit the highest predictive performance among the different techniques that have been applied. Nevertheless, the steady increase in spatial resolution (and, correspondingly, the increase in the number of recording sites) not only

1) This work was supported by the Deutsche Forschungsgemeinschaft (SFB-TR3 sub-project A2 and LE-660/4-1).

renders the statistical evaluation of bivariate algorithms more and more difficult, but also results in EEG signals that are highly correlated and, consequently, highly redundant. Thus, one of the major challenges in the future will be the extraction of relevant information from multivariate recordings. This can be achieved through multivariate time series analysis techniques. As in the bivariate case, multivariate techniques aim at investigating relationships between different recording sites. However, these relationships are characterized through a simultaneous analysis of all recording sites, in contrast to the pairwise analysis in the bivariate case. In light of the hypothesis of an epileptic network [34, 35], which enters pathological states and generates seizure activity, multivariate time series analysis techniques aiming at the identification and characterization of interaction networks become more and more attractive.

In the following, we present exemplary bivariate and multivariate measures for the quantification of interactions between dynamical (sub)-systems. The bivariate measures are derived from two different frameworks, namely, the synchronization of dynamical systems [36], and from information theoretical considerations [37]. We start with measures of synchronization that are derived from the concepts of phase synchronization [38] and of generalized synchronization [39]. We then present the transfer entropy [40] which is a modified version of the mutual information for conditional probabilities and thus can detect the directed exchange of information between two systems. Exemplary applications to EEG time series using some of these bivariate techniques will then be presented. Finally we give an overview about recently introduced multivariate time series analysis techniques which are based on two frameworks, namely Random Matrix Theory [41, 42] and Network Theory [43, 44].

15.2
Bivariate Time Series Analysis Techniques

15.2.1
Measures of Synchronization

When studying two interacting dynamical (sub-)systems, the investigation focuses on two aspects, namely, the *strength* and the *direction* of the interactions. In the field of seizure prediction, studies have so far been limited to the use of measures of the strength [23–25]. However, potential future developments are likely to involve directional measures. For this reason, this section includes a representative choice of bivariate measures for the strength of interaction that has already been used in different studies. In addition, we present exemplary directional approaches for each of these measures. From the literature, four different frameworks for the mathematical description of synchronization phenomena are well known. In the most simple case of *complete synchronization* the systems' states become identical. A natural extension of this case, *lag synchronization* [45], is reached when the states coincide if one of the systems is shifted in time. In terms of time series analysis, both synchronization forms can be regarded as special cases of the more general concepts of *phase synchronization* [38] and *generalized synchronization* [39]. In the former case, the difference between the phases (which have to be defined appropriately) of both systems are bounded, while in the latter case, the states of the systems can be mapped onto each other through some function. For these we present exemplary measures for the strength and direction of interactions.

15.2.2
Phase Synchronization

The investigation of both the strength and the direction of interaction requires an extraction of phase time series $\left(\phi_j^{(x)}\right)_{j=1\ldots N}$ and $\left(\phi_j^{(y)}\right)_{j=1\ldots N}$ from the measured time series $\left(x\left(j\delta t\right) = x_j\right)_{j=1\ldots N}$ and $\left(y\left(j\delta t\right) = y_j\right)_{j=1\ldots N}$. Depending on the properties of the signals as well as on the purpose of the investigation, several phase extraction techniques are available. These can be divided into adaptive and selective techniques. Phases obtained from adaptive techniques such as the analytical signal approach [46, 47] always relate to the predominant frequency in the Fourier spectrum [48]. In contrast, frequency selective techniques (e.g., based on the wavelet transform) allow investigations of interactions in specific frequency bands. For an overview of the different techniques we refer to [49].

In the noise-free case, phase-synchronized systems are related through the phase-locking-condition

$$\alpha\phi_j^{(x)} - \beta\phi_j^{(y)} = \text{const}, \quad \alpha, \beta \in \mathbb{N}. \tag{15.1}$$

We here restrict ourselves to the case $\alpha = \beta = 1$; more general cases can be treated in an almost identical manner. A well known measure for the strength of synchronization between the systems is the mean phase coherence $R\left(X|Y\right)$ [24]:

$$R\left(X|Y\right) = \frac{1}{N}\left|\sum_{j=1}^N \exp i\left(\phi_j^{(x)} - \phi_j^{(y)}\right)\right|. \tag{15.2}$$

In the case of independent systems, the phase differences are randomly distributed on the unit circle, yielding $R\left(X|Y\right) \to 0$ for $N \to \infty$, while in the synchronized case, the phase-locking condition implies that the phase differences are peaked around some constant value, yielding $R\left(X|Y\right) \to 1$.

As a measure for directionality using phase time series, we consider the evolution map approach proposed in [50]. The phase increments $\Omega_j^{(x,y)} = \phi_{j+\tau}^{(x,y)} - \phi_j^{(x,y)}$ over some fixed time τ are assumed to be generated by two-dimensional noisy maps

$$\Omega_j^{(x,y)} = \mathbb{F}^{(x,y)}\left[\phi_j^{(x)}, \phi_j^{(y)}\right] + \eta^{(x,y)}, \tag{15.3}$$

with the random terms $\eta^{(x,y)}$ representing noisy perturbations. The maps are then approximated using finite Fourier series

$$F^{(x,y)}\left(\phi^{(x)}, \phi^{(y)}\right) = \sum_{k,l} A_{k,l}^{(x,y)} \exp\left(ik\phi^{(x,y)} + il\phi^{(y,x)}\right). \tag{15.4}$$

Solving the linear least squares problem $F \approx \mathbb{F}$ provides estimates for the deterministic parts of the maps. The accuracy of approximation is determined by the order of the Fourier series (in both dimensions), e.g., by the choice of the pairs (k, l), which have to be selected *a priori*. On the one hand, high-order terms increase the quality of the fit. On the other hand, they lead to an approximation of the non-deterministic parts of the dynamics, which is not desired. It should be pointed out that the choice of the optimal set of pairs (k, l) is non-trivial and highly application-specific.

Given the function $F^{(x,y)}$, the mutual influence of the systems can be quantified as

$$\left(c^{(x,y)}\right)^2 = \int_0^{2\pi} \int_0^{2\pi} \left(\frac{\partial F^{(x,y)}}{\partial \phi^{(y,x)}} \right)^2 d\phi^{(x)} d\phi^{(y)} \tag{15.5}$$

since the Fourier series are 2π-periodic, by definition. Analytical integration [51] yields

$$c^{(x,y)} = \sqrt{\sum_{k,l} l^2 A_{k,l}^{(x,y)}}, \tag{15.6}$$

and the directionality of the coupling can be quantified by

$$d\,(X|Y) = \frac{c^{(y)} - c^{(x)}}{c^{(y)} + c^{(x)}}, \tag{15.7}$$

where $d\,(X|Y) \in [-1, 1]$ with properties

$$d\,(X|Y) \begin{cases} > 0 & \text{X drives Y} \\ = 0 & \text{symmetric bidirectional coupling} \\ < 0 & \text{Y drives X.} \end{cases} \tag{15.8}$$

Note, that a vanishing coupling between the systems is a special case of symmetric bidirectional coupling.

15.2.3
Generalized Synchronization

The concept of generalized synchronization applies in the state spaces of the investigated systems, which initially have to be reconstructed from time series using, e.g., delay embedding [52]:

$$\vec{x}_j = (x_j, x_{j+T}, \ldots, x_{j+(m-1)T}),$$
$$\vec{y}_j = (y_j, y_{j+T}, \ldots, y_{j+(m-1)T}), \quad j = 1, \ldots, M = N - (m-1)T. \tag{15.9}$$

The embedding dimension m and the time delay T have to be chosen appropriately. A reliable unfolding of the dynamics of a system with dimension d is given if $m > 2d$ [53,54]. While the topological properties of the reconstructed attractors are independent of the time delay, a proper detection of the coupling direction [55] between dynamical systems requires an appropriate choice of T. This relates to the fact that the unfolding of the attractor is optimal with linearly independent vector components. For this reason, the decorrelation time of the system or the delay corresponding to the first minimum of the *time delayed mutual information* [56] often turns out to be a good choice.

Generalized synchronization is now said to occur if a function $\mathcal{F}$ exists that maps the states of the systems onto each other:

$$\vec{y} = \mathcal{F}(\vec{x}). \tag{15.10}$$

While $\mathcal{F}$ was initially introduced as being completely arbitrary [57], the concept of mutual neighborhoods in state space, which is related to the idea of $\mathcal{F}$ being smooth, turned out to be appropriate when investigating interactions between the systems in their respective state spaces. Measures derived from this concepts mostly exploit the mutual neighborhoods either directly [58, 59] or by predicting the future states of the systems [60]. The prediction of future states of a system with the inclusion of the knowledge about the interacting second systems is closely related to the concept of *Granger causality* [61]. Other methods, which also rely on mutual neighborhoods, use joint recurrences for the detection of strength [62, 63] and direction [64] of interactions.

We here resort to the concept of *nonlinear interdependence* [23], which does not require the existence of a function $\mathcal{F}$. Consider two state space vectors $\vec{x}_j$ and $\vec{y}_j$ at time j. Let the r nearest neighbors of $\vec{x}_j$ be denoted by $\vec{x}_{s_j(i)}$, $i = 1 \ldots r$. The systems are considered to be interdependent if, for any j, the mean squared distance between $\vec{y}_j$ and its r *conditioned neighbors* $\vec{y}_{s_j(i)}$, $i = 1 \ldots r$ is small compared to the mean squared distance

$$Q_j^r(Y) = \frac{1}{M-1} \sum_{k \neq j} (\vec{y}_j - \vec{y}_k)^2 \tag{15.11}$$

between $\vec{y}_j$ and all remaining state space vectors. Given the mean squared distance between $\vec{y}_j$ and its conditioned neighbors

$$Q_j^r(X|Y) = \frac{1}{r} \sum_{k=1}^{r} \left(\vec{y}_j - \vec{y}_{s_j(i)}\right)^2 \tag{15.12}$$

one now defines

$$H_r(X|Y) = \frac{1}{M} \sum_{k=1}^{M} \log\left(\frac{Q_j^r(Y)}{Q_j^r(X|Y)}\right) \tag{15.13}$$

as a measure of interdependence. $H_r(Y|X)$ is defined in complete analogy. Measures for strength and direction of interaction are defined as:

$$H_r^+(X|Y) = H_r^+(Y|X) = \frac{H_r(X|Y) + H_r(Y|X)}{2}$$

$$H_r^-(X|Y) = -H_r^-(Y|X) = \frac{H_r(X|Y) - H_r(Y|X)}{2}. \tag{15.14}$$

As a measure for the strength of interaction, $H_r^+(X|Y) \in \mathbb{R}_0^+$ and increases with an increasing coupling between the systems. The directional measure $H_r^-(X|Y) \in \mathbb{R}$ attains positive values when X is driving Y, and negative values in the opposite case.

15.3
Information Theoretic Measures

Basic principles of information theory and the concept of entropy were derived almost 60 years ago [37, 65]. Entropy is a measure quantifying the order of a time series by mapping a probability function onto some scalar. Suppose the state x_j of process X is occupied with uncertainty $\log 1/p(x_j)$. The Shannon entropy is then defined as the average over all possible states:

$$S(X) = \sum_j p(x_j) \log \frac{1}{p(x_j)} = -\sum_j p(x_j) \log p(x_j).$$ (15.15)

In time series analysis, the Shannon entropy can be estimated via a box counting approach to quantify the average information gained by a measurement of X. A partition size θ is chosen and the data x_j are discretized into symbols depending on what bin of size θ they fall into. Up to a factor, the Shannon entropy is the only functional, which is continuous in the $p(x_j)$, additive, and increases with the number of possible symbols (size of alphabet A) [66].

Since $p(x_j)$ is measured from a signal that is generated by the process X, it is an estimate of the true distribution $q(x_j)$ only, which has to be assumed *a priori*. Averaging the difference in uncertainty, $\log 1/q(x_j) - \log 1/p(x_j)$, yields the Kullback–Leibler entropy [67] as a measure of the difference between two probability distributions

$$K(X) = \sum_{x_j \in A} p(x_j) \log \frac{p(x_j)}{q(x_j)}.$$ (15.16)

If we observe two processes X and Y that we assume to be independent, we can set

$$q(x_j, y_k) = p(x_j)p(y_k).$$ (15.17)

The Kullback–Leibler entropy (15.16) then reads

$$M(X|Y) = \sum_{j,k} p(x_j, y_k) \log \frac{p(x_j, y_k)}{p(x_j)p(y_k)}.$$ (15.18)

$M(X|Y)$ quantifies the deviation from the assumption that the processes X and Y are independent and is thus called mutual information [68, 69]. In order to analyze time-delayed dependencies, $j = k - T$ can be considered, with T denoting the time lag [70]. The mutual information is symmetric and zero for independent processes and can be expressed through Shannon entropies:

$$M(X|Y) = S(X) + S(Y) - S(X, Y).$$ (15.19)

The analysis of the *dynamical* structure of a process requires the consideration of transition probabilities $p(x_{j+1}|\mathbf{x}_j^{(l)})$, where p denotes the probability of a transition into a new state x_{j+1} given the last l states of X. If x_{j+1} depends on the l past states of X, but is independent of the last m states of Y, then

$$p(x_{j+1}|\mathbf{x}_j^{(l)}, \mathbf{y}_k^{(m)}) = p(x_{j+1}|\mathbf{x}_j^{(l)}). \qquad (15.20)$$

Transfer entropy [40, 71] quantifies the incorrectness of this assumption and is formulated as the Kullback–Leibler entropy between $p(x_{j+1}|\mathbf{x}_j^{(l)}, \mathbf{y}_k^{(m)})$ and $p(x_{j+1}|\mathbf{x}_j^{(l)})$:

$$T(Y|X) = \sum_{x_{j+1}, \mathbf{x}_j^{(l)}, \mathbf{y}_k^{(m)}} p(x_{j+1}, \mathbf{x}_j^{(l)}, \mathbf{y}_k^{(m)}) \frac{p(x_{j+1}|\mathbf{x}_j^{(l)}, \mathbf{y}_k^{(m)})}{p(x_{j+1}|\mathbf{x}_j^{(l)})}. \qquad (15.21)$$

Thus, transfer entropy specifies the deviation from the generalized Markov property (15.20). $T(Y|X)$ quantifies the degree of dependence of X on Y, and vice versa. In order to quantify the preferred direction of flow, a directionality index based on the transfer entropy can be introduced as

$$\hat{T}(X|Y) = \frac{T(X|Y) - T(Y|X)}{T(X|Y) + T(Y|X)}. \qquad (15.22)$$

$\hat{T}(X|Y)$ is expected to vary between 1 for unidirectional coupling with X as the driver and -1 for Y driving X. For symmetric bidirectional coupling $\hat{T}(X|Y) = 0$ is expected. The transfer entropy is an antisymmetric measure, i.e., $\hat{T}(X|Y) = -\hat{T}(Y|X)$.

15.4
Exemplary Applications

The mean phase coherence $R(X|Y)$ has repeatedly been applied to EEG time series as a potential measure for epileptic seizure prediction [72]. We here present an application where this measure was used to detect synchronization in different frequency bands in long-term multichannel intracranial EEG recordings from a patient suffering from left frontal lobe epilepsy. The EEG was sampled over a period of one week at 200 Hz, during which the patient had four seizures. The mean phase coherence was calculated for consecutive, non-overlapping segments of 4096 data points (corresponding to 20.48 s). In addition to a phase estimation using the Hilbert transform, we here filtered the EEG data using a morlet wavelet adapted to the frequency bands 0.5–4 Hz (δ-band), 4–8 Hz (θ-band), 8–13 Hz (α-band), 13–20 Hz (β1-band), and 20–30 Hz (β2-band) at full width half maximum.

In order to investigate the ability of the broad-band and frequency-selective mean phase coherence to detect changes predictive of an impending seizure, we here assumed a preictal period of four hours (cf. [29]). In the upper part of Figure 15.1 we present an exemplary time course of the mean phase coherence for EEG data filtered in the β2-band. In the middle and lower part of Figure 15.1 and in Figure 15.2 we

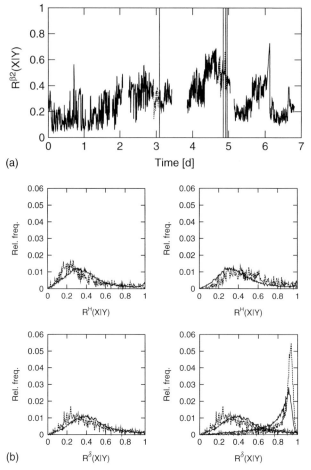

(a)

(b)

Fig. 15.1 (a) Exemplary time course of the mean phase coherence $R^{\beta 2}$ calculated from a combination of electrodes located in brain regions distant to seizure onset zone. Discontinuities in the time profile are due to recording gaps. Seizure onsets are marked by vertical lines, and the assumed preictal states are plotted with dashed lines. (b) Distributions of interictal (solid lines) and preictal (dashed lines) of the mean phase coherence obtained from estimating the phases using the Hilbert transform (upper plots) and from estimating band-limited phases (δ-band) using the wavelet transform (lower plots). Diagrams shown on the left refer to the channel combination which exhibited the most pronounced preictal decrease. Diagrams shown on the right refer to the channel combination that exhibited the most pronounced preictal increase.

show the distributions of values of the mean phase coherence for interictal and preictal periods assuming either a preictal increase or decrease in this measure. We here restrict ourselves to recording sites that exhibited the most pronounced differentiability (using receiver-operating-characteristic statistics) between interictal and preictal periods. For this patient, the mean phase coherence calculated from EEG data filtered in the upper frequency bands (α- and β-bands) allowed a better

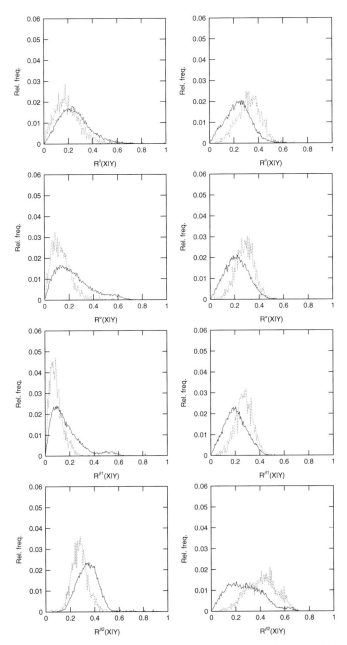

Fig. 15.2 Continuation of Figure 15.1. Top to bottom: Distributions of interictal (solid lines) and preictal (dashed lines) of the mean phase coherence obtained from estimating band-limited phases (θ-, α-, β1-, and β2-bands) using the wavelet transform. Diagrams shown on the left refer to the channel combination which exhibited the most pronounced preictal decrease. Diagrams shown on the right refer to the channel combination which exhibited the most pronounced preictal increase.

discrimination between interictal and preictal periods than phase coherence values from the lower frequency bands and from broad-band signals (cf. [73]). We observed both a preictal increase and decrease. However, these phenomena were not confined to a specific brain region but could be observed in the interactions between different neocortical structures covered by a 64-contact grid electrode.

As an example for the detection of directionality in EEG time series, we here present an application of the transfer entropy to long-term, multichannel EEG recordings from a patient suffering from right medial temporal lobe epilepsy. We here restrict ourselves to the analysis of spatial interactions between different brain regions during the interictal state (we excluded data that were recorded less than four hours prior to a seizure, during a seizure and less than one hour after any seizure). The EEG was sampled at 200 Hz and recorded via intrahippocampal depth electrodes, each equipped with 10 cylindrical contacts (diameter: 2.5 mm, intercontact distance: 4 mm) (cf. Figure 15.3a). $\hat{T}(X|Y)$ was calculated for consecutive, non-overlapping segments of 4096 data points for all possible combinations of electrode contacts.

Figure 15.3(b) shows the temporal average of $\hat{T}(X|Y)$ for each combination of electrode contacts in the form of a directionality matrix $M(\hat{T})$. $M(\hat{T})$ is antisymmetric since $\hat{T}(X|Y) = -\hat{T}(Y|X)$. Matrix entries $M(\hat{T})_{ij}$ attain positive values if the system corresponding to the row index i drives the system corresponding to the column index j. The structures recorded at the seventh and eighth electrode contacts in the right hemisphere (TR07,TR08) appear to drive both the left and right medial temporal regions of the brain. Visual inspection of the raw EEG data including seizure activity by a clinical expert confirmed that initial signs of seizure activity were always restricted to electrode contacts (TR06,TR07,TR08,TR09). Apparently, the seizure onset zone is driving remote brain areas during the interictal state (cf. [74]).

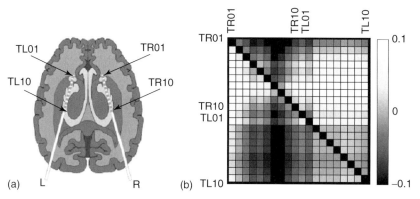

Fig. 15.3 (a) Implantation scheme for intrahippocampal electrodes. (b) Directionality matrix $M(\hat{T})$ obtained from the transfer entropy. Matrix entries are color coded and represent the temporal average of $\hat{T}$ for each combination of electrode contacts. (Please find a color version of this figure on the color plates.)

15.5
Multivariate Time Series Analysis Techniques

While, in the past, epilepsy has mainly been studied using a reductionistic approach on the level of ion channels, single neurons, or small neuronal networks *in vitro*, the present view is that epileptic seizures are produced by the collective electrical activity in large, i.e., spatially extended and interacting neuronal networks [34, 35, 75]. Furthermore, the occurrence of epileptic seizures is often related to specific behavioral states, which affect the global state of the brain. Improving our understanding of the pathophysiological basis of epilepsy thus requires an *in vivo* assessment of the changes of the collective spatio-temporal neuronal activity before seizures, at the onset of seizures, during seizure propagation, and when seizures terminate. Two powerful approaches for the quantitative assessment of large systems of interacting units are Random Matrix Theory and Network (or Graph) Theory, which will be briefly discussed in the following.

15.5.1
Approaches Based on Random Matrix Theory

In the 20th century the development of Random Matrix Theory (RMT) was initiated by Wigner in order to describe statistical properties of many-body quantum systems, which represent complex dynamical systems with a large number of degrees of freedom. Together with Dyson, Mehta, and others RMT was further developed, extended, and successfully applied in various physical contexts, e.g., to describe statistical properties of atomic nuclei, atoms, and complex molecules (see [41, 42] for an overview).

Recently, concepts from RMT have been adopted in the context of multivariate time series analysis. Matrices arise naturally when computing a bivariate measure of signal interdependence for all pairs of multi-channel time series. A comparison of spectral properties of these empirically obtained matrices with spectral properties of an appropriate ensemble of random matrices can help in distinguishing 'real' from random correlations and thus can enable one to extract the relevant information hidden in a multivariate dataset. As prominent examples for RMT-based measures we mention the *spectral density* ρ, the *nearest-neighbor distribution* P_{nn}, and the *number variance* Σ^2, which are based on the eigenvalue spectrum and are typically compared with their analytical counterparts for random matrix ensembles. Deviations between empirically obtained matrix ensembles and random matrix ensembles are usually interpreted as evidence for non-random correlations. For a detailed introduction to RMT and discussion of some of its measures in the context of time series analysis we refer the reader to Chapter 16 in this book. In the following, we give a brief overview of applications of RMT-based time series analysis techniques to human EEG and MEG data.

Kwapien and co-workers [76] studied sequences of evoked magnetic fields from the human auditory cortex during the delivery of acoustic stimuli. Correlation matrices were constructed by treating the time series of each single trial as a separate system. The authors stated that the observed nearest-neighbor distribution P_{nn}, which only accounts for correlations between neighboring eigenvalues, were largely consistent with the nearest-neighbor distribution of the *Gaussian Orthogonal Ensemble (GOE)* of random matrices but deviated for larger eigenvalue spacings. This supported earlier findings indicating the presence of a background state (associated with a GOE-like behavior here), which is slightly perturbed by the sound stimuli. The authors hypothesized that the observed separation of the eigenvalue spectrum into the background of 'noise' eigenvalues and a group of eigenstates with large eigenvalues possibly reflects different anatomical routes from the cochlea to the auditory cortex.

Seba [77] investigated scalp EEG data recorded from 90 healthy subjects under resting conditions and under visual stimulation. After artifact removal, the equal time (=zero-lag) correlation matrices were computed from non-overlapping moving windows and ρ, P_{nn}, and the number variance Σ^2, which takes into account simultaneous correlations between a group of subsequent eigenvalues, were studied and compared to the predictions made by RMT. The author reported that ρ and P_{nn} did not change during visual stimulations and, in addition, appeared to be subject-independent and in agreement with RMT predictions, thereby reflecting generic features of correlation matrices obtained from human EEG data. In contrast, Σ^2 was found to deviate from RMT predictions when subjects were visually stimulated, which was attributed to altered correlations between the visual and other cortical areas.

In order to detect phase-shape correlations in multivariate time series, Müller and co-workers [78] proposed a method based on the eigenvalues and eigenvectors of the equal time correlation matrix. The authors investigated the temporal evolution of eigenvalues obtained from correlation matrices of time series generated by model systems. They reported that the increase of phase-shape correlations in a number of time series caused a *level repulsion* of a corresponding number of eigenstates. Level repulsion is a combined increase and decrease of different eigenvalues, such that their sum remains constant. In addition, the authors analyzed a scalp EEG recording from a patient suffering from absence epilepsy. Using a moving window technique, equal time correlation matrices were computed, and the temporal evolution of their largest and smallest eigenvalues were investigated. At the onset of the absence seizure a sudden increase of the largest eigenvalue and a decrease of almost the entire rest of the spectrum was observed, which was interpreted as a possible indication for a phase transition (see [78] and references therein). Extending this analysis, Baier and colleagues [79] studied the temporal evolution of eigenvalues obtained from scalp EEG recordings of five patients with absence epilepsy. Investigating the transition from 'normal' background activity to seizures, the authors observed significant changes of the eigenvalues a few seconds prior to the occurrence of spike-wave complexes. The smallest eigenvalue increased, which was interpreted as a loss of correlation confined to a small number of electrodes. After the onset of the spike-wave complexes the largest eigenvalues were increased while the smallest eigenvalues were decreased, thereby reflecting an overall increase of cross-correlation during the seizures.

Schindler and co-workers [80] assessed the temporal evolution of the correlation structure in the course of 100 focal onset seizures from 60 patients recorded by intracranial, multi-channel EEG by computing the time-resolved eigenvalue spectrum of the equal time correlation matrix of a moving window. The results demonstrated statistically significant changes in the correlation structure. Specifically, the changes in the eigenvalue spectrum indicated that the zero-lag correlation of multi-channel EEG either remained approximately unchanged or – especially in the case of secondary generalization – even decreased during the first half of the seizures. Then correlation gradually increased again before the seizures terminated. This development was qualitatively independent of the anatomical location of the seizure onset zone and may thus be a generic property of focal onset seizures. It was suggested that the decorrelation of EEG activity might be due to different propagation delays of locally synchronous ictal discharges from the seizure onset zone to other brain areas. Furthermore, it was proposed that the increase in the correlation during the second half of the seizures could be causally related to seizure termination. This hypothesis was further tested in another study [81], by investigating the correlation dynamics during status epilepticus, i.e., epileptic seizures that last more than five minutes [82]. It was found that correlation was relatively low during ongoing status epilepticus and only persistently increased toward its end. Interestingly, the increase of correlation often followed the application of drugs, indicating that RMT-based methods might be helpful to also assess pharmacological effects on a network level. The findings of this study thus support the hypothesis that increasing correlation of neuronal activity may be considered as an emergent self-regulatory mechanism for seizure termination, which could also further our knowledge about seizure initiation. They furthermore challenge the traditional concept that increasing neuronal synchronization during epileptic seizures is always pathological and should be suppressed.

Carrying over concepts developed in [78] Allefeld and co-workers proposed a method to detect synchronization clusters in multivariate time series [83]. Based on the observation that the eigenvectors that are involved strongly in the level repulsion process contain information about the correlated groups of time series, the authors defined the *participation index p*. This index is determined by the eigenvalues and eigenvectors of a synchronization matrix, which contains the mean phase coherence (see Section 15.2.1) between each channel combination as its entries. Bialonski and Lehnertz [84] explored the limitations of this index by analyzing time series generated by coupled model systems. By applying the method to long-term intracranial EEG recordings from patients with focal epilepsy, they observed synchronization clusters, whose spatial extent and location coincided with the seizure onset zone as determined by the presurgical workup and the post-operative outcome. In addition, the method was shown to be sensitive to short-term changes of synchronization phenomena, which could be associated with physiological processes (language processing) in the brain.

15.5.2
Approaches Based on Network Theory

Network theory has received rapidly growing interest from various fields of science during the last decade (see [43, 44] for an overview). Inspired by this theory, multivariate time series analysis techniques have been developed to gain more insights into dependencies between structure and dynamics of complex systems. A network can be formally described by a graph consisting of a set of nodes (vertices) and a set of links (edges) connecting them. *Unweighted networks* are characterized by links that only can exist (1) or not exist (0) between nodes (binary network). If links carry a numerical value (indicating the strength of the connection), the network is called a *weighted network*. Unlike links of *undirected networks*, links of *directed networks* describe connections between nodes that are associated with a direction. A network can be represented by an *adjacency matrix A*, where the entries a_{jk} indicate the strength (or existence) of a link from node j to node k. Thus, undirected networks are described by symmetric adjacency matrices. By interpreting bivariate time series analysis techniques as measures of link strengths within an interaction network, the analysis can be carried over into the domain of network theory, where a plethora of methods can be applied to further investigate the obtained structure. In the following, we report on measures that have been widely used in the literature to distinguish between different classes of networks.

The *average shortest path length L* of a network quantifies how efficiently information can propagate over the network. Since L relies on the overall integrity of the network it characterizes the network on a *global scale*. Let N be the number of nodes and let d_{jk} denote the shortest path length between nodes j and k, which corresponds to the smallest number of links needed to travel from j to k in unweighted networks. Then L can be defined by

$$
L = \frac{1}{N(N-1)} \sum_{j \neq k} d_{jk}.
\tag{15.23}
$$

In the case of disconnected components in the network L can diverge, since for some j and k no path exists and hence d_{jk} becomes infinite. Other measures have been proposed to circumvent this problem, e.g.,

$$
\tilde{L} = E^{-1} = \left(\frac{1}{N(N-1)} \sum_{j \neq k} \frac{1}{d_{jk}} \right)^{-1},
\tag{15.24}
$$

where E is called the *efficiency* and considers the harmonic mean of the shortest path lengths.

The *clustering coefficient C* [85] quantifies how well connected the nodes in a network are on a *local scale*. Let k_j be the number of nodes that are connected to node j (neighbors of j). For unweighted networks, C is then defined as

$$C = \frac{1}{N}\sum_{j=1}^{N} C_j = \frac{1}{N}\sum_{j=1}^{N} \left(\frac{\sum\limits_{k,m} a_{jk}a_{km}a_{mj}}{k_j(k_j-1)} \right), \qquad (15.25)$$

where C_j denotes the fraction of existing links among all possible links between the neighbors of node j. If a node with highly connected neighbors is removed, the information can most likely still travel over the network without interference. Thus C has sometimes been interpreted as a measure of the robustness of the network against random failures. Note that alternative measures for weighted networks have been proposed (e.g. see [43] and references therein).

Based on L and C, Watts and Strogatz [85] distinguished between three classes of networks, namely regular networks (e.g., lattices), small-world networks, and random networks, whose links are randomly assigned to nodes. Whereas regular networks are characterized by high clustering coefficients and long average shortest path lengths, random networks display short path lengths and low clustering coefficients. Exploring the regime between both network types the authors used a random rewiring procedure to generate artificial networks with varying degree p of randomness (starting from $p = 0$ (regular network) to $p = 1$ (random network)). They found that for a slight increase in p the clustering coefficient C remained almost constant, while L decreased rapidly. Networks of this regime, where a high C and a low L prevail, are called small-world networks. Because of these features, they can efficiently transfer information, while at the same time being quite robust against random errors.

Measures motivated by network theory have found various applications in different sciences. In the following, we briefly summarize research done in the context of epilepsy. For a general overview about network analyses of the brain we refer the reader to two recently published reviews [86, 87].

Ponten and colleagues [88] analyzed intracranial EEG recordings from seven patients suffering from medial temporal lobe epilepsy. The authors used the synchronization likelihood index [62] to detect interdependencies between pairs of channels for different frequency bands and to construct networks. The clustering coefficient C and a differently normalized version of $\tilde{L}$ were computed for the obtained networks as well as for random networks with the same degree distribution (C_r and $\tilde{L}_r$). The authors reported that the mean values C/C_r and $\tilde{L}/\tilde{L}_r$ of all patients increased during and after seizures when compared to the interictal period. This suggests a 'movement' from random interictal to small-world like and therefore more regular ictal configurations. In this context, we mention a case study by Wu and co-workers [89] who analyzed an intracranial EEG recording of ten minutes duration from an epilepsy patient before and during a seizure. Bispectral analysis techniques were applied to detect interdependencies between channels, and adjacency matrices of unweighted networks were obtained in the way described above. The authors reported that the small-world property (high C values, low L values) was more pronounced in networks obtained during the seizure than in the seizure-free time interval, which supports the findings by Ponten and colleagues.

Furthermore, recent theoretical studies [90] indicate that random networks are more synchronizable than small-world networks. These results might indicate that epilepsy comes along with interictal networks that possess a random structure and thus a lower threshold for seizure generation [86].

15.6
Conclusions

We have presented bivariate and multivariate EEG analysis techniques and have discussed their potential impact for seizure prediction. Research over recent years has demonstrated in particular that bivariate EEG analysis techniques – ones which allow one to estimate the strength of interactions between different brain regions – can be regarded as superior for the detection of long-lasting preictal phenomena. Nevertheless, the performance of seizure prediction approaches based on bivariate EEG analysis techniques cannot yet be regarded as sufficient for clinical applications. Our preliminary findings obtained from analyses involving different methods that allow one to measure coupling asymmetries and thus the direction of interactions indicate that these approaches can be regarded as helpful in expanding our knowledge of how seizures are generated in the brain and of the intermittent dysfunctioning of the brain between seizures.

Multivariate EEG analysis techniques are attractive because of their ability to simultaneously assess and quantify the collective dynamics of the epileptic brain. These techniques might help to gain deeper insights into the components of the epileptic network that are necessary for clinical events to occur and into the temporal and spatial distribution of seizure precursors. The multitude of new insights that can already be achieved within a short time period after introducing these techniques clearly demonstrates the power of these concepts. Multivariate techniques form a natural generalization of bivariate techniques, and further extensions and new developments can be expected within the near future. Given these developments that are also fertilized by research into the prediction of extreme events arising in other scientific areas, we expect increasing progress in the field of seizure prediction.

References

1 B. Litt and J. Echauz. Prediction of epileptic seizures. *Lancet Neurol.*, **1**, 22–30 (2002).

2 B. Litt and K. Lehnertz. Seizure prediction and the preseizure period. *Curr. Opin. Neurol.*, **15**, 173–7 (2002).

3 L. D. Iasemidis. Epileptic seizure prediction and control. *IEEE Trans. Biomed. Eng.*, **50**, 549–58 (2003).

4 F. Mormann, C. E. Elger and K. Lehnertz. Seizure anticipation: from algorithms to clinical practice.

Curr. Opin. Neurol., **19**, 187–93 **(2006)**.

5 F. Mormann, R. G. Andrzejak, C. E. Elger and K. Lehnertz. Seizure prediction: the long and winding road. *Brain*, **130**, 314–33 **(2007)**.

6 K. Lehnertz, M. Le Van Quyen and B. Litt. Seizure prediction. In J. Engel Jr and T. A. Pedley, editors, *Epilepsy: A Comprehensive Textbook*, pp 1011–24. Lippincott Williams & Wilkins, Philadelphia, second edition **(2007)**.

7 K. Lehnertz, F. Mormann, H. Osterhage, A. Müller, A. Chernihovskyi, M. Staniek, J. Prusseit, D. Krug, S. Bialonski and C. E. Elger. State-of-the-art of seizure prediction. *J. Clin. Neurophysiol.*, **24**, 147–53 **(2007)**.

8 Z. Rogowski, I. Gath and E. Bental. On the prediction of epileptic seizures. *Biol. Cybern.*, **42**, 9 **(1981)**.

9 A. Siegel, C. L. Grady and A. F. Mirsky. Prediction of spike wave–bursts in absence epilepsy by EEG power–spectrum signals. *Epilepsia*, **23**, 47 **(1982)**.

10 H. H. Lange, J. P. Lieb, J. Engel Jr and P. H. Crandall. Temporo–spatial patterns of pre-ictal spike activity in human temporal lobe. *Electroencephalogr. Clin. Neurophysiol.*, **56**, 543 **(1983)**.

11 A. Katz, D. A. Marks, G. McCarthy and S. S. Spencer. Does interictal spiking rate change before seizures? *Electroencephalogr. Clin. Neurophysiol.*, **79**, 153 **(1991)**.

12 L. D. Iasemidis, J. C. Sackellares, H. P. Zaveri and W. J. Williams. Phase space topography and the Lyapunov exponent of electrocorticograms in partial seizures. *Brain Topography*, **2**, 187–201 **(1990)**.

13 L. D. Iasemidis, L. D. Olson, R. S. Savit and J. C. Sackellares. Time dependencies in the occurrences of epileptic seizures. *Epilepsy Research*, **17**, 81–94 **(1994)**.

14 K. Lehnertz and C. E. Elger. Spatiotemporal dynamics of the primary epileptogenic area in temporal lobe epilepsy characterized by neuronal complexity loss. *Electroencephalogr. Clin. Neurophysiol.*, **95**, 108 **(1995)**.

15 K. Lehnertz and C. E. Elger. Can epileptic seizures be predicted? Evidence from nonlinear time series analysis of brain electrical activity. *Phys. Rev. Lett.*, **80**, 5019–22 **(1998)**.

16 C. Elger and K. Lehnertz. Seizure prediction by nonlinear time series analysis of brain electrical activity. *Europ. J. Neurosc.*, **10**, 786 **(1998)**.

17 J. Martinerie, C. Adam, M. Le Van Quyen, M. Baulac, S. Clémenceau, B. Renault and F. J. Varela. Epileptic seizures can be anticipated by nonlinear analysis. *Nature Medicine*, **4**, 1173 **(1998)**.

18 M. Le Van Quyen, J. Martinerie, M. Baulac and F. J. Varela. Anticipating epileptic seizures in real time by a non-linear analysis of similarity between EEG recordings. *Neuroreport*, **10**, 2149 **(1999)**.

19 M. Le Van Quyen, C. Adam, J. Martinerie, M. Baulac, S. Clémenceau and F. Varela. Spatio-temporal characterization of non-linear changes in intracranial activities prior to human temporal lobe seizures. *Eur. J. Neurosci.*, **12**, 2124 **(2000)**.

20 M. Le Van Quyen, J. Martinerie, V. Navarro, P. Boon, M D'Havé, C. Adam, B. Renault, F. Varela and M. Baulac. Anticipation of epileptic seizures from standard EEG recordings. *Lancet*, **357**, 183 **(2001)**.

21 V. Navarro, J. Martinerie, M. Le Van Quyen, S. Clémenceau, C. Adam, M. Baulac and F. Varela. Seizure anticipation in human neocortical partial epilepsy. *Brain*, **125**, 640 **(2002)**.

22 L. D. Iasemidis, P. Pardalos, J. C. Sackellares and D. S. Shiau. Quadratic binary programming and dynamical system approach to determine the predictability of epileptic seizures. *J. Comb. Optim.*, **5**, 9–26 **(2001)**.

23 J. Arnhold, P. Grassberger, K. Lehnertz and C. E. Elger. A robust method for detecting interdependences: application to intracranially recorded EEG. *Physica D*, **134**, 419–30 **(1999)**.

24 F. Mormann, K. Lehnertz, P. David and C. E. Elger. Mean phase coherence as a measure for phase synchronization and its application to the EEG of epilepsy patients. *Physica D*, **144**, 358–69 (**2000**).

25 M. Le Van Quyen, J. Martinerie, V. Navarro, M. Baulac and F. J. Varela. Characterizing neurodynamic changes before seizures. *J. Clin. Neurophysiol.*, **18**, 191 (**2001**).

26 R. G. Andrzejak, F. Mormann, T. Kreuz, C. Rieke, A. Kraskov, C. E. Elger and K. Lehnertz. Testing the null hypothesis of the nonexistence of a preseizure state. *Phys. Rev. E*, **67**, 010901(R) (**2003**).

27 T. Kreuz, R. G. Andrzejak, F. Mormann, A. Kraskov, H. Stögbauer, C. E. Elger, K. Lehnertz and P. Grassberger. Measure profile surrogates: A method to validate the performance of epileptic seizure prediction algorithms. *Phys. Rev. E*, **69**, 061915 (**2004**).

28 B. Schelter, T. Maiwald, A. Brandt, A. Schad, A. Schulze-Bonhage and J. Timmer. Testing statistical significance of multivariate time series analysis techniques for epileptic seizure prediction. *Chaos*, **16**, 013108 (**2006**).

29 F. Mormann, T. Kreuz, C. Rieke, R. G. Andrzejak, A. Kraskov, P. David, C. E. Elger and K. Lehnertz. On the predictability of epileptic seizures. *Clin. Neurophysiol.*, **116**, 569 (**2005**).

30 M. Le Van Quyen, J. Soss, V. Navarro, R. Robertson, M. Chavez, M. Baulac and J. Martinerie. Preictal state identification by synchronization changes in long-term intracranial eeg recordings. *Clin. Neurophysiol.*, **116**, 559–68 (**2005**).

31 M. Winterhalder, B. Schelter, T. Maiwald, A. Brandt, A. Schad, A. Schulze-Bonhage and J. Timmer. Spatio-temporal patient-individual assessment of synchronization changes for epileptic seizure prediction. *Clin. Neurophysiol.*, **117**, 2399–413 (**2006**).

32 B. Schelter, M. Winterhalder, H. Drentrup, J. Wohlmuth, J. Nawrath, A. Brandt, A. Schulze-Bonhage and J. Timmer. Seizure prediction: the impact of long prediction horizons. *Epilepsy Research*, **73**, 213–17 (**2007**).

33 H. Osterhage, F. Mormann, M. Staniek and K. Lehnertz. Measuring synchronization in the epileptic brain: a comparison of different approaches. *Int. J. Bifurcat. Chaos*, **17**, 3539–44 (**2007**).

34 D. A. McCormick and D. Contreras. On the cellular and network bases of epileptic seizures. *Annu. Rev. Physiol.*, **63**, 815–46 (**2001**).

35 S. S. Spencer. Neural networks in human epilepsy: Evidence of and implications for treatment. *Epilepsia*, **43**, 219–27 (**2002**).

36 A. S. Pikovsky, M. Rosenblum and J. Kurths. *Synchronization: A Universal Concept in Nonlinear Sciences*. Cambride Nonlinear Science Series 12. Cambridge University Press, Cambridge, UK (**2001**).

37 C. E. Shannon. A mathematical theory of communication. *Bell Sys. Tech. J.*, **27**, 379 (**1948**).

38 M. G. Rosenblum, A. S. Pikovsky and J. Kurths. Phase synchronization of chaotic oscillators. *Phys. Rev. Lett.*, **76**, 1804–7 (**1996**).

39 V. S. Afraimovich, N. Verichev and M. I. Rabinovich. General synchronization. *Izv. Vyssh. Uch. Zav. Radiofizika*, **29**, 795–803 (**1986**).

40 T. Schreiber and A. Schmitz. Surrogate time series. *Physica D*, **142**, 346–82 (**2000**).

41 T. Guhr, A. Müller-Groeling and H. A. Weidenmüller. Random-matrix theories in quantum physics: Common concepts. *Phys. Rep.*, **29**, 190–425 (**1998**).

42 T. A. Brody, J. Flores, J. B. French, P. A. Mello, A. Pandey and S. S. M. Wong. Random-matrix physics: Spectrum and strength fluctuations. *Rev. Mod. Phys*, **53**, 385–479 (**1981**).

43 S. Boccaletti, V. Latora, Y. Moreno, M. Chavez and D.-U. Hwang. Complex networks: Structure and dynamics. *Phys. Rep.*, **424**, 175–308 (**2006**).

44 R. Albert and A.-L. Barabási. Statistical mechanics of complex networks. *Rev. Mod. Phys.*, **74**, 47–97 (**2002**).

45 A. S. Pikovsky and J. Kurths. Coherence resonance in a noise-driven excitable system. *Phys. Rev. Lett.*, **78**, 775–8 (**1997**).

46 D. Gabor. Theory of communication. *J. IEE (London)*, **93**, 429–57 (**1946**).

47 P. Panter. *Modulation, Noise and Spectral Analysis*. New York. McGraw–Hill (**1965**).

48 B. Boashash. *Time-frequency Signal Analysis: Methods and Applications*. Longman Cheshire, Melbourne (**1992**).

49 A. S. Pikovsky. Statistics of trajectory separation in noisy dynamical systems. *Phys. Lett. A*, **165**, 33–6 (**1992**).

50 M. G. Rosenblum and A. S. Pikovsky. Detecting direction of coupling in interaction oscillators. *Phys. Rev. E*, **64**, 045202 (**2001**).

51 D. A. Smirnov and B. P. Bezruchko. Estimation of interaction strength from short and noisy time series. *Phys. Rev. E*, **68**, 046209 (**2003**).

52 F. Takens. Detecting strange attractors in turbulence. In D. A. Rand and L.-S. Young, editors, *Dynamical Systems and Turbulence (Warwick 1980)*, volume 898 of Lecture Notes in Mathematics, page 366. Springer-Verlag, Berlin (**1981**).

53 H. Whitney. Differentiable manifolds. *Ann. Math.*, **37**, 645 (**1936**).

54 T. Sauer, J. A. Yorke and M. Casdagli. Embedology. *J. Stat. Phys.*, **65**, 579–616 (**1991**).

55 H. Osterhage, F. Mormann, T. Wagner and K. Lehnertz. Measuring the directionality of coupling: phase versus state space dynamics and application to EEG time series. *Int. J. Neur. Sys.*, **17**, 139 (**2007**).

56 H. Kantz and Th. Schreiber. *Nonlinear Time Series Analysis*. Cambridge Univ. Press, Cambridge, UK, second edition (**2003**).

57 V. S. Afraimovich, N. N. Verichev and M. I. Rabinovich. General

synchronization. *Radiophys. Quantum Electron.*, **29**, 795 (**1986**).

58 N. F. Rulkov, M. M. Sushchik and L. S. Tsimring. Generalized synchronization of chaos in directionally coupled chaotic systems. *Phys. Rev. E*, **51**, 980 (**1995**).

59 J. Arnhold, P. Grassberger, K. Lehnertz and C. E. Elger. A robust method for detecting interdependences: application to intracranially recorded EEG. *Physica D*, **134**, 419 (**1999**).

60 S. J. Schiff, P. So, T. Chang, R. E. Burke and T. Sauer. Detecting dynamical interdependence and generalized synchrony through mutual prediction in a neural ensemble. *Phys. Rev. E*, **54**, 6708 (**1996**).

61 C. W. J. Granger. Investigating causal relations by econometric models and cross–spectral methods. *Econometrica*, **37**, 424 (**1969**).

62 C. J. Stam and B. W. Dijk. Synchronization likelihood: An unbiased measure of generalized synchronization in multivariate data sets. *Physica D*, **163**, 236–51 (**2002**).

63 M. C. Romano, M. Thiel, J. Kurths and W. von Bloh. How much information is contained in a recurrence plot? *Phys. Lett. A*, **330**, 343 (**2004**).

64 M. C. Romano, M. Thiel, J. Kurths and C. Grebogi. Estimation of the direction of the coupling by conditional probabilities of recurrence. *Phys. Rev. E*, **76**, 036211 (**2007**).

65 W. Weaver and C. E. Shannon. *The Mathematical Theory of Communication*. University of Illinois Press (**1949**).

66 C. S. Daw, C. E. A. Finney and E. R. Tracy. A review of symbolic analysis of experimental data. *Review of Scientific Instruments*, **74**, 916 (**2003**).

67 S. Kullback. *Information Theory and Statistics*. Dover, New York (**1959**).

68 A. Fraser and H. Swinney. Interdependent coordinates for strange attractors from mutual information. *Phys. Rev. A*, **33**, 1134 (**1986**).

69 T. M. Cover and J. A. Thomas. *Elements of Information Theory*. Wiley, New York (**1991**).

70 J. A. Vastano and H. L. Swinney. Information transport in spatiotemporal systems. *Phys. Rev. Lett.*, **60**, 1773–6 (**1988**).

71 A. Kaiser and T. Schreiber. Information transfer in continuous processes. *Physica D*, **166**, 43–62 (**2002**).

72 F. Mormann, R. Andrzejak, T. Kreuz, C. Rieke, P. David, C. Elger and K. Lehnertz. Automated detection of a preseizure state based on a decrease in synchronization in intracranial electroencephalogram recordings from epilepsy patients. *Phys. Rev. E*, **67**, 021912 (**2003**).

73 M. Le Van Quyen, J. Martinerie, V. Navarro, M. Baulac and F. J. Varela. Characterizing neurodynamic changes before seizures. *J. Clin. Neurophysiol.*, **18**, 191–208 (**2001**).

74 M. Paluš, V. Komàrek, T. Procházka, Z. Hrnčíř and K. Štěrbová. Synchronization and information flows in EEGs of epileptic patients. *IEEE Eng. Med. Biol. Mag.*, **20**, 65–71 (**2001**).

75 B. J. Weder, K. Schindler, T. J. Loher, R. Wiest, M. Wissmeyer, P. Ritter, K. Lovblad, F. Donati and J. Missimer. Brain areas involved in medial temporal lobe seizures: a principal component analysis of ictal SPECT data. *Hum. Brain. Mapp.*, **27**, 520–34 (**2006**).

76 J. Kwapien, S. Drozdz and A. A. Ioannides. Temporal correlations versus noise in the correlation matrix formalism: An example of the brain auditory response. *Phys. Rev. E*, **62**, 5557–64 (**2000**).

77 P. Seba. Random matrix analysis of human EEG data. *Phys. Rev. Lett.*, **91**, 198104 (**2003**).

78 M. Müller, G. Baier, A. Galka, U. Stephani and H. Muhle. Detection and characterization of changes of the correlation structure in multivariate time series. *Phys. Rev. E*, **71**, 046116 (**2005**).

79 G. Baier, M. Müller, U. Stephani and H. Muhle. Characterizing correlation changes of complex pattern transitions: The

80 K. Schindler, H. Leung, C. E. Elger and K. Lehnertz. Assessing seizure dynamics by analysing the correlation structure of multichannel intracranial EEG. *Brain*, **130**, 65–77 (**2007**).

81 K. Schindler, C. E. Elger and K. Lehnertz. Increasing synchronization may promote seizure termination: Evidence from status epilepticus. *Clin. Neurophysiol.*, **118**, 1955–68 (**2007**).

82 W. Y. Chen and C. G. Wasterlain. Status epilepticus: pathophysiology and management in adults. *Lancet Neurol.*, **5**, 246–56 (**2006**).

83 C. Allefeld, M. Müller and J. Kurths. Eigenvalue decomposition as a generalized synchronization cluster analysis. *Int. J. Bifurcat. Chaos*, **17**, 3493–7 (**2007**).

84 S. Bialonski and K. Lehnertz. Identifying phase synchronization clusters in spatially extended dynamical systems. *Phys. Rev. E*, **74**, 051909 (**2006**).

85 D. J. Watts and S. H. Strogatz. Collective dynamics of 'small-world' networks. *Nature*, **393**, 440–2 (**1998**).

86 C. J. Stam and J. C. Reijneveld. Graph theoretical analysis of complex networks in the brain. *Nonlinear Biomed. Phys.*, **1**, 3 (**2007**).

87 J. C. Reijneveld, S. C. Ponten, H. W. Berendse and C. J. Stam. The application of graph theoretical analysis to complex networks in the brain. *Clin. Neurophysiol.*, **118**, 2317–31 (**2007**).

88 S. C. Ponten, F. Bartolomei and C. J. Stam. Small-world networks and epilepsy: Graph theoretical analysis of intracerebrally recorded mesial temporal lobe seizures. *Clin. Neurophysiol.*, **118**, 918–27 (**2006**).

89 H. Wu, X. Li and X. Guan. Networking property during epileptic seizure with multi-channel EEG recording. *Lecture Notes in Computer Science*, **3976**, 573–8 (**2006**).

90 M. Chavez, D.-U. Hwang, A. Amann and S. Boccaletti. Synchronizing weighted complex networks. *Chaos*, **16**, 015106 (**2006**).

case of epileptic activity. *Phys. Lett. A*, **363**, 290–6 (**2007**).

16
A Multivariate Approach to Correlation Analysis Based on Random Matrix Theory

Markus Müller, Gerold Baier, Christian Rummel, Kaspar Schindler, Ulrich Stephani

16.1
Introduction

In this contribution we present a multivariate method for the analysis of inter-relations between data channels of an M-dimensional recording. We describe in detail how and in which sense genuine *multivariate* features of the data set are extracted and illustrate the performance of the method with the help of numerical examples. As tools known from Random Matrix Theory are used, a brief overview of the origin of this field is given and some technical aspects of the calculation of RMT-measures are discussed. Finally we discuss several examples where this method has been applied successfully to the analysis of electroencephalographic recordings of epileptic patients.

For the analysis of complex systems like the human brain the development and application of sophisticated tools of time series analysis is indispensable. During the last decade, the development and application of nonlinear measures became particularly popular, mostly because it is supposed that the underlying mechanism generating EEG signals is nonlinear or even chaotic [1]. Hence a multitude of different techniques such as estimating the largest Lyapunov exponents [2], the correlation dimension [3], generalized correlations via Mutual Information [4], nonlinear regression [5] or synchronization measures [6] have been applied. All these methods are uni- or bivariate evaluating properties of single or the interplay of pairs of data channels of a multivariate EEG recording. Although spectacular results, in particular in the field of seizure prediction, were published [3, 7, 8] a certain disenchantment for nonlinear techniques appeared. Comparing the performance of different linear and nonlinear measures within controlled testframes, it turned out that simple linear measures like, e.g., the Pearson correlation coefficient, do not perform worse or even better than sophisticated nonlinear techniques [9, 10]. In these studies, the sensitivity of a measure in detecting the (nonlinear and linear) coupling between units has been tested. For the particular application of characterizing preseizure dynamics, linear as well as nonlinear measures have been applied to EEG recordings in [11, 12]. Also there the conclusion has been drawn that linear approaches are highly competitive

Seizure Prediction in Epilepsy. Edited by Björn Schelter, Jens Timmer and Andreas Schulze-Bonhage
Copyright © 2008 WILEY-VCH Verlag GmbH & Co. KGaA, Weinheim
ISBN: 978-3-527-40756-9

with fashionable nonlinear ones, and that bivariate techniques in general deliver more favorable results than do univariate approaches. However, if one likes to detect relationships between $m > 2$ data channels and to quantify the strength of interrelation, a multivariate approach seems most promising. This conjecture has its foundation in the typical network structure of mammalian brains where interconnections lead to more complex structures than given by the sum of bivariate relations.

Motivated by these findings we present in this work a linear but multivariate approach based on the zero-lag correlation matrix and techniques known from Random Matrix Theory (RMT). RMT tools have been successfully applied to time series stemming from stock markets [13], electroencephalographic as well a magnetoencephalographic recordings [14, 15], climate data [16] and data derived from the internet traffic [17]. However, we proceed differently in several aspects from these authors in order to increase the sensitivity of the approach in detecting and characterizing spatio-temporal correlation patterns. Although in the present contribution the Pearson coefficient is used as a basic measure in order to construct an interdependence matrix, we emphasize that also nonlinear measures like mutual information could be used instead (work in this direction is underway). In this contribution we describe in detail the method and illustrate its performance with the help of test systems. We will give evidence, that although the Pearson coefficient is a bivariate measure, we will be able to estimate the interrelation between m data channels, for any $m \geq 2$. Finally, we report in the last two chapters successful applications of our approach to EEG data of epilepsy patients. In two, recently published contributions [18, 19] the evolution of the spatio-temporal correlation pattern during focal onset seizures as well as status epilepticus is described while in a third publication [20] also the transition towards spontaneous epileptiform activity is characterized for the case of primary generalized absence seizures.

16.2
The Equal-time Correlation Matrix

Let us start with some basic definitions and properties of the equal-time correlation matrix. For a measured multivariate time series $X_i(t)$ ($i = 1 \ldots M$) the equal-time correlation matrix $\mathbf{C}$ (see, e.g., [21]) is constructed by first normalizing

$$\tilde{X}_i(t) = \frac{X_i(t) - \langle X_i \rangle_t}{\sigma_i^t}, \tag{16.1}$$

and then evaluating the matrix elements as Pearson's correlation coefficients

$$C_{ij}(t) = \frac{1}{T} \sum_{t=1}^{T} \tilde{X}_i(t)\tilde{X}_j(t) = \langle \tilde{X}_i \tilde{X}_j \rangle_t. \tag{16.2}$$

In the last two equations averages are denoted by $\langle \cdot \rangle$ and the standard deviations by σ. They are calculated over all time steps $t \in [1, T]$. Equation (16.2) can equivalently be written as

$$C = \frac{1}{T} \tilde{X}\tilde{X}^t. \tag{16.3}$$

Here a matrix notation using the $M \times T$-data-matrix with elements $\tilde{X}_{it} = \tilde{X}_i(t)$ has been introduced and the symbol ^{t} denotes the transposition of the data matrix $\tilde{X}$.

The normalization eq. (16.1) removes the amplitude information of the data and implies for the average $\langle \tilde{X}_i \rangle_t = 0$ and for the variance $\langle \tilde{X}_i^2 \rangle_t = 1$. Hence, only the shape but not the variance or the offset of the signals influence the value of the correlation coefficients. Let us resume the following important properties of the C-matrix:

$$C_{ii}(t) \equiv 1 \tag{16.4}$$

$$C_{ij}(t) = C_{ji}(t) \tag{16.5}$$

$$-1 \leq C_{ij}(t) \leq 1. \tag{16.6}$$

The first equation implies simply that each signal is perfectly correlated to itself. The symmetry property (16.5) takes into account that the correlation between signal $X_i(t)$ and $X_j(t)$ is the same as that between signal $X_j(t)$ and $X_i(t)$ and finally, (16.6) indicates that the correlation coefficients vary between ± 1, the completely correlated or anti-correlated cases. Furthermore, each row (or equivalently column) of C displays the cross correlations of a specific data channel with all others, i.e., the matrix (16.2) is written in the 'channel basis'.

16.3
Eigenvalues, Eigenvectors and Interrelations between Data Channels

The discussion of the M eigenvalues λ_i and eigenvectors $\vec{v}_i$ of C, which can be obtained by solving the equation

$$C\vec{v}_i = \lambda_i \vec{v}_i, \tag{16.7}$$

will be a central part of this contribution. Together with (16.4) one obtains an additional constraint for the eigenvalues which is given by the trace of C:

$$\text{Tr } C = \sum_{i=1}^{M} C_{ii} = \sum_{i=1}^{M} \lambda_i = M. \tag{16.8}$$

If some of the eigenvalues λ_i increase, this change has to be compensated by a corresponding decrease in some of the remaining ones such that the sum (16.8) is equal to the number of data channels.

In fact, all the information about the linear two-point correlation structure of two time series $X_i(t)$ and $X_j(t)$ is contained in the *bivariate* measures represented by the $M(M-1)/2$ independent matrix elements C_{ij}. The advantage of the linear correlation coefficient is that it is easy to calculate and provides a straightforward interpretation. However, there are two serious deficiencies in these quantities. First, particularly for large M, it is difficult to follow all matrix elements in the course of time. Second, due to the finite value of T there is a certain number of random correlations of order $\sim 1/\sqrt{T}$ hidden in the C_{ij}. That is to say, for finite T the sum (16.2) results in a non-zero value even in the case of completely uncorrelated signals $X_i(t)$ and $X_j(t)$. On the contrary, with the help of the eigenvalues and eigenvectors it is not only possible to separate random and non-random contributions, but in addition, to capture interrelations between $2 \leq m \leq M$ signals. In principle, the Pearson coefficient quantifies the 'similarity' between two time series. It turns out that the eigenvalues and eigenvectors provide a kind of sorting of the data channels due to their similarity. Independently whether two or more data channels are simultaneously 'similar' the groups of interrelated signals can be identified. For this reason these quantities are the central focus of this contribution.

Let us illustrate this aspect by the following geometrical argument. For this purpose we consider a multivariate random (not necessarily uncorrelated) process described by $\vec{X} = (X_1, \ldots, X_M)$. In this representation the $X_i = \tilde{X}_i(t)$ at a fixed time step such that each $\vec{X}$ represents a single point in the M-dimensional phase space spanned by the data channels, i.e., $\vec{X}$ is one column of the $M \times T$ data matrix $\mathbf{X}$ from (16.3). In the special case where the process which generates the X_i is Gaussian and properly normalized, the joint probability distribution of the events X_i in the M-dimensional phase space can be described by [22]:

$$G(\vec{X}) = G(X_1, \ldots, X_M) = \frac{1}{\sqrt{(2\pi)^M \det \mathbf{C}}} \exp\left\{ -\frac{1}{2} \vec{X}^t \mathbf{C}^{-1} \vec{X} \right\}. \tag{16.9}$$

where $\vec{X}^t$ is the transposed M-dimensional data vector and $\mathbf{C}^{-1}$ is the inverse correlation matrix, which is symmetric as $\mathbf{C}$ is symmetric. The reason why the inverse of the correlation matrix appears in the last formula can intuitively be understood by considering the probability density of a univariate (i.e., one-dimensional) Gaussian process which is given by

$$G(x) = \frac{1}{\sqrt{2\pi\alpha}} \exp\left(-\frac{1}{2}(x-\mu)\alpha^{-1}(x-\mu) \right), \tag{16.10}$$

where μ is the center of the distribution and $\alpha = \sigma^2$ the square of the standard deviation or variance. In the multivariate case, the scalar value μ has to be replaced by the vector of mean values $\vec{\mu}$ and the variance α by the covariance matrix $\Sigma = \frac{1}{T} \mathbf{X}\mathbf{X}^t$. If additionally the random process is normalized to zero mean and unit variance as in (16.1), the vector $\vec{\mu}$ will be equal to the null vector and the covariance matrix converts to the correlation matrix (16.3).

The contour where the exponential (16.9) takes $1/e$th of its maximum value is given by the $(M-1)$-dimensional set of points satisfying

$$\frac{1}{2} \vec{X}^t \mathbf{C}^{-1} \vec{X} = 1. \tag{16.11}$$

The last equation defines an ellipsoid in the M-dimensional phase space, whose symmetry axes (the principal axes) are in general not parallel to the chosen coordinate system. This can be achieved by introducing some orthogonal transformation $\mathbf{D}$

$$\frac{1}{2}\vec{X}^t\mathbf{D}^t\mathbf{D}\mathbf{C}^{-1}\mathbf{D}^t\mathbf{D}\vec{X} = \frac{1}{2}\vec{X}^t\mathbf{D}^t\mathrm{diag}(\mathbf{C}^{-1})\mathbf{D}\vec{X} = 1. \tag{16.12}$$

$\mathbf{D}$ provides a rotation of the original coordinate system in which the correlation matrix is written (the channel basis) to its principal axes, i.e., a coordinate system which is parallel to the symmetry axes of the point set defined by (16.11).

In (16.12) $\mathrm{diag}(\mathbf{C}^{-1})$ denotes the diagonalized form of the inverse of the correlation matrix. Hence, the directions of the symmetry axes are given by the orthogonal transformation performing the rotation. As the eigensystem of the inverse of a matrix is identical to the eigensystem of the matrix itself, the columns of the orthogonal transformation $\mathbf{D}$ are the eigenvectors of the correlation matrix. As the eigenvalues of $\mathrm{diag}(\mathbf{C}^{-1})$ are equal to $1/\lambda_i$; $i = 1, \ldots, M$, it turns out that half of the lengths of the M symmetry axes of the characteristic ellipsoid of the joint probability distribution d_i are equal to the square root of the eigenvalues of the correlation matrix: $d_i = \sqrt{\lambda_i}$. For the uncorrelated case, the ellipsoid is the M-dimensional unit sphere. In that case, all eigenvalues are degenerate and equal to one. For any non-zero correlations the sphere deforms to some ellipsoid whose half-axes length satisfy the trace condition (16.8) as

$$\sum_{i=1}^{M}\lambda_i = \sum_{i=1}^{M}d_i^2 = M. \tag{16.13}$$

For the Gaussian processes considered above, the joint probability distribution is completely determined if one knows the direction and length of the symmetry axes of the ellipsoid, which is by means of the normalization (16.1) centered at the origin. Hence, for such processes, the solution of (16.7) provides the complete information. For non- or almost Gaussian processes, the knowledge of the eigenvalues and eigenvectors might give at least a first approximation of the shape of the joint probability distribution. The quality of this approximation depends essentially on the higher moments of the joint distribution such as skewness, kurtosis, etc. However, in any case it remains clear that the set of eigenvalues and eigenvectors present *multivariate* quantities, which provides information about the characteristics of the joint probability distribution of the underlying process. As we will argue in the next chapter, they measure the degree of 'similarity' between $2 \leq m \leq M$ data channels.

16.4
Random and Non-random Level Repulsion

In this section we discuss in which manner two point correlations between data channels provoke non-random repulsions between the eigenvalues of $\mathbf{C}$. Most of the arguments given in this section are equally valid if, e.g., mutual information instead of the Pearson coefficient is used in order to create some interdependence matrix. The only necessary requirement is that the chosen bivariate measure is real and symmetric in order to provide real eigenvalues and, in order to guarantee the validity of the trace condition (16.8), the measure should be appropriately normalizable.

A qualitative understanding of how the eigenvalues react under a given correlation pattern can be obtained by considering the following simple cases. If, for example, all signals $X_i(t)$ are completely independent, all non-diagonal elements of **C** will be equal to zero, at least for stationary, infinitely long time series. In that case, **C** is already diagonal and all M eigenvalues will be degenerate (i.e., they have the same value) and will be equal to one. Consequently, the ellipsoid in phase space turns to an M-dimensional unit sphere as mentioned above. For finite but long data segments $X_i(t)$, all non-diagonal elements of **C** will take random values close to zero which lifts the degeneration and the eigenvalues will be randomly distributed around unity, see e.g. [23].

If, on the other hand, all signals $X_i(t)$ are identical, all entries of the C-matrix will be equal to one. As a consequence there remains only one non-zero eigenvalue $\lambda_M = M$ in order to satisfy the conservation of the trace (16.8) and the prolate ellipsoid collapses to a one-dimensional subset given by a finite segment of length $2\sqrt{M}$ along the main diagonal in the M-dimensional phase space. For M highly correlated (but not identical) time series, all non-diagonal elements will take values close to unity and a large gap between the largest eigenvalue and all others occurs. In such a situation λ_M is close to the maximum value M while all others take values close to zero [23] and the ellipsoid also shows a well pronounced prolate shape.

Considering a more complicated example, we suppose that the dynamical state of the system in question can be characterized by K groups of correlated data channels ('correlation clusters') where, for the moment, interrelations between these groups are of negligible strength. If one choses an adequate indexing of the data channels, the corresponding correlation matrix will have a block-diagonal structure. For each group containing m_k correlated signals an $m_k \times m_k$-dimensional block of elevated correlation coefficients occurs, while all other entries of **C**, including the components connecting the different blocks, will be given by random numbers whose magnitude (e.g., the standard deviation of its distribution function) is determined by the amount of random correlations, i.e., the length T of the data segment. Note, eigenvalues and eigenvectors are invariant under any permutation of the index set of the data channels, but only with an adequate choice of the indices is the block-diagonal structure visible. By diagonalizing the correlation matrix, for each of those blocks, exactly one large eigenvalue and $m_k - 1$ small ones appear. If there are additionally signals belonging to none of these blocks, there will be a corresponding number of eigenvalues randomly distributed around unity. Again, the widths of this randomly distributed eigenvalues are determined by the amount of random correlations and noise. In the sequel, this group of exclusively randomly correlated eigenvalues will be denoted as the 'bulk' [13]. Geometrically, in phase space K prolate ellipsoids of dimension m_k appear additionally to a unit sphere whose dimension is given by the number of data channels belonging to the bulk. If all matrix elements of **C** connecting the blocks and those connecting the blocks with the bulk are exactly zero they are located in different orthogonal subspaces of the whole phase space.

In general, one can summarize the reaction of the eigenvalues on a given correlation structure as follows [23]:

1. The number of displaced eigenvalues equals the total number of correlated data channels. This displacement is generated by a repulsion between eigenvalues located at *both* edges of the spectrum in order to satisfy (16.8).

2. The relevant information about the correlation structure of a multivariate data set is imprinted *locally*, in small parts of the entire eigenvalue spectrum.

3. For each cluster of correlated signals one large and a certain number of small eigenvalues are repelled.

4. The strength of the repulsion is determined by the amount of correlation and the size of the cluster.

5. In principle, the structure of the corresponding eigenvectors contains information about which of the data channels belong to which correlated group. However, in the case of non-stationary time series it is hard to extract this information from the eigenvector components, see [24, 25] and Section 16.7.

16.5
RMT Measures: Motivation and Definition

In principle, the scheme described in the last section suffices to separate random and genuine correlations and gives a method of extracting information from the eigenvalue spectrum of **C**. This concept has been used for multivariate EEG analysis in [18] where a systematic study of the ictal activity of 100 focal onset seizures has been presented in order to characterize the spatio-temporal correlation dynamics of these extreme events. Using a running-window approach, the characteristics of the spatio-temporal correlation pattern can be revealed via the time evolution of the spectrum of eigenvalues and eigenvectors of the correlation matrix. Once the gaps in the spectrum provoked by the (non-random) repulsion of eigenvalues are identified, one can, in principle, extract the complete information about the correlation structure. On the other hand, not only are the eigenvalues λ_i of the bulk affected by random fluctuations, but to a certain extent the repelled ones are also. Hence, if the amount of random correlations is large in comparison to the genuine correlations, it is difficult to identify the repelled eigenvalues by visual inspection. Therefore, quantitative measures are required, which are able to detect even tiny non-random displacements of the eigenvalues.

Such tools are provided by Random Matrix Theory (RMT) which originated from the field of nuclear physics in the early fifties [26]. In general, the description of many-body systems is quite cumbersome and computationally time consuming. On the other hand, the nuclear system has a rich structure whose signatures have been found experimentally and its theoretical description presented a real challenge. As a prominent example we mention the so-called 'neutron resonances' [27], with lifetimes which are about five to six orders of magnitude longer than those expected for nuclei. A detailed microscopic description seemed hopeless. Instead Wigner proposed a statistical treatment of these compound nuclear states, by using (almost) random Hamiltonians, which respect only general symmetry properties such as e.g., time reversal invariance. This approach turned out to be very successful for the statistical description of nuclear spectra [28, 29] and provided the theoretical basis for a new research field, the physics of quantum chaos [30]. Using the mathematical apparatus of RMT [31], universal statistical properties of a variety of physical systems could be compared and classified. The key point of all these studies was to apply sensitive measures, able to detect correlations within the eigenvalue spectra of (Hamiltonian) matrices [29]. In the context of the correlation matrix we aim to distinguish random from non-random repulsions of eigenvalues of $\mathbf{C}$. Therefore, it seems most appropriate to apply RMT correlation measures to the eigenvalue spectra of the empirical correlation matrix.

One of the most popular measures is the so-called nearest-neighbor distribution $P(s)$ [29, 30, 32] which measures the amount of fluctuations of the spacings s between neighboring eigenvalues. Suppose the size of an eigenvalue λ_i is known, $P(s)$ quantifies the probability of finding the next eigenvalue at a distance s. As a matter of fact, this probability depends crucially on the density function of the eigenvalues

$$\rho(\lambda) = \frac{1}{N} \sum_{i=1}^{N} \delta(\lambda - \lambda_i), \tag{16.14}$$

i.e., the average distance between adjacent λ_i is, in general, different close to the borders of a spectrum and in the central part. If one wishes to compare results obtained from different systems such as compound nuclear states [28], a hydrogen atom in a magnetic field [33], eigenmodes of a microwave cavity [34] or elastic waves within a crystal of irregular shape [35] one has to put the eigenvalue spectra on the same footing. This is done via the so-called 'unfolding procedure' where the spectra of $\mathbf{C}$ are transformed such that the distance between adjacent λ_i becomes unity on average and one is left exclusively with the fluctuations around this mean value.

To this end, one usually calculates the so-called accumulated level density

$$N(\lambda) = \int_{-\infty}^{\lambda} \rho(\lambda') \, d\lambda' \tag{16.15}$$

which counts the number of states in the interval $[-\infty, \lambda]$. It can be split into a smooth and a fluctuating part

$$N(\lambda) = N_{\text{smooth}}(\lambda) + N_{\text{fluct}}(\lambda). \tag{16.16}$$

As the fluctuating part

$$\rho_{\text{fluct}}(\lambda) = \frac{dN_{\text{fluct}}(\lambda)}{d\lambda} \qquad (16.17)$$

is zero on average, the mean level density is solely given by

$$\rho_{\text{smooth}}(\lambda) = \frac{dN_{\text{smooth}}(\lambda)}{d\lambda}. \qquad (16.18)$$

If the smooth part of the accumulated level density $N_{\text{smooth}}(\lambda)$ is known, the eigenvalues Λ_i of the unfolded spectrum are obtained by

$$\Lambda_i = N_{\text{smooth}}(\lambda_i). \qquad (16.19)$$

The crucial problem of unfolding a spectrum in this way is to find the correct form of $\rho_{\text{smooth}}(\lambda)$ or $N_{\text{smooth}}(\lambda)$, respectively. In particular, if the analytical formula for the smooth part of the level density is unknown, one has to perform a fit to the numerical probability distribution of the eigenvalues. To this end different strategies can be used. A common procedure is a polynomial fit to the numerically obtained accumulated level density $N(\lambda)$. This fit function is then used to perform the unfolding transformation equation (16.19), see, e.g., [32]. Another technique, the so-called Gaussian broadening, approximates the delta distributions in (16.14) by Gaussian distributions centered at each eigenvalue λ_i. The width of the Gaussians is adjusted such that the resulting curve describes $N_{\text{smooth}}(\lambda)$.

However, independent of which kind of procedure is used, at the edges of the spectrum the description of the experimental data is not satisfactory. Therefore, in any practical application, a certain percentage of the spectrum at both edges has to be neglected (usually about 20 to 30%). Additionally, a fit to the accumulated level density always implies a certain mixture of bulk and edge properties (spectral average). As we are aiming at an extraction of the fluctuation properties:

(a) *locally* for small parts of the eigenvalue spectrum (in extreme cases between two eigenvalues only);

(b) along the whole spectrum *including* the edges,

it is preferable to proceed differently in this step.

Under the assumption that one can create a (possibly small) ensemble of correlation matrices over a certain interval of the data set, we simply normalize each distance by its own ensemble average [36]:

$$s_i^{(n)} = \frac{\lambda_{i+1}^{(n)} - \lambda_i^{(n)}}{\langle \lambda_{i+1} - \lambda_i \rangle_n} \qquad \text{where} \quad i = 1 \ldots M - 1 \quad \text{and} \quad n = 1 \ldots N_{\text{ens}} \qquad (16.20)$$

Here $\langle \ldots \rangle_n$ denotes the average over an ensemble of size N_{ens} and we naturally obtain $\langle s_i \rangle_n = 1$ for all i. Using (16.20) we can calculate the nearest-neighbor distribution $P(s_i)$ for each of the distances s_i *individually*.

However, the application of these concepts to EEG recordings bears the additional difficulty that the experimental data do not only contain the correlations produced by the electrical brain dynamics but will be contaminated by correlations uniquely produced by the chosen EEG reference. Each reference point influences the result of a measurement in a particular way such that different parts of the spectrum of eigenvalues (and eigenvectors) is modified by it. This distortion of the eigenvalues spectrum can be sufficiently strong such that possible changes in the correlation pattern caused by the brain dynamics are completely blurred [37]. Hence a systematic analysis is required in order to determine: (i) how a given EEG reference influences the results obtained by measures derived from Random Matrix Theory; and (ii) if it is possible to correct for this influence, at least to a certain extent. Such a study is presented in [37] where six commonly used EEG references have been considered, their influence on the results has been quantified and the performance of a correction method evaluated.

Finally the question remains as to which distributions can be used to test the null hypothesis of only randomly correlated data. The natural choice in our context will be correlation matrices derived from independent Gaussian white noise, the so-called Wishart Ensemble (WE) [21,38]. Unfortunately, for this ensemble no exact formula for the nearest-neighbor distribution of the WE is available. Nevertheless, empirically it is well established that at least for the case $M \ll T$ the $P(s)$ of the WE is not distinguishable from the Gaussian Orthogonal Ensemble (GOE), for which a multitude of analytical results is published, see e.g., [31]. The GOE consists in the set of matrices that are invariant under orthogonal transformations and having random elements which are drawn from a Gaussian probability distribution.

16.6
Application to a Test System

In order to check the validity of these concepts with some concrete examples we generate data from the $M = 20$ Rössler system in the chaotic regime:

$$\dot{X}_i = -\omega Y_i - Z_i + \eta \sum_{k=1}^{K} \sum_{n \in N(k)} (X_n - X_i)$$

$$\dot{Y}_i = \omega X_i + \beta Y_i + \eta \sum_{k=1}^{K} \sum_{n \in N(k)} (Y_n - Y_i) \tag{16.21}$$

$$\dot{Z}_i = 0.1 + Z_i(X_i - 8.5)$$

with $\omega = 0.98$ and $\beta = 0.28$. The coupling strength between units is controlled via the parameter η, which is non-zero only if unit i belongs to one of the K correlation clusters. $N(k)$ denotes the index set of those oscillators belonging to cluster k. The equations (16.21) are integrated via a fourth-order Runge–Kutta algorithm with step width 0.1 and the time series are sampled from the X-coordinates. In a first example only two units are coupled with $\eta = 0.044$. We generated $N_{ens} = 500$ and 10^5 data segments of length $T = 8000$ sampling points and

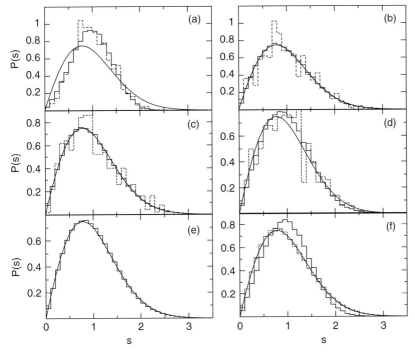

Fig. 16.1 (a) to (d) Nearest-neighbor spacing distribution of the system (16.21) when two of 20 units are coupled. The dashed lines indicate a statistics with $N_{ens} = 500$ correlation matrices and the solid line histograms are obtained for $N_{ens} = 10^5$ matrices. (a) to (d) shows the results for s_1, s_2, s_{18} and s_{19}, respectively. In (e) the spectrally averaged result is shown, while (f) gives the nearest-neighbor distribution for the system (16.21) for the case of two mutually independent correlation clusters. Shown are the results for s_{17} and s_{19} as dashed and dotted histograms. The spacing distribution for distance s_{18} is drawn as a solid line. In all panels the solid continuous line indicates the theoretical result for the GOE.

calculated the correlation matrix as well as its eigenvalues. Then the unfolding procedure (16.20) for each of the $M - 1$ distances was performed and the nearest-neighbor distributions $P(s_i), i = 1, \ldots, M - 1$ were calculated. The results are summarized in Figure 16.1, which shows the nearest-neighbor distribution for distance s_1, s_2, s_{18} and s_{19} in (a) to (d) respectively, in comparison with the theoretical result for the GOE. Figure 16.1(e) shows the spectrally averaged result.

For the given coupling scheme one expects a repulsion between the largest and lowest eigenvalue. Consequently the distribution of distances $s_1 = \lambda_2 - \lambda_1$ and $s_{19} = \lambda_{20} - \lambda_{19}$ are more narrow than the theoretical result for the GOE which reflects the situation of purely random correlations, or in the present context, purely random repulsions between the corresponding eigenvalues. The non-random repulsion between λ_1 and λ_{20} leads to a decrease in the fluctuations of s_1 and s_{19}. However, this effect is more pronounced at the lower edge than for the distance between the two largest eigenvalues. This is remarkable in the sense that the lowest eigenvalues (and their corresponding eigenvectors) are commonly considered as strongly noise contaminated and therefore disregarded; a point of view which is supported not only by recent publications applying RMT methods to time series analysis [13] but is also an essential aspect of the theoretical background of Principal Component Analysis where only a few of the largest eigenvalues are taken into account. In the present example, the opposite is true. The changes observed for the distribution of s_1 are more significant than those for s_{19}. Even for a comparably small ensemble of $N_{ens} = 500$ the difference between $P(s_1)$ and the GOE is significant while for the same statistical quality $P(s_{19})$ is hardly distinguishable from the universal RMT result (compare Figures 16.1(a) and (d)). This is at first glance a surprising result which finds an explanation with the help of (16.8). When the repulsion occurs only between a single large and a small number $m \ll M$ of small eigenvalues, it is evident that the relative change of λ_M is smaller than that of the lowest eigenvalues in order to satisfy the trace condition (16.8). Consequently, the changes at the lower border of the spectrum are more significant than those for the largest eigenvalues. This is also imprinted in the distribution of level spacings as illustrated by this example.

Furthermore, we find it remarkable that the information about the correlation structure is not smeared out along the spectrum but is induced locally. Only the distances s_1 and s_{19} show a reduced amount of fluctuations, while the adjacent distances s_2 and s_{18} already follow the GOE distribution precisely. Finally, we emphasize that spectral averages wipe out the effect. If, as in the present example, deviations from the universal RMT behavior are only visible at the edges of the spectrum or locally in a few level spacings, correlations are hardly detected when spectral averages are considered. This is exemplified in Figure 16.1(e), where the spectrally averaged nearest-neighbor distribution is presented. Note, in the average the distributions of distances s_1 and s_{19} are also included, which are usually neglected due to a poor performance of a fit at the borders. However, deviations from the GOE curve are only significantly visible when the ensemble size N_{ens} is quite large. Otherwise, the erroneous conclusion from this result would be that the system does not contain any genuine correlations.

Next we illustrate the results when $K = 2$ small mutually independent correlation clusters are formed. For each cluster $m_k = 2$ Rössler oscillators are coupled with $\eta = 0.048$ such that one expects repulsions between two eigenvalues at each border of the eigenvalue spectrum. In this way two elevated eigenvalues at the upper edge and two decreased eigenvalues at the lower border occur. As the coupling between the clusters is zero, the spacing distribution between the two largest, and correspondingly between the two lowest, eigenvalues should follow the random matrix prediction, while a significant deviation should appear for the distribution of distances s_{18} and s_2. In Figure 16.1(f) only the spacing distributions between the four largest eigenvalues are shown. As a similar behavior can be seen for the spacings of the corresponding eigenvalues at the lower border, these figures are disregarded. The numerical results confirm our supposition nicely. While $P(s_{18})$ is more narrow than the GOE, both, $P(s_{17})$ as well as $P(s_{19})$ follow the RMT prediction with high accuracy. Note, in the case of any non-zero interrelation between the two clusters a deviation for $P(s_{19})$ (and correspondingly for $P(s_1)$) has also to be expected. Hence, the identification of all non-random repulsions between eigenvalues provides important information about the number of correlation clusters and the total number of correlated signals can be deduced. Furthermore, the distribution of spacings between the enlarged eigenvalues gives an additional hint to possible inter-cluster relations.

16.7
Cluster Detection based on Eigenvectors

A complete picture of the correlation structure; namely, which data channel correspond to which of the correlation clusters, requires a reliable clustering algorithm. In principle, this information is encoded in the structure of the eigenvectors corresponding to the repelled eigenvalues. However, inter-cluster correlations and random correlations will cause a non-trivial mixing of the eigenvectors such that the cluster structure cannot be seen directly from their large components (Figure 16.2(a)). Therefore, in [24, 25] an algorithm is proposed which not only allows one to identify the members of the correlation clusters but additionally provides an estimate of how strongly involved a certain data channel is in a given correlation cluster, and how strong the inter-cluster correlations are.

The scheme of this algorithm can be sketched as follows. The number K of clusters is given by the number of enlarged eigenvalues. Their corresponding eigenvectors will be denoted by $\vec{v}_i$. The basic idea of this technique is to construct suitable linear combinations of the $\vec{v}_i$ such that each for the new vectors (denoted as *Cluster Participation Vectors* $\vec{w}_i$) only those components which correspond to data channels belonging to one of the clusters have a prominent size. This is achieved by maximizing the distance measure

$$D_{ln} = D(\vec{w}_l, \vec{w}_n) = \sum_{i=1}^{M} |b_{il}^2 - b_{in}^2| \qquad (16.22)$$

where the b_{il} are the components of the vector $\vec{w}_l$. From the b_{ij}^2 the data channels belonging to each of the K clusters can be identified, see Figure 16.2(b). Furthermore, with the help of the so-called Cluster Participation Coefficients

$$\Lambda_l = \sum_{k=1}^{K} |\langle \vec{v}_k | \vec{w}_l \rangle|^2 \lambda_k \qquad (16.23)$$

it is possible to estimate the average strength of the intra-cluster correlations

$$C_k \approx \frac{\Lambda_k - 1}{m_k - 1} \qquad (16.24)$$

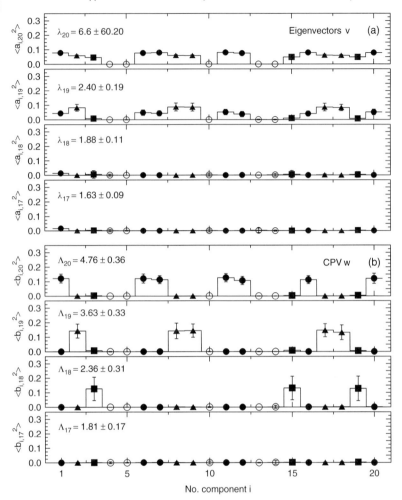

Fig. 16.2 Comparison of the average of the squared components of the four largest eigenvectors $\vec{v}_l$ of a correlation matrix of dimension $M = 20$ with those of their linear combinations $\vec{w}_l$ for $N_{ens} = 100$ independent realizations. The error bars are often the same size as the symbols. The channels that contribute to the $K = 3$ clusters are marked by full symbols (●, ▲, ■) and the uncorrelated ones by open circles (○).

as well as the inter-cluster correlation strength

$$D_{kl} \approx \langle \vec{w}_l | \vec{v}_k \rangle \frac{\Lambda_k - \Lambda_l}{\sqrt{m_k m_l}}. \tag{16.25}$$

In these equations $\langle \cdot | \cdot \rangle$ denotes the scalar product of two vectors, m_l the number of data channels belonging to cluster l, C_k the average correlation strength of cluster k and finally D_{kl} the average strength of the inter-cluster correlation between cluster k and l. Note that the statistical uncertainty of the relations (16.24) and (16.25) are smaller than for simple averages over pre-identified blocks of the correlation matrix.

A detailed description of the method as well as the discussion of several applications to artificially created and experimental recordings can be found in [24, 25].

16.8
Application to EEG Recordings

The method described above has so far found several applications to electroencephalographic recordings. In [18] the evolution of the correlation pattern during epileptic seizures was studied. To this end, 100 focal onset seizures (in particular 49 secondarily generalized and 51 complex partial seizures) of 60 patients were analyzed, i.e., the relative eigenvalues $\lambda_i^{rel} = \frac{\lambda_i - \bar{\lambda}_i}{\sigma_i}, i = 1, \ldots, M$ are calculated, where $\bar{\lambda}_i$ and σ_i are the mean and standard deviation of the eigenvalue λ_i evaluated over an adequately chosen reference interval. By using a sliding window the relative changes of the eigenvalues and consequently the development of the correlation structure could be traced. As the number of electrodes of the recordings vary from patient to patient and, in order to make the results more transparent, only the average of a certain number of large as well as small eigenvalues has been considered. Contrary to the general belief that epileptic seizures present a kind of static hypersynchronous state, the authors of [18] found that the brain dynamics at seizure onset is *not* characterized by a drastic increase in synchronized, namely correlated, activity. Instead they report for secondary generalized seizures a small but significant correlation loss during the first half of the seizure. For each of the 100 recordings they found a substantial increase of correlations at seizure termination. They consider this change of correlation structure as 'generic' because it was observed in all seizures independently of the anatomical location of the seizure onset zone or of seizure duration, number of channels, etc. The authors argued that the decorrelation of EEG activity might be caused by different propagation times of locally synchronous ictal discharges from the seizure onset zone to other brain areas. At the same time they speculate that the gradual increase of correlated EEG activity before seizure end could be an active seizure termination mechanism. These results could be confirmed by a subsequent study of six status epilepticus EEG recordings from six patients [19] where a similar analysis to that in [18] was performed. In all six recordings the amount of correlation fluctuates around some relatively low level during ongoing epileptiform activity and only persistently increased before, or in one case, at the end of status epilepticus. These findings support the hypothesis that increasing synchronization of neuronal activity may be considered as an emergent self-regulatory mechanism for seizure termination.

In [20] a study of surface EEG recordings of nine primarily generalized absence seizures from five patients was presented. There, the temporal evolution of the absolute and relative eigenvalues was analyzed (similar to [18, 19]). In addition, the components of the largest eigenvectors shown for preictal and ictal activity were compared, and the nearest-neighbor distribution was calculated individually for each distance between eigenvalues (as described above). The objective of this work was two-fold. It was aimed at characterizing the temporal evolution of the correlation pattern *during* the seizure as well as the *transition* to the seizure.

As a result it was found that in six of nine seizures, the two largest eigenvalues separate significantly from all others while almost all of the remaining eigenvalues decrease. In one patient a strong separation of only the largest eigenvalue was observed. Evaluating an ensemble of correlation matrices for the ictal period it turned out that the spacing distribution of the distance between the two largest eigenvalues deviated only slightly from the GOE while the distribution of the distance between the second and third largest eigenvalue was significantly narrower (Figure 2 in [20]). This situation is qualitatively similar to that shown in Figure 16.1(f). In terms of zero lag correlations this behavior can be interpreted as the formation of two weakly connected correlation clusters during the seizure.

The calculation of the relative eigenvalues as a running window revealed interesting changes of the correlation pattern at the transition to the seizure. A detailed inspection of the time course of the smallest relative eigenvalue reveals a significant correlation loss just *before* seizure onset. A few seconds before the first spike-wave activity is observed, the smallest relative eigenvalue increases drastically, while during spike-wave activity it strongly decreases. In fact, the increase prior to the seizure could be observed in all nine seizures and although the recordings were contaminated by (occasionally strong) artifacts, the increase in λ_1^{rel} was more pronounced than many of the correlation changes due to muscle artefacts. Often this decorrelation was manifested only in the evolution of the smallest eigenvalue. For this reason it is considered the result of a correlation change where probably only two data channels participate. Repeating the analysis with subsets of electrodes in one patient showed that the observed correlation loss was restricted to electrodes positioned in the frontopolar, frontal right region of the scalp.

16.9
Conclusions

In this contribution a truly multivariate method for the analysis of the spatio-temporal evolution of inter-relation patterns measured in extended systems is proposed. Here the term *multivariate* does not refer to a simple accumulation of bivariate measures but to a characterization of inter-relations between *all* data channels. The information contained in the data is neither frozen in a single quantity nor are M^2 quantities produced from an M-dimensional data set.

The keypoint of the method is the distinction of random and non-random repulsion of the eigenvalues of the interdependence matrix by using techniques known from Random Matrix Theory. The method indicates, in a self-contained manner, which part of the spectrum of eigenvalues and eigenvectors carry genuine information of the interdependence structure of the system. In addition it is possible to deduce how many and which channels are inter-related.

Although in principle the method can be generalized to symmetric and normal-izable nonlinear measures like, e.g., mutual information, so far results have been obtained mainly for the linear Pearson's correlation coefficient. Using this measure for Gaussian processes the joint probability distribution in the M-dimensional phase space is completely determined, while for other processes a good approximation of linear inter-relations between all data channels is obtained.

The comparably small computational effort of the method allows for a real-time application to surface as well as intracranial EEG recordings. In several recent studies it was confirmed that the approach presented in this contribution provides a powerful tool for EEG analysis.

References

1 K. Lehnertz, L. Arnhold, P. Grassberger and C. E. Elger: *Chaos in Brain?*, World Scientific, Singapore (**2000**).

2 I. Iasemides *et al.*, *IEEE Trans. Biomed. Engr.* **50**, 616 (**2003**).

3 K. Lehnertz and C.E. Elger, *Phys. Rev. Lett.* **80**, 5019 (**1998**).

4 N. J. Mars and F. H. Lopez da Silva, *Electroencephalogr. Clin. Neurophysiol.* **56**, 194 (**1983**).

5 J. P. Pijn and F. H. Lopes da Silva, *Basic Mechanisms of the EEG Brain Dynamics*, edited by S. Zschocke and E. J. Speckmann Birkhauser, Boston, 1993 pp. 41–61. F. Wendling, F. Bartolomei, J. J. Bellanger and P. Chauvel *Clin. Neurophysiol.* 112, 1201 (**2001**).

6 M. Rosenblum, A. Pikovsky and J. Kurths, *Fluct. Noise Lett.* **4**, L53 (**2004**). J. Bhattacharya *Acta Neurobiol. Exp. Warsz* **61**, 309 (**2001**). J. Arnhold, P. Grassberger, K. Lehnertz and C. E. Elger, *Physica D* **134**, 419 (**1999**). C. J. Stam and B. W. van Dijk, *Physica D* **163**, 236 (**2002**).

7 E. Rodriguez, N. George, J.-P. Lachaux, Jacques Martinerie, B. Renault and F. Varela, *Nature* **397**, 430 (**1999**).

8 M. Le Van Quyen, J. Martinerie, V. Navarro, P. Boon, M. D' Havre, C. Adam, B. Renault, F. Varela and M. Baulac, *Lancet* **357**, 183 (**2001**).

9 K. Ansari-Asl, L. Senhadji, J.-J. Bellanger and F. Wendling, *Phys. Rev. E* **74**, 031916 (**2006**).

10 T. Kreuz, F. Mormann, R. G. Andrzejak, A. Kraskov, K. Lehnertz and P. Grassberger, *Physica D* **225**, 29 (**2007**).

11 F. Mormann, T. Kreuz, C. Rieke, R. G. Andrzejak, A. Kraskov, P. David, C. E. Elger and K. Lehnertz, *Clin. Neurophysiol.* **116**, 569 (**2005**).

12 K. K. Jerger, T. I. Nethoff, J. T. Francis, T. Sauer, L. Pecora, S. L. Weinstein and S. J. Schiff, 'Comparison of Methods for Seizures Detection' in *Epilepsy as a Dynamic Disease*, J. Milton and P. Jung (eds), Springer-Verlag, Berlin (**2002**).

13 L. Laloux, P. Cizeau, J. P. Bouchaud and M. Potters, *Phys. Rev. Lett.* **83**, 1467 (**1999**), V. Plerou, P. Gopikrishnan, B. Rosenow, L. A. Nunes Amaral and H. E. Stanley, *Phys. Rev. Lett.* **83**, 1471 (**1999**), V. Plerou, P. Gopikrishnan, B. Rosenow, L. A. Nunes Amaral, T. Guhr and H. E. Stanley, *Phys. Rev. E* **65**, 066126 (**2002**).

14 P. Seba, *Phys. Rev. Lett.* **91**, 198104 (**2003**).

15 J. Kwapien, S. Drozdz, L. Liu and A. Ioannides, *Phys. Rev. E* **58**, 6359 (**1998**).

16 M. S. Santhanam and P. K. Patra, *Phys. Rev. E* **64**, 016102 (**2001**).

17 M. Barthélemy, B. Gondran and E. Guichard, *Phys. Rev. E* **66**, 056110 (**2002**).

18 K. Schindler, H. Leung, C. E. Elger and K. Lehnertz, *Brain* **130**, 65 (**2007**).

19 K. Schindler, C. E. Elger and
K. Lehnertz, *Clin. Neurophysiol.* **118**,
1955 (**2007**).

20 G. Baier, M. Müller, U. Stephani
and H. Muhle, *Phys. Lett. A* **363**, 290
(**2007**).

21 R. J. Muirhead, *Aspects of Multivariate Statistical Theory*, John Wiley & Sons ltd, New York (**1982**).

22 T. W. Anderson, *An Introduction
to Multivariate Statistical Analysis*, Chapter 2.3 of the Third edition, John Wiley & Sons Ltd, Inc.
Hoboken, New Jersey (**2003**).

23 M. Müller, G. Baier, A. Galka,
U. Stephani and H. Muhle,
Phys. Rev. E **71**, 046116 (**2005**).

24 C. Rummel, G. Baier and M. Müller,
Eur. Phys. Lett. **80**, 68004 (**2007**).

25 C. Rummel, *Phys. Rev. E* **77**, 016708
(**2008**).

26 E. P. Wigner, *Ann. Math.* **53**, 36
(**1951**), E. P. Wigner, *Proc. Cambridge Philos. Soc.* **47**, 790 (**1951**).

27 A. Bohr and B. R. Mottelson,
Nuclear Structure, (Benjamin,
New York, **1969**), Vols. 1 and 2.

28 R. U. Haq, A. Pandey and
O. Bohidas, in: K. H. Böchhoff (Ed.),

*Nuclear Data for Science and Technol-
ogy*, Reidel, Dordrecht, p 809 (**1983**).

29 T. Brody, J. Flores, J. French,
P. Mello, A. Pandey and S. Wong,
Rev. Mod. Phys. **53**, 385 (**1981**).

30 F. Haake, *'Quantum Signatures of Chaos'*, 2nd edition,
Springer-Verlag Berlin, (**2004**).

31 M. L. Mehta, *Random Matrices*, 3rd Edition, Elsevier Academic Press, Amsterdam, (**2004**).

32 J. Flores, M. Horoi, M. Müller and
T. H. Seligman, *Phys. Rev. E* **63**,
026204 (**2000**).

33 D. Wintgen and H. Friedrich,
Phys. Rev. A **35**, 1464 (**1987**).

34 O. Bohigas, M. J. Giannoni and
C. Schmit, *Phys. Rev. Lett.* **52**, 1
(**1984**).

35 C. Ellegard, T. Guhr, K. Lindemann,
J. Nygard and M. Oxborrow,
Phys. Rev. Lett. **77**, 4918 (**1996**).

36 M. Müller, Y. López Jiménez,
C. Rummel, G. Baier, A. Galka,
U. Stephani and H. Muhle,
Phys. Rev. E **74**, 041119 (**2006**).

37 C. Rummel, G. Baier and M. Müller,
J. Neurosc. Meth. **166**, 138 (**2007**).

38 J. Wishart, *Biometrika* **20A**, 3 (**1928**).

17
Seizure Prediction in Epilepsy: Does a Combination of Methods Help?

Hinnerk Feldwisch genannt Drentrup, Michael Jachan, Björn Schelter

17.1
Introduction

Epilepsy, as one of the most common neurological diseases, affects up to 1 % of the world population. Due to the unforseeable nature of epileptic seizures, many patients suffer from severe impairments (see Chapter 1). In addition to established therapies like antiepileptic medication or surgical resections, closed-loop prediction and intervention methods could open up a new window of treatment strategies. For this prospect, several methods have been proposed for the prediction of epileptic seizures (see Chapter 2). By means of linear and nonlinear time series analysis, predictive features have been found in several studies (for a review see [1]). If preictal changes in the EEG could be detected early and with a sufficient sensitivity and specificity, alarms could be raised to warn the patients. Furthermore, automatically triggered interventions like focal drug applications or electrical stimulations would be feasable [2, 3].

Several prediction methods which were proposed up to now show a statistically significant prediction performance, but are hardly sufficient for clinical applications [1, 4]. As a promising approach to improve prediction results, a combination of two prediction methods is investigated in this study. If it would be possible to combine the predictive power of different methods in a complementary manner, a considerable increase in sensitivity and/or specificity could be expected. Especially a combination of a bivariate and a univariate prediction method has some prospect of success, as also concluded by Mormann et al. [5]. Here, we analyze the bivariate mean phase coherence and the univariate dynamical similarity index. For both methods, statistically significant prediction performances have been reported in earlier studies [5,6]. The mean phase coherence is a measure based on the concept of phase synchronization which was successfully applied to detect preictal changes in synchronization of intracranial EEG recordings [7,8]. For one channel of the EEG, the dynamical similarity index continuously compares the reconstructed dynamics of a sliding window of the signal to a reference window, which contains no seizure activity [9].

The seizure prediction characteristic (SPC, [10, 11]) provides a general methodology for a systematic analysis of the prediction power of these methods. It allows the assessment of the sensitivities of prediction methods dependent on their specificity, such that the performance and applicability of therapeutic intervention strategies can be evaluated. A comparison with a random predictor raising alarms just by chance and without utilizing any EEG data enables the assessment of statistical significance of each method [12].

For the combination of prediction methods, two different kinds are conceivable following Boolean logics. A logical 'OR' combination would consider each alarm of any of the individual methods as an alarm of the combined system, whereas for a combination using a logical 'AND' all individual methods have to trigger an alarm during a given time window to cause an alarm of the combined system.

For a statistical analysis of the combined prediction systems, the seizure prediction characteristic is adapted to this paradigm. Besides the intervention time and the seizure occurrence period, a new time window has to be defined for the 'AND' combination to reflect the time window during which all methods must trigger an alarm to cause an alarm of the prediction system. Since the degree of freedom is increased for the combined methods, the concept of the random predictor is extended accordingly.

17.2
Materials and Methods

In this chapter, we propose two different kinds of combination of two individual prediction methods: the mean phase coherence and the dynamical similarity index. Both the prediction performances for the individual methods and the performances of the combinations are assessed. The seizure prediction characteristic is used as the conceptual framework of the prediction and for statistical validation of these methods.

17.2.1
The Seizure-prediction Characteristic

For a thorough assessment of the predictive power of seizure-prediction methods, the seizure-prediction characteristic was introduced (see [10–12]). The essence of a clinically applicable 'prediction' is covered by the concept of the intervention time (IT), which is defined as the minimum period of time between a triggered alarm and the earliest possible occurrence of a subsequent seizure. This time window can be used in clinical applications, e.g., allowing an intervention method to take effect. To restrict the subsequent interval during which the seizure is expected to occur, the seizure-occurrence period (SOP) is defined. A short SOP limits the time a patient has to be under alert for a predicted seizure. However, prediction performance will likely decrease with shorter SOPs due to the immanent variability in seizure occurrence. Therefore, a trade-off between a short SOP and sufficient sensitivities may be chosen dependent on the clinical need.

Apart from the sensitivity as the ratio of correctly predicted seizures to the number of seizures investigated, assessment of the specificity of a certain prediction method by evaluating interictal data is also crucial [1,12]. For this purpose, a maximum rate of false predictions is predefined in order to limit the number of false predictions to clinically reasonable values.

To assess the statistical significance of prediction performance, based on the seizure prediction characteristic, an analytical random predictor can be utilized. If the performance of a specific prediction method is higher than that of a random predictor which does not exploit any information contained in the EEG data, it can be regarded as statistically significant. The random predictor raises alarms just by chance in accordance with a Poisson distribution resulting in inter-alarm intervals that are exponentially distributed and limited by the same maximum rate of false alarms FPRmax. In dependence on FPRmax and the length of SOP, there is a certain opportunity to raise an alarm preceding a seizure, such that it will be regarded as a correct prediction. The probability to predict k out of K seizures correctly can be calculated by a binomial distribution (see Chapter 18). Furthermore, a correction for multiple testing is possible to take the retrospective selection of a best feature into account.

17.2.2
Combination of Individual Prediction Methods

In order to merge the predictive power of individual prediction methods, several ways to combine the alarms, which are triggered by the methods, are conceivable. Following Boolean logics, both an 'AND' combination and an 'OR' combination are studied in this manuscript. For the 'AND' combination, all individual methods have to trigger an alarm during a certain time window, the so-called combination window (CW). If this is the case, the combined system triggers an alarm, which is followed by the intervention time and the seizure occurrence period (see Figure 17.1). This alarm is a correct prediction if a seizure occurs during this time window, otherwise it is a false positive prediction. For the 'OR' combination, each alarm of each respective method triggers an alarm of the combined system. However, if two alarms follow each other closer than within IT + SOP, only the first alarm will be taken into account.

Depending on the coincidence of alarms triggered by the individual methods, either of the two types of combination could be advantageous. Two methods with low thresholds of triggering a prediction, i.e., causing a large number of false predictions if applied individually, could be linked using the 'AND' combination to select those predictions which are generated contemporaneously by both methods. This procedure is expected to reduce the number of false predictions considerably while preserving the sensitivity of the individual methods. The 'OR' combination results in a unification of individual prediction methods as all alarms of individual methods lead to alarms of the combined system. If, for example, each of the

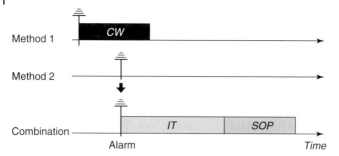

Fig. 17.1 Combination of two prediction methods following a logical 'AND': An alarm of one of the methods is followed by a combination window (CW). If the other method also triggers an alarm during CW, a prediction of the combination is triggered. In accordance with the seizure-prediction characteristic, the intervention time (IT) has to be seizure-free to allow a successful intervention. The seizure is predicted to occur during the seizure occurrence period (SOP).

methods is able to predict a specific seizure type, a combination using a logical 'OR' would match the correct predictions of both methods together. But the false alarms also add up. Hence, the 'OR' combination would only be useful if the benefits of the combination of the true predictions exceed the disadvantages regarding the increase in false alarms.

To investigate these dependencies, we apply the mean-phase coherence and the dynamical similarity index to an intracranial long-term recording of the EEG of one patient, and combine them with both types of combination. In order to reduce the number of channels and channel combinations of both methods, three focal and three extra-focal electrode contacts are preselected by a certified epileptologist, resulting in 15 features for the mean phase coherence and in six features for the dynamical similarity index. Furthermore, a best feature of both the mean phase coherence and the dynamical similarity index is selected in the first part of the data. For this purpose, both methods are applied individually to this data, and the features which achieve the best sensitivities given the maximum false prediction rate are used for the combination in the second part of the data. For the combination, both thresholds of the individual methods are chosen such that the sensitivity of the combination is optimal (see Figures 17.2 and 17.3). For the 'AND' combination, several false alarms (marked by dash-dotted black vertical lines) of one method were not followed by alarms of the other methods during the combination window. Therefore, for the combination, less false alarms arise. Threshold crossings, which triggered false alarms of the individual methods, led to correct alarms (solid black vertical lines) from the combination. In this example, an improved sensitivity is observed for the 'AND' combination, while fewer false predictions were made. By applying the 'OR' combination, both methods are joined in a complementary manner. Seizures, which can be predicted correctly only by one method, can be predicted by the 'OR' combination. But also the false alarms of both methods are joined together.

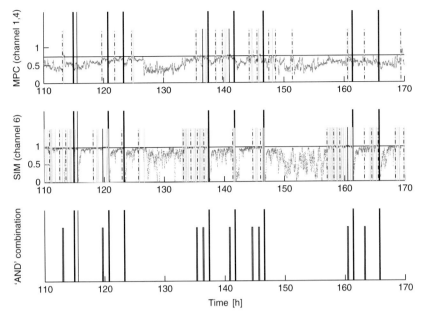

Fig. 17.2 Results of the 'AND' combination. The time course of the features is shown for the mean phase coherence (MPC) in the top row and for the dynamical similarity index (SIM) in the middle row. The thresholds (gray horizontal line) for both methods were chosen such that the sensitivities of the combinations (bottom rows) were optimal. Seizures less than 1 h apart from the previous one are marked by bold black lines and threshold crossings by gray black lines mark threshold crossings causing false (correct) alarms. After an alarm, no further alarms are raised during the duration of IT + SOP. Furthermore, the combination windows of the 'AND' combination are shown in gray. Its length and also SOP was set to 10 min, FPRmax to 0.15 false predictions per hour. An optimal value for IT of 50 min was chosen.

17.2.3
Patient Characteristics

For an analysis of seizure prediction methods and their combination, a continuous registration of intracranial EEG of one patient was used, which was recorded during presurgical epilepsy monitoring at the Epilepsy Center of the University Hospital Freiburg, Germany. The patient suffered from pharmacoresistant epilepsy including simple and complex partial seizures of hippocampal and neocortical origin. After the resection the patient became seizure-free (Engel Ia [13], Wieser 1 [14]). The retrospective evaluation of the data received prior approval by the Ethics Committee, Medical Faculty, University of Freiburg, Germany. Informed consent was obtained from the patient. The EEG data were recorded by strip and depth electrodes using a Neurofile NT digital video EEG system with a sampling rate of 1024 Hz. Digitized with a 16 bit analogue-to-digital converter, the data were high-pass filtered at 0.5 Hz and anti-aliasing filtered. The EEG channels were

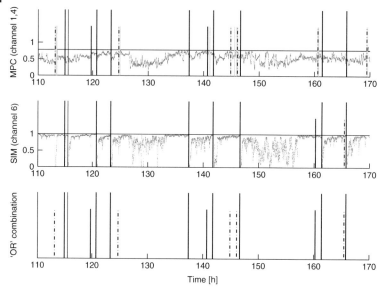

Fig. 17.3 Results of the 'OR' combination. The time course of the features is shown for the mean phase coherence (MPC) in the top row and for the dynamical similarity index (SIM) in the middle row. The thresholds (gray horizontal line) for both methods were chosen such that the sensitivities of the combinations (bottom rows) were optimal. Seizures less than 1 h apart from the previous one are marked by bold black lines and threshold crossings by gray vertical lines. Dash-dotted (solid) black lines mark threshold crossings causing false (correct) alarms. After an alarm, no further alarms are raised during the duration of IT + SOP. SOP was set to 30 min, FPRmax to 0.15 false predictions per hour. An optimal value for IT of 60 min was chosen.

referenced to the channel displaying lowest epileptic activity. A 50 Hz notch filter was used to eliminate possible line noise.

The recording included 6.8 days and 22 seizures. For the selection of best features of both individual methods, 36 hours of recording including 12 seizures were used.

17.3
Results

In order to test both the 'AND' and 'OR' combination, they are applied to the mean phase coherence and the dynamical similarity index of one patient as described in the previous section. For this study the false prediction rate for each individual algorithm was restricted to FPRmax = 0.15 false predictions per hour, and the duration of IT was optimized between 10 min and 60 min. In Figure 17.4, the prediction performance of the methods based on the mean phase coherence (MPC) and the dynamical similarity index (SIM) and their combinations are shown dependent on the duration of the seizure occurrence period. For the 'AND' combination, the duration of the combination window was set to the duration of SOP. In addition to the observed sensitivity, the lower and upper critical sensitivity

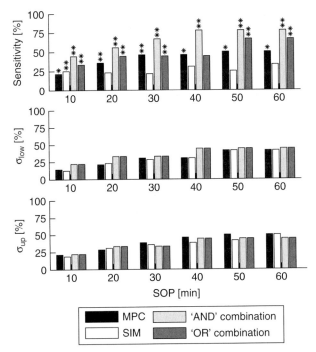

Fig. 17.4 Results of the individual methods mean phase coherence (MPC, black) and dynamical similarity index (SIM, white) in comparison with the optimized combinations following a logical 'AND' (light gray) and 'OR' (dark gray) applied on selected features dependent on the seizure occurrence period. Upper row: optimal sensitivities; middle and lower row: lower and upper critical sensitivity values of the random predictor. Sensitivity values which exceed the corresponding lower/upper critical values are marked by one/two asterisks. A maximum false prediction rate of 0.15 false predictions per hour was used. The intervention time was optimized between 10 min and 60 min.

values of the random predictor are displayed. The lower critical value corresponds to the assumption of complete dependence on the information contained in all features of one method, whereas the upper critical value corresponds to complete independence of these.

On the one hand, it can be observed that with longer seizure occurrence periods the sensitivity increases, since the probability increases to correctly predict the seizures. On the other hand, for the same reason the critical values of the random predictor also increase. The sensitivities of the phase synchronization index exceed the upper critical value for SOP = 20 min and 30 min. The same is true for the dynamical similarity index for SOP = 10 min. For the 'AND' combination, a doubled sensitivity of 44.4 % to 77.8 % can be observed in comparison with the average of the individual methods. Due to the fact that only one feature is analyzed for each method, the upper and lower critical sensitivity values are identical. For the 'AND' combination, the sensitivity exceeds the critical value for all durations of SOP. For the 'OR' combination, slightly lower sensitivities of 33.3 % to 66.7 %

are observed, which exceed the critical value for SOP = 10 min, 20 min, 30 min, 50 min and 60 min.

17.4
Discussion

Precursors of imminent epileptic seizures have been found in a number of studies. Whereas some of them have proved their predictive value by also showing statistical significance when rigorous testing methods have been applied, the relationship of sensitivity and specificity is currently insufficient for clinical applications [4, 5].

In this manuscript, we investigate the possibility of combining different prediction methods to improve their prediction performances. By using exemplary, intracranial recordings of the EEG of one patient, two different kinds of combinations are tested. For a logical 'AND' combination, both methods have to raise alarms during a specified time window, whereas each single alarm of one method leads to an alarm of the combined system following a logical 'OR' combination. For a maximum false prediction rate of 0.15 false predictions per hour and intervention times between 10 min and 60 min, an increase in sensitivity is observed. For the 'AND' combination, the sensitivity doubles in comparison with the average of the individual methods; for the 'OR' combination it increases slightly less. In comparison with critical sensitivity values of an analytical random predictor, the combinations show statistically significant prediction performances.

These findings demonstrate that the combination of prediction methods is a promising novel approach in order to increase the performance of seizure-prediction methods. The extraction of different aspects of chances in preictal dynamics can be combined and advantageous characteristics of various methods can be joined. This is of particular value, as a number of individual prediction algorithms have proven statistically significant performances, but an insufficient relation between sensitivity and specificity to be useful in a clinical setting [15, 16]. The improvements in sensitivity by the combination of two methods could represent a relevant step in the direction of future clinical applications.

In this study, the combination of a univariate and a bivariate measure was applied. This can be transferred to other algorithms. It is easily conceivable that prediction methods could be designed focusing on a certain aspect of brain dynamics with high sensitivities not covered by others. A combination could then improve the predictive power.

These results show that combinations can improve seizure-prediction performance as compared with an application of individual prediction methods alone. Using the Boolean operations introduced here, two scenarios are conceivable. If, on the one hand, the applied individual prediction methods are known to produce independently occurring false alarms, the 'AND' combination is a promising approach to reduce the number of false alarms considerably. Particularly with regard to future applications, this is advantageous especially if a warning is to be given to patients, or if an intervention method to be applied critically depends on a low false prediction rate to be tolerable for clinical applications. Based on the results presented here, further studies of the occurrence of false alarms should be undertaken for currently available and future seizure-prediction methods.

If, on the other hand, prediction methods are available which complement each other with respect to correct predictions, the 'OR' combination may offer advantages by combining the predictive power of the individual methods. In situations when the specificity is less relevant than the sensitivity, this can help to increase the number of correctly predicted seizures.

The optimization of the combination of prediction methods allows one to suit the prediction method to the specific requirements of clinical settings. Due to the optimization, best sensitivities can be achieved for a given maximum false prediction rate. For intervention systems with negligible side effects, like pharmacological or electrical interventions at the focus site, a high sensitivity is desirable, whereas false alarms are of less importance in this scenario. If, on the other hand, alarms are raised to warn the patient, an optimal specificity has to be achieved, such that the patient is not impaired by false alarms. By optimizing the combinations, both can be achieved such that the combinations behaves as a single-prediction system. The observed improvements suggest that this could help to enhance prediction performances considerably.

17.5
Acknowledgments

This work was supported by the German Federal Ministry of Education and Research (BMBF grant 01GQ0420) and the German Science Foundation (DFG grant Ti315/2-1).

References

1 F. Mormann, R. G. Andrzejak, C. E. Elger and K. Lehnertz, *Brain* **130**, 314 (**2007**).

2 A. G. Stein, H. G. Eder, D. E. Blum, A. Drachev and R. S. Fisher, *Epilepsy Res.* **39**, 103 (**2000**).

3 Y. Li and D. J. Mogul, *J Clin. Neurophysiol.* **24**, 197 (**2007**).

4 J. S. Ebersole, *Clin. Neurophysiol.* **116**, 489 (**2005**).

5 F. Mormann, T. Kreuz, C. Rieke, R. Andrzejak, A. Kraskov, P. David, C. Elger and K. Lehnertz, *Clin. Neurophysiol.* **116**, 569 (**2005**).

6 B. Schelter, M. Winterhalder, T. Maiwald, A. Brandt, A. Schad, J. Timmer and A. Schulze-Bonhage, *Epilepsia* **47**, 2058 (**2006**).

7 M. G. Rosenblum, A. S. Pikovsky and J. Kurths, *Phys. Rev. Lett.* **76**, 1804 (**1996**).

8 F. Mormann, K. Lehnertz, P. David and C. Elger, *Physica D* **144**, 358 (**2000**).

9 M. Le van Quyen, J. Martinerie, M. Baulac and F. Varela, *Neuroreport* **10**, 2149 (**1999**).

10 M. Winterhalder, T. Maiwald, H. U. Voss, R. Aschenbrenner-Scheibe, J. Timmer and A. Schulze-Bonhage, *Epilepsy Behav.* **4**, 318 (**2003**).

11 T. Maiwald, M. Winterhalder, H. U. Voss, R. Aschenbrenner-Scheibe, A. Schulze-Bonhage and J. Timmer, In P. Pardalos, J. Sackellares, P. Carney and L. Iasemidis (editors) *Quantitative Neuroscience* (Kluwer Academic Publishers, **2004**), Biocomputing Vol. 2.

12 B. Schelter, M. Winterhalder, T. Maiwald, A. Brandt, A. Schad, A. Schulze-Bonhage and J. Timmer, *Chaos* **16**, 013108 (**2006**).

13 J. Engel and T. Rasmussen, *In*: J. Engel (Ed), *Surgical Treatment of Epilepsies* (New York, Raven Press, pp 609–12, **1993**).

14 H. G. Wieser, W. T. Blume, D. Fish, E. Goldensohn, A. Hufnagel, D. King, M. R. Sperling, H. Lüders, T. A. Pedley and Commission on Neurosurgery of the International League Against Epilepsy (ILAE), *Epilepsia* **42**, 282 (**2001**).

15 F. Mormann, C. Elger and K. Lehnertz, *Curr. Opin. Neurol.* **19**, 187 (**2006**).

16 B. Schelter, M. Winterhalder, H. Feldwisch genannt Drentrup, J. Wohlmuth, J. Nawrath, A. Brandt, A. Schulze-Bonhage and J. Timmer, *Epilepsy Res.* **73**, 213 (**2007**).

18

Can Your Prediction Algorithm Beat a Random Predictor?

Björn Schelter, Ralph G. Andrzejak, Florian Mormann

18.1
Introduction

To date, approximately one in four epilepsy patients cannot be treated successfully by common therapeutic strategies, so they continue to suffer from unforeseen seizures. A precise prediction at a sufficiently early stage before seizure onset would offer new therapeutic options such as warning devices or even seizure-prevention techniques, e.g., by applying electric stimuli [1]. For this purpose, several time series analysis techniques based on the theory of linear and nonlinear dynamics have been applied to intracranial and scalp EEG data. For an overview of these studies see, e.g., [2, 3]. Significant changes in the EEG dynamics in a range from seconds up to hours in advance of seizure onsets have been reported. These studies have strengthened the hope that not only *interictal* states between seizures but also specific *preictal* states exist preceding seizures. The existence of preictal periods is the basic requirement for genuine seizure prediction.

When a focal seizure is generated, synchronized epileptic brain activity is initially observed only in a small area of the brain. From this focus, the activity spreads out to other brain areas. Provided that there is information about an impending seizure contained in the EEG data in advance of the seizure onset, time series analysis techniques are supposed to detect such changes. Visual inspection of the EEG data has not yet led to the detection of any characteristic changes preceding seizure onsets. In recent years, several seizure-prediction algorithms have been claimed to be capable of detecting a pre-seizure state.

Many of the studies addressing the challenge of predicting seizures lack methodological rigor. For this reason four major methodological guidelines have been proposed in a recent review of the field [3] to ensure a basic methodological standard. First, seizure-prediction techniques should be evaluated on long-term continuous data. No preselection of the data should be performed. Several physiological as well as pathological states of the patients should be covered by the data. Secondly, if training of a prediction algorithm is necessary, results should be reported separately for training and testing data. Thirdly, when assessing seizure-prediction performance, the assessment has to be based on sensitivity as

Seizure Prediction in Epilepsy. Edited by Björn Schelter, Jens Timmer and Andreas Schulze-Bonhage
Copyright © 2008 WILEY-VCH Verlag GmbH & Co. KGaA, Weinheim
ISBN: 978-3-527-40756-9

well as specificity. The procedure of how the specificity is assessed should be clearly reported. A natural, straightforward measure for specificity is one minus the fraction of time under false warning. Fourthly, a proper statistical validation of seizure-prediction performance is mandatory. To this end, either analytical or bootstrap-based techniques are available in the literature.

In this chapter, we concentrate on the last two points, namely the assessment of performance and statistical significance. The remaining two points require a rather huge amount of data that is not available to several groups. But ensuring a rigorous assessment, including a sound statistical analysis, are the crucial steps in the field of seizure prediction for the near future. Comparisons of published seizure-prediction performances also become possible especially by following the two guidelines 3 and 4.

The chapter is structured as follows. First, for the performance assessment, the ROC statistics and the seizure-prediction characteristic are briefly summarized. Subsequently, for the statistical evaluation three different random predictors are reviewed. First, an analytic random predictor is presented followed by two bootstrap-based techniques. The two bootstrap-based techniques are the seizure time surrogates and the measure profile surrogates. The chapter concludes with some remarks about the advantages and disadvantages of the respective techniques.

18.2
Performance Assessment

Briefly, we review a general methodology used to predict epileptic seizures before we address the two assessment methodologies, namely, the receiver operating characteristic (ROC) and the seizure-prediction characteristic.

18.2.1
General Methodology of Seizure Prediction

Seizure-prediction techniques generally utilize electroencephalography (EEG) data to predict epileptic seizures. Although other signals than the EEG are also conceivable for seizure prediction, we explain the basic methodology on EEG data. Time series analysis techniques are applied to EEG data, which may or may not be preprocessed by applying for example filters. Using a moving-window technique, the output of the time series analysis techniques is also a time series, the so-called *measure profile* or *feature* of the seizure-prediction technique. The measure profile can also be post-processed, for instance by smoothing. For example, a crossing of a certain threshold of the actual measure profile can be considered as the predictive alarm for the upcoming seizure. Since it would be useless to indicate only that there will be a seizure some time after the alarm, a certain time window in which the seizure is predicted to occur is needed. This time interval is the so-called prediction horizon or seizure-occurrence period *SOP*.

To analyze seizure-prediction performance, a study design which does not rely on the notion of an alarm issued by a quasi-prospective algorithm is also conceivable. In a retrospective, statistical (as opposed to algorithmic) study design, certain statistical characteristics (e.g., the amplitude distributions) of a certain feature during a pre-seizure period of the same length as the SOP and prior to this period, are compared (e.g. [4]). A certain difference between the feature characteristics of these two periods indicates a prediction performance which subsequently has to be tested for statistical significance. Alternatively, the time a feature is above a certain threshold in the pre-seizure period can be compared with the time the feature is above the threshold during the interictal epochs. The former would be a marker of the sensitivity while the latter quantifies the specificity.

18.2.2
The ROC Curve

The basic idea of the receiver-operating characteristic originates from signal-detection theory. A receiver-operating characteristic (ROC) or simply ROC curve or ROC statistics is a graphical plot of the sensitivity versus one minus the specificity of a binary classifier system as its discrimination threshold is varied.

Adjusted to seizure prediction, the sensitivity is the number of true positive predictions divided by the total number of seizures analyzed. A prediction is considered to be a true one, if the seizure onset falls into the occurrence period. The specificity is the fraction of true predictions divided by the total number of predictions. In terms of false predictions the specificity is identical to one minus the number of false predictions divided by the total number of predictions made.

Considering the consequences of actions to be taken, a statistical evaluation of the relative number of both correct and false predictions as measures of sensitivity and specificity, has to be made. Sensitivity and specificity, however, are not independent. It is always possible to obtain a high sensitivity at the expense of a low specificity, a problem which becomes particularly relevant if the costs of erroneous actions are high. In addition, patients' trust and willingness to follow instructions based on such predictions critically depend on their specificity.

In Figure 18.1 a schematic drawing of a ROC curve is shown. The dashed line indicates the performance achieved by chance in the ROC statistics. To quantify the prediction performance, the area under the curve should be used. The obtained area should be compared with the area under the diagonal, which is identical to 0.5. Since the ROC curve can take values of one for a 'perfect' performance (star in Figure 18.1) or zero indicating no performance at all, the difference between the area under the curve and the one under the diagonal takes values between 0.5 and −0.5. To interpret these values, a statistical evaluation procedure is, however, needed.

In a retrospective statistical analysis that compares predefined preictal and inter-ictal periods, definitions of sensitivity and specificity are quite straightforward [4]. In the case of a quasi-prospective algorithm, however, each false alarm forces the patient to wait for the full duration of the SOP before he can determine that this was

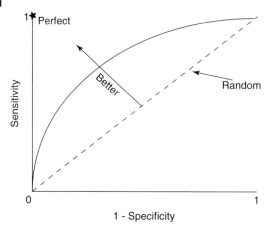

Fig. 18.1 Example of an ROC curve (solid line). The dashed main diagonal line represents the performance which can be achieved by chance. The star denotes the best possible ROC statistics as both the sensitivity and the specificity are identical to one. Values of the ROC above the diagonal line indicate better results, although they are not yet statistically compared to a random predictor. A natural choice to quantify the performance of the ROC curve; the area under the curve, is compared to 0.5, which corresponds to the area under the diagonal line.

indeed a false alarm. During this time, the patient must assume a seizure is impending and if another (false) alarm occurs within this period, the patient needs to 'restart' the SOP. A natural measure of specificity, or rather of 1-specificity, for quasi-prospective algorithms is therefore to quantify the fraction of time spent under false warning. This also allows for an easy performance evaluation since the time under (false) warning is also the expected value for 'accidently' predicting seizures, i.e., the 'random' sensitivity corresponding to the dashed diagonal line in Figure 18.1.

18.2.3
The Seizure-prediction Characteristic

To base actions on predictions, the above mentioned requirements have to be fulfiled. Additionally, in order to react to a prediction, a time period has to be specified between the time of prediction and the beginning of the occurrence period; the intervention time. Depending on the desired interactions, a spectrum of time periods may be conceivable and reasonable. In the case of warning a patient, an intervention time of minutes to hours is necessary.

In contrast to the ROC statistics, the seizure-prediction characteristic (SPC) [2,5] evaluates the seizure-prediction performance with respect to the false-prediction rate FPR, i.e., the number of false predictions per certain time window, and the two time intervals, the intervention time and the occurrence period.

Although the two approaches ROC and SPC are basically the same, in certain cases one might be favorable to the other. As described above, the time under false warning is a superior quantity used to measure the burden which the prediction technique puts on the patient. Having the false prediction rate at hand when answering this question is rather straightforward, as the false-prediction rate has simply to be multiplied by the seizure-occurrence period. But this holds only under some assumptions. For instance, two false predictions should be separated by a time interval larger than the occurrence period. Moreover, the occurrence period has to be constant for all predictions.

In Figure 18.2 the seizure-prediction characteristic describing the sensitivity as a function of intervention time and the occurrence period is shown. Figure 18.2(a) depicts the intervention time and occurrence period, Figure 18.2(b) exemplifies the prediction characteristic for different prediction methods with respect to the occurence period.

18.3
Statistical Validation

For both methodologies given above, an evaluation procedure is needed to determine if the seizure-prediction performance is significantly different from the chance level. Besides the number of seizures investigated and the duration of the interictal period, the number of EEG channels utilized by a seizure-prediction method has to be taken into account. It ranges from a few to more than a hundred. Using, for instance, symmetric and bivariate synchronization quantities between each pair of electrode contacts, leads to a high number of possible combinations. Assuming that there is no predictive information in the EEG data and thus in the feature time series, the probability of predicting at least some of the seizure onsets correctly by chance increases with increasing number of electrode contacts. To assess the superiority of a prediction method over a random predictor, the same number of electrode contacts has to be considered. There are several possibilities for the choice of a random predictor. Here, the notion random predictor covers analytic as well as bootstrap-based techniques. We present an overview of the random predictors currently used in prediction studies.

18.3.1
The Analytic Random Predictor

A test to decide on the statistical significance of a given value of the seizure-prediction characteristic is defined by the 'prediction performance' of an unspecific random prediction. For an unspecific random prediction, alarms are triggered completely at random without using any information contained in the EEG. The random predictor is based on a homogenous Poisson process for the false predictions.

At any single sampling point of a feature extracted from a time series, the probability of raising an alarm is

$$P_{\text{Poiss}} = \frac{\text{FP}}{N} \tag{18.1}$$

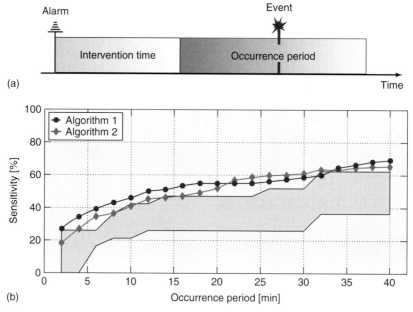

Fig. 18.2 (a) Schematic drawing of the time relationships between an alarm based on a prediction algorithm, the period preceding an expected event (intervention time) and the uncertainty of the exact time point of occurrence (occurrence period). Only events occurring within the occurrence period are considered as correct predictions in an analysis of the performance of the algorithm. (b) Seizure-prediction Characteristic for a fixed intervention time of 10 minutes and false prediction rate of 0.15 false predictions per hour depending on the duration of the occurrence period evaluated for two exemplary seizure-prediction algorithms based on intracranial long-term electroencephalographic data and a range of sensitivities for a random predictor. It is possible to compare different algorithms and to clarify their superiority over random predictors. Note the increase in correct predictions for the random algorithm with increasing occurrence periods.

where FP denotes the number of false predictions and N the number of samples. Consider a time period of duration equal to SOP. If the product of the maximal allowed false prediction rate and the occurrence period, $FPR_{max} \cdot SOP$, is considerably smaller than one, which is reasonable to ensure that the patient is not under continuous warning, the probability of raising at least one alarm within SOP for a given value of FPR_{max} can be approximated by [7]

$$P \approx 1 - e^{-FPR_{max} \cdot SOP} \approx FPR_{max} \cdot SOP. \qquad (18.2)$$

The probability P forms the basis for a significance level to test whether the sensitivity $S(FPR_{max}, SOP, IT)$ of a prediction method is higher than the 'sensitivity' of a random predictor.

The intervention time, while being essential for the concept of prediction, has no influence on the random predictor; only the duration of SOP is contained in (18.2) for the probability P. This is intuitively understandable. Although the intervention time allows an intervention, it is only restricted by an upper limit demanded by the seizure-prediction technique. Assume that a prediction of epileptic seizures is possible with an intervention time of one hour, but the intervention needs only minutes to become effective, one can always wait until the very last moment before applying the intervention. However, if the desired intervention time is one hour, and imagine that this intervention time is really needed, all patients having an optimal intervention time of less than one hour cannot be treated by this intervention. Here, optimal means that a highly significant seizure-prediction performance could be obtained.

One should note that the duration of the intervention time will, in practice, always influence the random predictor. This is because, for long intervention times, it is not guaranteed that there is enough time between two consecutive seizures to allow for such a long intervention time. Therefore the number of seizures is indirectly decreased, which influences the random predictor as discussed in the following.

A significance level should take into account that more than one seizure is being investigated. Furthermore, usually a set of electrode contacts is analyzed which is associated with a multiple testing problem. For instance, applying the bivariate and symmetric measure mean phase coherence [6] to EEG time series recorded from six electrode contacts, 15 different time courses of the mean phase coherence are extracted simultaneously. Increasing the number of electrode contacts to a value of ten, for instance, would result in 45 different pairs of electrodes with corresponding time courses of a possibly bivariate seizure-prediction technique. Therefore, the probability of changes in any of these time courses leading to a prediction of seizures by chance, strongly increases with the number of electrode contacts investigated. Both aspects are discussed in the following. The probability of predicting at least k out of K seizures by chance follows a binomial distribution with probability P [7]

$$P_{\text{binom}}(k, K, P) = \sum_{j \geq k} \binom{K}{j} P^j (1 - P)^{K-j} \tag{18.3}$$

Furthermore, the dimension d of the feature vector has a direct influence on the significance level. Increasing the number of features also increases the probability of predicting seizures by chance. The maximum number of independent features for an r-variate, symmetric feature extraction using n electrode contacts is given by

$$d_{\text{max},r} = \binom{n}{r}. \tag{18.4}$$

The dimension d of the extracted feature vector has to be included in the probability of predicting seizures by chance. The probability of predicting at least k of K seizures using d measure profiles is

$$P_{\text{binom},d}(k, K, P) = 1 - \left[\sum_{j \leq k} \binom{K}{j} P^j (1 - P)^{K-j} \right]^d. \tag{18.5}$$

This equation is based on the assumption that the *P*-values are identical for all measure profiles. In applications, the seizure-occurrence period is likely constant, but the false-prediction rate is usually not the same for all channels. This is the reason for using the concept of the maximum false-prediction rate. It provides an upper bound for all features. Therefore (18.5) yields a conservative approximation for the true probability if the *P*-values are considerably different.

The number of independent degrees of freedom spanned by the measure profiles is usually unknown. The effective value $d_{\mathrm{eff}}(n)$ of d might be smaller than $d_{\mathrm{max},r}$. For instance, signals from neighboring electrode contacts may be correlated, leading to a lower effective dimension of the extracted feature vector. Therefore, two critical values have to be taken into account to test for statistical significance. For a significance level α, the lower critical value is given by

$$\sigma_{\mathrm{low}} = \frac{\mathrm{argmax}_k \left\{ P_{\mathrm{binom},1}(k, K, P) > \alpha \right\}}{K} \cdot 100\% \tag{18.6}$$

for $d = 1$. For an *r*-variate, symmetric feature extraction and n electrode contacts investigated, the upper critical value

$$\sigma_{\mathrm{low}} = \frac{\mathrm{argmax}_k \left\{ P_{\mathrm{binom},d_{\mathrm{max},r}}(k, K, P) > \alpha \right\}}{K} \cdot 100\% \tag{18.7}$$

is obtained for independent features [7].

The above-mentioned assessment methodology is also applicable to the ROC statistics after some minor adaptations. Alternatively, two different data-driven assessment concepts; so-called bootstrap techniques, have been suggested [8, 9]. The advantage of the analytic random predictor lies in the analytic expression for its 'sensitivity'. Thus, the design of studies is possible. Assume that a given prediction performance will be verified based on certain parameters for the intervention time, the occurrence period, and maximum false-prediction rate based on a *d*-dimensional feature. The above equations would provide the information of the minimal number of seizures that have to be contained in the data to ensure that a performance above chance level can be shown.

The bootstrap-based techniques show superior characteristics, for instance, when assumptions made by the random predictor are not reasonable. For example, the analytic random predictor is based on a homogeneous Poisson distribution for the false predictions. This might not be the case or one might want to consider different distributions for certain reasons. In those cases the analytic random predictor is expected to be anticonservative. The bootstrap techniques reviewed in the following sections present a way out.

18.3.2
Bootstrapping Techniques

An alternative approach to analytic random predictors is the concept of numerical Monte Carlo simulations based on surrogate seizure predictors which are constructed from constrained randomizations of the original seizure predictor

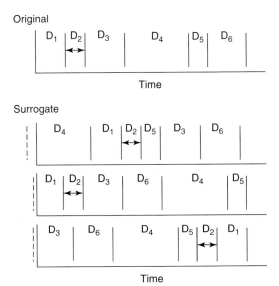

Fig. 18.3 Original seizure times and the surrogate times bootstrapped from the inter-seizure intervals. The arbitrary onset times for the surrogates are obtained from a uniform distribution and are indicated by the dashed vertical lines. Note that, by randomly selecting the offset of the starting point (compared to the original one), the end point is different from the original one for all surrogates.

(bootstrapping). These seizure-predictor surrogates are constrained to share specified properties with the original seizure predictor, but are otherwise random. This approach offers a greater flexibility than analytical random predictors since it allows one to test different null hypotheses by composing appropriate sets of assumptions and constraints. Specifically, a certain assumption about the original seizure predictor can be translated into a corresponding randomization constraint. For example, if one assumes that the alarms are raised at a time-independent mean rate, the predictor surrogates must be constrained to be time-independent, regardless of potential time-dependencies of the original seizure predictor. If alarms are assumed to be generated by a Poisson process, the predictor surrogate must have an exponential inter-alarm-interval distribution, regardless of the original distribution. If no assumptions about a potential time-dependence of the predictor or the inter-alarm-interval distribution are intended, the predictor surrogate must be constrained to share any time-dependence and the inter-alarm-interval distribution with the original predictor. Except for these constraints, the surrogate seizure predictor must be random. The assessed performance value for the original predictor is then compared with the predictive performance obtained for an ensemble of predictor surrogates. If the performance of the original predictor is significantly higher than the performance of the predictor surrogates, the respective underlying null hypothesis can be rejected, i.e., the prediction algorithm performs better than chance with respect to the assumptions described above.

Seizure-times Surrogates

As a first Monte Carlo approach for the evaluation of seizure-prediction statistics, the technique of seizure-times surrogates was proposed [8]. In this approach the intervals between seizures are shuffled while the original measure profiles are kept unchanged (random permutation of inter-seizure-intervals). Accordingly, the original seizure times are replaced with random times under the constraint that the inter-seizure-interval distribution is maintained (Figure 18.3).

Let $D_1, \ldots, D_n$ denote the intervals between consecutive seizures, where D_1 is the interval from the first seizure back to an arbitrarily defined starting point T_0. The seizure-times surrogates are now generated by the following steps. First, a new starting point is defined as $T_0^* = T_0 - \epsilon(1h)$, with ϵ being a random number uniformly distributed in $[0, a]$ with a some arbitrary number. This is necessary as the sum of all D_i is constant. If the starting point was kept constant the very last seizure would occur at the very same time for both the original seizure times as well as the seizure-times surrogates, which would definitely influence the statistical assessment. Thereby, starting at T_0^*, surrogate seizure onset times $T_1^*, \ldots, T_n^*$ are produced by random permutations of $D_1, \ldots, D_n$ (Figure 18.3).

The predictive performance is then calculated again from the unchanged measure profile but now with regard to the pseudo seizure times of the surrogates. If the predictive performance obtained for the original seizure times is higher than the one for the seizure-times surrogates the underlying null hypothesis, which is that the seizure predictor is not only naïve and unspecific but also stationary, can be rejected (see, e.g., [4]). While conceptually straightforward, the practical application of seizure-times surrogates can be problematic. Sometimes only a few seizures are included in the recordings, and recordings can be interrupted by gaps. Both problems can make the generation of a sufficient number of independent realizations of seizure-times surrogates impossible.

Measure-profile Surrogates

As an alternative approach to seizure-times surrogates, the technique of measure-profile surrogates was suggested [9]. These surrogates are generated by randomizing the original measure profiles, while keeping the original seizure times unchanged. In particular, the randomization has to be constrained in such a way that any feature of the measure profile which is evidently not related to a true predictive power of the seizure predictor, but which might influence the predictive performance, should be mimicked by the measure-profile surrogates. This constrained randomization can readily be carried out using the technique of simulated annealing. In the original publication [9], this technique was illustrated by using the autocorrelation function and amplitude distribution of the measure profile as constraints for the measure-profile surrogates. This approach offers a much higher flexibility than seizure-times surrogates since any time-dependence of the measure-profile statistics could readily be used as a constraint for the generation of measure-profile surrogates. A drawback of this approach is that simulated annealing can be computationally expensive.

18.4
Conclusion

In this chapter the receiver-operating characteristic and also the seizure-prediction characteristic have been discussed. Both assessment methodologies are inconclusive without a reasonable comparison to a prediction performance which can be achieved by chance. Two basic concepts of this aim have been suggested and used. Although the analytic random predictor has been developed in the framework of the seizure-prediction characteristic, it can be readily applied to the ROC statistics. Similarly, the bootstrap-based random predictors can be readily applied to both the seizure-prediction characteristic and the ROC curve, as they are extremely flexible in this respect.

Nevertheless, the random predictors have certain advantages and disadvantages some of which will be mentioned here. The most obvious advantage of the analytic random predictor is that it can be calculated in almost no time and is therefore suitable for study design. In contrast, bootstrap-based techniques are comparably slow. Albeit that bootstrap techniques are also subject to certain assumptions, the assumptions made by the random predictor are quite crucial. If for some reason the inter-alarm interval is not exponentially distributed as assumed by the analytic random predictor, bootstrap-based techniques are favorable. Moreover, if one wants to distinguish a seizure-prediction performance achieved by the evaluation of circadian dependencies alone, the measure-profile surrogates might be the method of choice. In contrast, for the seizure-times surrogates it is feasible to keep part of the inter-seizure intervals preserved, which might become important for clustered seizures. While bootstrapping techniques in principle have the advantage of being more flexible, great care must be taken to implement them according to the correct null hypothesis in order to avoid hidden bias.

A detailed and sound comparison of the different approaches is, however, an important task for future research. In the meantime one might consider it reasonable to use all three complementary approaches. It would be worse to use none of the random predictors. This would inevitably render a conclusive evaluation of seizure-prediction performance impossible.

In conclusion, the methodological framework of the seizure-prediction characteristic or ROC statistics, together with random predictors, allows for a scientific evaluation of the prediction performance. In particular, it enables one to assess prediction methods with predefined ranges for occurrence periods and parameter ranges which are suitable for rationally based decisions.

18.5
Acknowledgments

This work was supported by the German Science Foundation (DFG grant Ti315/2-1) and the German Federal Ministry of Education and Research (BMBF grant 01GQ0420). F.M. acknowledges support from the 6th Framework Programme of the European Commission (Marie Curie OIF 40445).

References

1 J. Milton and P. Jung, in *Epilepsy as a Dynamic Disease*, edited by J. Milton and P. Jung (Springer, **2003**), pp. 341–52.

2 T. Maiwald, M. Winterhalder, R. Aschenbrenner-Scheibe, H. Voss, A. Schulze-Bonhage and J. Timmer, *Physica D* **194**, 357 (**2004**).

3 F. Mormann, R. G. Andrzejak, C. Elger and K. Lehnertz, *Brain* **130**, 314 (**2007**).

4 F. Mormann, T. Kreuz, C. Rieke, R. G. Andrzejak, A. Kraskov, P. David, C. Elger and K. Lehnertz, *Clin. Neurophysiol.* **116**, 569 (**2005**).

5 M. Winterhalder, T. Maiwald, H. U. Voss, R. Aschenbrenner-Scheibe, J. Timmer and A. Schulze-Bonhage, *Epilepsy Behav.* **4**, 318 (**2003**).

6 F. Mormann, K. Lehnertz, P. David and C. Elger, *Physica D* **144**, 358 (**2000**).

7 B. Schelter, M. Winterhalder, T. Maiwald, A. Brandt, A. Schad, A. Schulze-Bonhage and J. Timmer, *Chaos* **6**, 013108 (**2006**).

8 R. G. Andrzejak, F. Mormann, T. Kreuz, C. Rieke, C. E. Elger and K. Lehnertz, *Phys. Rev. E* **67**, 010901 (**2003**).

9 T. Kreuz, R. G. Andrzejak, F. Mormann, A. Kraskov, H. Stögbauer, C. E. Elger, K. Lehnertz and P. Grassberger, *Phys. Rev. E* **69**, 061915 (**2004**).

19

Testing a Prediction Algorithm: Assessment of Performance

J. Chris Sackellares, Deng-Shan Shiau, Kevin M. Kelly, Sandeep P. Nair

19.1
Introduction

Early investigations of seizure-prediction research were aimed at finding characteristic and consistent changes in EEG prior to seizure onset. Evidence of seizure precursors in the EEG were reported as early as the 1970s when investigators, using linear signal processing methods [1–3], reported changes in EEG characteristics beginning minutes prior to the onset of seizures. Some investigators found changes in interictal spike distribution or incidence approaching seizure onset [4,5], whereas others found no consistent changes in spike patterns [6–8]. These observations were made by analysis of relatively brief EEG samples in a limited number of patients. Even at that time, the researchers realized that the presence of preictal changes in the EEG raised the possibility that seizures could be predicted.

With the introduction of faster computers with larger storage capacity, pioneers in the field of seizure prediction began systematically to analyze longer segments of EEG preceding and following seizures from a larger number of patients in epilepsy monitoring units and to use novel and more sophisticated approaches to signal processing. Motivated by theories that seizures may result from spontaneous state transitions in a chaotic nonlinear system [9–18], some investigators began to apply mathematical techniques developed for the study of complex nonlinear systems to analyze EEG signals for characteristics unique to the transitions into and out of seizures [18–34]. In addition, using linear measures, other investigators reported new evidence of measurable preictal EEG changes [35] and their ability to predict clinical seizure onset [36]. Some researchers also began to question the usefulness of nonlinear measures, using similar methods to those previously described [37–39]. These investigators raised questions about methods that employed the similarity index, the correlation dimension, the correlation integral, and the Lyapunov exponent. In most instances, the investigators did not precisely duplicate the methods they challenged. However, their reports served to temper initial enthusiasm and confidence in finding clinically useful seizure-prediction algorithms. Recently, the ability of a number of linear/nonlinear and univariate/bivariate measures to distinguish the preictal from the interictal state have

Seizure Prediction in Epilepsy. Edited by Björn Schelter, Jens Timmer and Andreas Schulze-Bonhage
Copyright © 2008 WILEY-VCH Verlag GmbH & Co. KGaA, Weinheim
ISBN: 978-3-527-40756-9

Table 19.1 Some Reported EEG Seizure Precursors.

Report	Author(s)
Increased spatial distribution of interictal spikes	[4]
Increased incidence of interictal spikes	[8]
Spatial convergence of the short-term largest Lyapunov exponent	[19]
Reduction in signal complexity	[25]
Reduction of signal similarity between segments	[28]
Increased incidence of power bursts	[35]
Changes in marginal predictabilities between regions	[34]
Reduced synchronization	[41]

provided further evidence of significant differences in EEG characteristics between the two periods [40]. While several measures showed significant differences, bivariate measures were generally more effective than univariate ones. A summary of some the most highly cited of these reports is provided in Table 19.1.

Based on these findings, many researchers have concluded that it may be possible to predict seizures with automated algorithms based on quantitative EEG analysis [41–45]; see [46] for a detailed review. These algorithms are computer programs that read the raw EEG signal, calculate measures of specific signal characteristics, compare these measures to threshold values, and generate seizure warnings when pre-established criteria are met. The parameters of the algorithms can be set to alter the sensitivity/specificity ratio. In most cases, specificity is expressed as the false positive rate (i.e., the number of false warnings per unit of time). In general, the higher the sensitivity obtained (i.e., the percentage of seizures correctly predicted, by a preset definition), the greater the false positive rate observed. Initial tests involved EEG recordings from a small number of patients. While some of the results looked promising, the algorithms were not validated statistically.

Although there remains a debate as to what constitutes an appropriate experimental design, what statistical comparisons are optimal, and what benchmarks constitute a good algorithm performance, several studies have been designed to assess the performance of a seizure-prediction algorithm [45,47–52]. Some of these studies assessed the test algorithm by reproducing the results in limited datasets, and others re-evaluated the performance reported in the literature using the data characteristics (seizure intervals, number of seizures, etc.) provided in the papers. Unfortunately, most of these studies suggested that none of the reported seizure-prediction algorithms performed at a level better than chance, and therefore the algorithms were considered not useful for clinical applications. Because the results from such assessments will have a significant impact on this research field, it is essential that future studies be well designed with clearly defined research questions.

While seizure-prediction research has significant scientific implications regarding the basic mechanisms of seizure generation, the potential impact on clinical care is of equal importance. For this reason, we wish to consider some of the issues pertinent to testing seizure-prediction algorithms for clinical applications.

Up to this time, the focus of seizure-prediction research has been on developing quantitative descriptors for analyzing EEG during clinically relevant states (interictal, ictal and postictal) and descriptors useful in the analysis of transitions between these states. The existing body of work in this research area has provided us with a rich variety of candidate measures for developing seizure-prediction algorithms. Subsequent investigations could best be categorized as investigations into the feasibility of seizure prediction with the algorithms incorporating these quantitative descriptors. To advance the field into clinical applications, the next phases of research will need to be preclinical testing and clinical trials. There is a need to establish scientific criteria for preclinical efficacy and safety standards and for clinical trial designs for subsequent human studies. Given the novelty of this field, these phases will require some creativity. To some extent, the preclinical testing must be driven by the clinical needs and the anticipated clinical evaluations. Therefore, we will confine the remainder of our remarks to issues pertaining to the clinical evaluation of seizure-prediction algorithms and devices.

Any clinical trial should include clearly stated clinical questions, well-defined statistical hypothesis, and statistical justification. More specifically, a sound study should be clear about the following questions: (1) What clinical application is being considered? (2) What are the statistical hypotheses? (3) Are the confidence levels sufficient for estimating the performance statistics, and for rejecting or accepting the null hypothesis?

19.2
Correlation between Study Design and Clinical Application

The study design for statistical evaluation of a seizure-prediction algorithm depends upon its intended clinical application. The definition of 'effectiveness' of the algorithm, as it pertains to the study will depend entirely upon the intended clinical application. For example, an implantable closed-loop seizure control device would require a near-perfect prediction rate (sensitivity) with a moderately good false prediction rate (specificity). One would want the device to be activated before every impending seizure, provided that the side effects due to unnecessary interventions are acceptable. In contrast, a monitoring device designed for use by nursing staff would require a high, but not near-perfect, sensitivity and a very high specificity. A few missed predictions is a great improvement on the current situation where seizures are entirely unpredictable, but too many false predictions will not be tolerated by busy monitoring unit staff personnel. In addition, different clinical applications have different (1) patient populations, (2) potential research subject pools (each with different clinical characteristics), and (3) types of EEG recordings (e.g., intracranial versus extracranial electrodes). All of these factors affect the design of the study.

Generally, the main clinical applications of a seizure-prediction algorithm are for inpatient monitoring and seizure-control devices. Inpatient monitoring can be done in an epilepsy monitoring unit, a general care unit, or an intensive care unit. The criteria for a clinically useful prediction algorithm are obviously different in different monitoring settings. For example, an epilepsy monitoring unit would require a seizure-prediction system providing moderate high sensitivity with low false-positive rate, whereas a general care or intensive care unit would require higher prediction sensitivity with a moderately low false-positive rate. In addition, EEG data characteristics are different among monitoring units. For example, the average number of seizures recorded for each patient in an epilepsy monitoring unit is between three and four. Therefore, it is not meaningful to determine whether the performance of a prediction algorithm for each individual patient is significantly better than chance (e.g., a random predictor). It would be more meaningful to evaluate the algorithm performance with respect to the proportion of test patients whose x % (e.g., two out of three) of seizures can be predicted by the algorithm within a pre-determined horizon. However, in the intensive care monitoring unit, where tens of critical events could be recorded from each individual patient, it is reasonable to evaluate the performance on each individual patient, although an overall assessment from a group of patients still gives more meaningful conclusions. Other factors, such as proper prediction horizon and cost/risk of a false positive or negative, which are different in various monitoring units, would also affect the study design.

Similarly, in the application for seizure-control devices, the satisfaction of a seizure-prediction performance depends on the mechanism of the intervention system, including the type, the effectiveness and the side effects of the controller. For example, the criteria for a prediction performance would be different for an electrical stimulation device compared with a drug-release control system.

19.3
Statistical Hypothesis

The first task of statistical design in a clinical trial is to translate clinical questions into testable statistical hypotheses. In assessment of a seizure-prediction algorithm, the underlying clinical question is whether the test algorithm is effective, or more specifically, performs better than a pre-determined performance level (e.g., chance level). However, the statistical hypotheses and the study design depend on how the pre-determined performance level is defined. Using the 'chance performance level' as an example, two different types of design have been reported. The first type is based on the comparisons of some performance statistic estimated from the test prediction algorithm against those from 'controlled algorithms' (e.g., random predictor, periodic predictor), using the same test datasets [45, 48–51]. In this case, the 'chance performance level' is defined as the prediction performance by some 'controlled prediction scheme' that does not use any information from the test EEG data. The statistical (null) hypothesis for this design (H_0) is: the mean performance statistic obtained from the test algorithm is the same as that from the 'controlled algorithm.' This design is straightforward because a 'controlled prediction method' is easy to design and can be justified statistically. One obvious choice is to simulate the predictions from a random Poisson process. Using the same prediction parameter settings (e.g., prediction horizon), the hypothesis can be tested by comparing the performance statistic between the test and the 'controlled algorithm' using a group of patients.

The second type of design is based on the comparison of a performance statistic estimated from the test datasets against those from 'controlled datasets,' using the same test algorithm. In this case, the 'chance performance level' is defined as the prediction performance of the test algorithm on the manipulated 'controlled data.' The (null) hypothesis for this design (H_0) is: the mean performance statistic obtained from the original datasets is the same as that from the 'controlled datasets.'

This type of design is less straightforward than the first type, mainly because there is a smaller consensus on what constitutes an appropriate 'controlled dataset' for testing a prediction algorithm. Andrzejak et al. (2003) [47] carefully constructed 'seizure-time surrogates' as 'controlled data.' However, not only is the statistical justification of 'seizure-time surrogates' in relation to the 'chance performance level' difficult to establish, but also the inter-dependency among the surrogates would confound the conclusions of the hypothesis testing.

So, what are the differences between the two types of assessment? Do they provide the same assessment of the test algorithm? Is it possible that, for the same test algorithm, one type of null hypothesis is rejected and the other is not? Are there other aspects of hypotheses that should be tested? These are important questions in the assessment of a prediction algorithm, and they need to be answered by a consensus of the community.

19.4
Statistical Justification

Besides the need for specification of the intended clinical application and the statistical hypotheses, another important issue is the choice of test datasets, in terms of their quality as well as the quantity, in order to have more reliable and confident estimates of the prediction-performance statistics. Inappropriate choice of the test datasets will not only waste the effort and resources spent in conducting the study, but, worse, may also result in generating a false conclusion.

The most commonly used performance statistics for evaluating a seizure-prediction algorithm are sensitivity and false prediction rate (per unit time), or their combination such as the receiver-operating characteristic (ROC) curve. The reliability of their estimations depends on the quality and quantity of the test datasets. Several review articles have given guidelines for selecting test datasets with respect to their qualitative characteristics (see [46] and the references therein). Here, we will provide guidelines with statistical justification for the minimum required number of seizures and the length of interictal recordings for reliable estimates of the performance statistics of a seizure-prediction algorithm.

19.4.1
Prediction Sensitivity

Sensitivity is estimated by dividing the number of correctly predicted seizures by the total number of seizures. Therefore, the accuracy and the confidence level of the estimated sensitivity depend on the total number of seizures included in the estimation. When the number of seizures included in the analysis is too small, the confidence level will be very low for a desired accuracy of the estimation. For example, if a prediction sensitivity of 80 % is estimated from the prediction of 10 seizures, there will be only a 20 % confidence level that the error of the estimation is less than 10 %. Stated differently, there is only a 20 % chance that the true sensitivity will be higher than 70 % (and lower than 87 %, but we are more concerned about over-estimation for sensitivity estimation). Hence, there is a great chance that the sensitivity is over-estimated. So, the question is: how many seizures are sufficient to claim a meaningful sensitivity?

The number of seizures required can be calculated based on the confidence limits for a binomial parameter [53] with the desired accuracy and the confidence level. Figure 19.1 shows the 95 % confidence limits as a function of the number of seizures included in the sensitivity estimation, with sensitivity estimated as 80 %. In this example, at least 80 seizures are required to ensure, with 95 % confidence, that the sensitivity is not over-estimated by more than 10 %. This suggests that claiming or evaluating prediction sensitivity for an individual patient is almost meaningless because of the lack of the required number of seizures. It is therefore

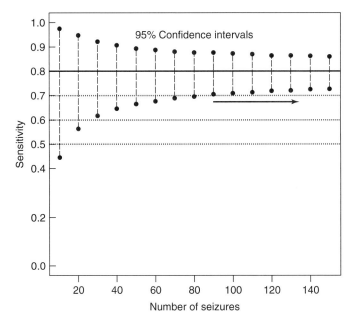

Fig. 19.1 95 % confidence interval as a function of number of seizures used to estimate prediction sensitivity as 80 %.

necessary to combine seizures from a group of patients for a more meaningful estimation. If one assumes that the average number of seizures recorded from each patient is four, at least 20 patients will be required. Similar analysis can be applied to determine the minimum number of seizures (or patients) required for different sensitivity estimations.

19.4.2
False-positive Rate

The false-positive rate is estimated by dividing the total number of false predictions by the total unit time (e.g., hours or days) that are outside the 'prediction zone' before each seizure (referred to as 'interictal data' hereafter). 'Prediction zone' of a seizure, by definition, means the time interval during which a true prediction for the seizure is possible, i.e., when a false warning cannot be produced. Obviously, the accuracy and the confidence level of false-positive-rate estimation depend on the length of the test 'interictal data.' If the interval is too short, then the confidence level for the estimation will be very low for a desired error. For example, assume that 24 hours of 'interictal data' are analyzed and a false-positive rate of three per day is estimated. In this case, the researcher will only have 10 % confidence level that the false-positive rate was not under-estimated by one per day. Hence, there is a great chance that the false-positive rate is under-estimated. So, the question is: how long must the cumulative interictal period be in a recording to claim a meaningful false-positive rate?

The length of 'interictal data' required for estimating a reliable false-positive rate can be calculated based on the confidence limits for a Poisson distribution [54] with the desired accuracy and the confidence level. Figure 19.2 shows the 95 % confidence limits as a function of the length of the 'interictal data' (in days), with a false-positive rate estimated as three per day. In this example, at least 15 days of 'interictal data' are required to have the accuracy of the estimation of less than one per day with 95 % confidence. This again suggests that claiming or evaluating a false-positive rate for an individual patient is not meaningful, especially for an inpatient monitoring setting. Similar analysis can be applied to determine the minimum number of days (or patients) required for different false-positive-rate estimation.

19.5
Discussion and Conclusion

While there is much ongoing debate among researchers in the field of seizure prediction, there is an approaching consensus that seizures are often preceded by measurable changes in the EEG and that it is possible to design automated algorithms to detect such changes. As long as we are in the phase of scientific discovery, it is acceptable, and even preferable, for researchers to use different approaches for hypothesis testing and algorithm development and testing. However,

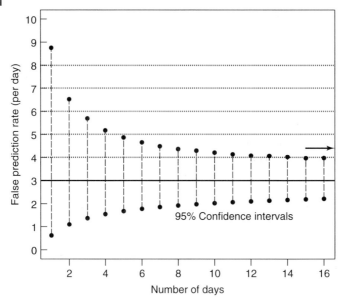

Fig. 19.2 95 % confidence interval as a function of number of days of 'interictal recording' used to estimate a false-positive rate of three per day.

as we enter into the phase of clinical translation, it is highly desirable to establish standards for clinical trials and standards of algorithm performance for specific clinical applications.

There is a need to establish standards for preclinical testing to determine whether or not there is sufficient evidence for 'efficacy' to support clinical trials. In addition, there is a need to establish standards of performance and experimental designs for clinical trials. It is clear that preclinical testing must be designed to predict the outcome of clinical trials. Therefore, a reasonable first step is to design clinical trials that will provide meaningful results in terms of predicting the utility of a given algorithm, or more precisely, devices incorporating that algorithm, for specific clinical applications. The performance requirements and the conditions under which the device must perform will differ for different clinical applications. Further, experimental designs will also be restricted by the research subject population. For example, if the intent is to test the seizure predictor in patients with primary generalized tonic-clonic seizures, it is likely that the research will require large-scale, long-duration outpatient trials using scalp electrodes. In fact, such studies may not be feasible. On the other hand, some implantable devices will require that research subjects be limited to patients undergoing electrode implantation for presurgical evaluation, thus markedly restricting the available patient population.

Given the clinical limitations, it is essential that experimental design and statistical analysis be well conceived. Further, it would be ideal for the community to reach consensus on standards of experimental design and statistical analysis as well as standards of acceptable performance for specific clinical applications.

One can look to the anticonvulsant drug experience for guidance. Unfortunately, modern standards of clinical research and experimental design were not applied to anticonvulsant drug testing until well after the introduction of many anticonvulsant drugs, some of which are still in use today.

In this chapter, we have not provided answers to the ideal experimental design and approach to statistical analysis for clinical trials of seizure-prediction devices, but have attempted to outline many of the clinical and statistical issues that need to be considered in designing such trials.

References

1 S. S. Viglione and G. O. Walsh, *Electroenceph. Clin. Neurophysiol.* **39**, 435–6 (**1975**).

2 Z. Rogowski, I. Gath and E. Bental, *Biol. Cybern.* **42**, 9–15 (**1981**).

3 A. Siegel, C. L. Grady and A. F. Mirsky, *Epilepsia*, **23**, 47–60 (**1982**).

4 H. H. Lange, J. P. Lieb, J. Engel Jr. and P. H. Crandall, *Electroenceph. Clin. Neurophysiol.* **56**, 543–55 (**1983**).

5 H. G. Wieser, *Epilepsia* **30**, 669 (**1989**).

6 J. Gotman and M. G. Marciani, *Ann. Neurol.* **17**, 597 (**1985**).

7 J. Gotman and D. J. Koffler, *Electroenceph. Clin. Neurophysiol.* **71**, 7–15 (**1989**).

8 A. Katz, D. A. Marks, G. McCarthy and S. S. Spencer, *Electroenceph. Clin. Neurophysiol.* **79**, 153–6 (**1991**).

9 J. G. Milton, A. Longtin, A. Beuter, M. C. Mackey and L. Glass, *J. Theoret. Biol.* **138**, 129–47 (**1989**).

10 J. G. Milton, In: J. G. Milton, P. Jung (Eds), *Epilepsy as a Dynamic Disease*, (Springer, Berlin, 1–14, **2003**).

11 M. C. Mackey and U. an der Heiden, *Funkt. Biol. Med.* **1**, 156–164 (**1982**).

12 G. Nicolis and I. Prigogine, *Self-Organization in Nonequilibrium Systems: From Dissipative Structures to Order through Fluctuations*, (John Wiley & Sons Ltd, **1977**).

13 P. E. Rapp, I. D. Zimmerman, A. M. Albano, G. C. deGuzman, N. N. Greenbaum and T. R. Bashore, In H. G. Othmer (Ed), *Nonlinear Oscillations in Biology and Chemistry*, (Springer, 175–805, **1986**).

14 P. E. Rapp, R. A. Latta and A. I. Mees, *Bull. Math. Biol.* **50**, 227–53 (**1988**).

15 L. D. Iasemidis, J. C. Sackellares, H. P. Zaveri and W. J. Williams, *Brain Topogr.* **2**, 187–201 (**1990**).

16 L. D. Iasemidis, *Ph.D. Dissertation*, University of Michigan (**1991**).

17 S. J. Schiff, K. Jerger, D. H. Duong, T. Chay, M. L. Spano and W. L. Ditto, *Nature* **370**, 615–20 (**1994**).

18 L. D. Iasemidis and J. C. Sackellares, *The Neuroscientist* **2**, 118–26 (**1996**).

19 L. D. Iasemidis and J. C. Sackellares, In D. W. Duke, W. S. Pritchard (Eds), *Measuring Chaos in the Human Brain*, (World Scientific, Singapore, 49–82, **1991**).

20 L. D. Iasemidis, J. C. Principe, J. M. Czaplewski, R. L. Gilmore, S. N. Roper and J. C. Sackellares, In F. Lopes da Silva, J. C. Principe, L. B. Almeida (Eds), *Spatiotemporal Models in Biological and Artificial Systems*, (IOS Press, Amsterdam, The Netherlands, 81–8, **1997**).

21 L. D. Iasemidis, P. M. Pardalos, J. C. Sackellares and D. S. Shiau, *J. Comb. Optim.* **5**, 9–26 (**2001**).

22 M. C. Casdagli, L. D. Iasemidis, J. C. Sackellares, S. N. Roper, R. L. Gilmore and R. S. Savit, *Physica D* **99**, 381–99 (**1996**).

23 M. C. Casdagli, L. D. Iasemidis, R. S. Savit, R. L. Gilmore, S. N. Roper and J. C. Sackellares, *Electroenceph. Clin. Neurophysiol.* **102**, 98–105 (**1997**).

24 C. E. Elger and K. Lehnertz, *Eur. J. Neurosci.* **10**, 786–9 (**1998**).

25 K. Lehnertz and C. E. Elger, *Electroenceph. Clin. Neurophysiol.* **95**, 108–17 (1995).

26 K. Lehnertz and C. E. Elger, *Phys. Rev. Letters.* **80**, 5019–23 (1998).

27 J. Martinerie, C. Adam, M. le Van Quyen, M. Baulac, S. Clemenceau, B. Renault and F. J. Varela, *Nat. Med.* **4**, 1173–6 (1998).

28 M. Le Van Quyen, J. Martinerie, M. Baulac and F. Varela, *NeuroReport* **10**, 2149–55 (1999).

29 M. Le Van Quyen, C. Adam, J. Martinerie, M. Baulac, *et al*, *European J. Neuroscience* **12**, 2124–34 (2000).

30 M. Le Van Quyen, J. Martinerie, V. Navarro, P. Boon, M. DŌHave, C. Adam, B. Renault, F. Varela and M. Baulac, *Lancet* **357**, 183–8 (2001).

31 V. Navarro, J. Martinerie, M. Le Van Quyen, S. Clemenceau, C. Adam, M. Baulac and F. Varela, *Brain* **125**, 640–55 (2002).

32 W. Van Drongelen, S. Nayak, D. M. Frim, M. H. Kohrman, V. L. Towle, H. C. Lee, A. B. McGee, M. S. Chico and K. E. Hecox, *Pediatr. Neurol.* **29**, 207–13 (2003).

33 H. R. Moser, B. Weber, H. G. Wieser and P. F. Meier, *Physica D* **130**, 291–305 (1999).

34 I. Drury, B. Smith, D. Li and R. Savit, *Exp. Neurol.* **184** (Suppl 1), S9–S18 (2003).

35 B. Litt, R. Esteller, J. Echauz, M. DŌAlessandro, *et al.*, *Neuron* **30** (Suppl 1), 51–64 (2001).

36 I. Osorio, M. G. Frei and S. B. Wilkinson, *Epilepsia* **39**, 615–27 (1998).

37 Y. C. Lai, M. A. F. Harrison, M. G. Frei and I. Osorio, *Phys. Rev. Lett.* **91**, 068102 (2003).

38 Y. C. Lai, M. A. F. Harrison, M. G. Frei and I. Osorio, *Chaos* **14**, 630–42 (2004).

39 W. DeClercq, P. Lemmerling, S. Van Huffel and W. Van Paesschen, *Lancet* **361**, 971 (2003).

40 F. Mormann, T. Kreuz, C. Rieke, R. G. Andrzejak, A. Kraskov, P. David, *et al.*, *Clin. Neurophysiol.* **116**, 569–87 (2005).

41 F. Mormann, R. G. Andrzejak, T. Kreuz, *et al.*, *Physical Review E* **67**, 021912-1–10 (2003).

42 L. D. Iasemidis, D. S. Shiau, W. Chaowolitwongse, *et al.*, *IEEE Trans. Biomed. Eng.* **15**, 616–27 (2003).

43 L. D. Iasemidis, D. S. Shiau, P. M. Pardalos, W. Chaovalitwongse, *et al.*, *Clinical Neurophysiol.* **116**, 532–44 (2005).

44 W. Chaovalitwongse, P. M. Pardalos, L. D. Iasemidis, P. R. Carney, D.-S. Shiau and J. C. Sackellares, *Epi. Res.* **64**, 93–113 (2005).

45 J. C. Sackellares, D. S. Shiau, J. C. Principe, M. C. K. Yang, L. K. Dance, W. Suharitdamrong, W. Chaovalitwongse, P. M. Pardalos and L. D. Iasemidis, *J. Clin. Neurophysiol.* **29**, 509–20 (2006).

46 F. Mormann, R. G. Andrzejak, C. E. Elger and K. Lehnertz, *Brain* **130**, 314–33 (2007).

47 R. Andrzejak, F. Mormann, T. Kreuz, *et al.*, *Phys. Rev. E.* **67**, 010901-1–4 (2003).

48 R. Aschenbrenner-Scheibe, T. Maiwald, M. Winterhalder, H. U. Voss, J. Timmer and A. Schulze-Bonhage, *Brain* **126**, 2616–26 (2003).

49 M. Winterhalder, T. Maiwald, H. U. Voss, R. Aschenbrenner-Scheibe, J. Timmer and A. Schulze-Bonhage, *Epilepsy Behav.* **4**, 318–25 (2003).

50 M. C. K. Yang, D. S. Shiau and J. C. Sackellares, In P. M. Pardalos, J. C. Sackellares, P. R. Carney, L. D. Iasemidis (Eds), *Biocomputing II: Quantitative Neuroscience*, (Kluwer Academic Publishers, Dordrecht, The Netherlands, 251–62, 2003).

51 T. Maiwald, M. Winterhalder, R. Aschenbrenner-Scheibe, H. U. Voss, *et al.*, *Physica D* **194**, 357–68 (2004).

52 B. Schelter, M. Winterhalder, T. Maiwald, A. Brandt, A. Schad, A. Schulze-Bonhage and J. Timmer, *Chaos* **16**, 013108 (2006).

53 R. A. Fisher and F. Yates, *Statistical Tables for Biological, Agricultural, and Medical Research*, (Hafner, New York, **1963**).

54 E. S. Pearson and H. O. Hartley, *Biometrika Tables for Statisticians*, (Cambridge University Press, Cambridge, England, **1966**).

20
Considerations on Database Requirements for Seizure Prediction

Carolin Gierschner, Andreas Schulze-Bonhage

20.1
Introduction

The need for extensive EEG databases for seizure prediction has increasingly become apparent to the prediction community since rigorous statistical analyses have been introduced for the assessment of prediction algorithms (e.g., [1, 2]). As epileptic seizures are relatively rare and short events compared to extended seizure-free periods, the importance of having long-term interictal data for an assessment of specificity is evident. Early reports on high sensitivities achieved by algorithms optimized to short preictal periods (e.g., [3–9]) could no longer be upheld when the relative specificity of predictions was simultaneously considered using long-term data.

Progress in the development of computer hardware, particularly the increases in affordable storage capacities, now allow us to store multichannel EEG signals even at high sampling frequencies, recorded continuously over periods of weeks. On the other hand, the number of epilepsy centers performing intracranial EEG (iEEG) recordings is to a relevant extent limited, and increasing economic pressure to restrain recording periods to a minimum in these centers (P. Carney, lecture held at the 3rd International Seizure Prediction Workshop in Freiburg) limits the availability of long-term iEEG data comprising high numbers of seizures distributed over long recording periods.

Currently, only a limited number of research groups have access to adequate EEG data. The few available datasets (e.g., http://epilepsy.uni-freiburg.de/freiburg-seizure-prediction-project/eeg-database) are being used by various research groups from many nations (see Figure 20.1), although they do not provide continuous long-term data and may thus be considered suboptimal, at least for the purpose of developing and analyzing prediction algorithms.

Here, we outline prerequisites for an EEG database which fulfils some minimal requirements for seizure prediction based on current practice at the Freiburg Epilepsy Center.

Seizure Prediction in Epilepsy. Edited by Björn Schelter, Jens Timmer and Andreas Schulze-Bonhage
Copyright © 2008 WILEY-VCH Verlag GmbH & Co. KGaA, Weinheim
ISBN: 978-3-527-40756-9

Fig. 20.1 Currently, 110 research groups from more than 20 countries (shown in black) use the Freiburg EEG database.

20.2
General Requirements for a Prediction Database

As pointed out above, a prediction database should contain biophysical raw data of sufficiently long duration, including a relevant number of events of interest for the purpose of analysis. In addition, the database structure should give access to metadata allowing for the identification of patients undergoing a specified type of telemetry, should provide access to information regarding the quality of EEG data, and should allow the correlation of EEG data with clinically important information.

20.3
Raw Data

Raw data may be an integral part of the database; due to the size of EEG and other data, these raw data may alternatively be stored independently and linked by pointers. For seizure prediction, key raw data include:

- EEG
- ECG
- other physiological parameters (e.g., EOG, EMG, polysomnographic data)
- complementary data relevant to localization of recording sites like 3D MRI and/or CT data sets
- video data on ictal and other events relevant to EEG interpretation.

As indicated above, the continuity of long-term EEG and other physiological data is of critical importance for statistical analyses. Given usual seizure frequencies, a duration of several days to weeks is necessary for statistical evaluation of the sensitivity and specificity of prediction algorithms. In addition, a minimum number of events (clinically manifest seizures; possibly also subclinical electrographic seizure patterns) must be present in the data; a lower limit may be 4–6. For the desirable separation of datasets into training and test data, long-term recordings would have to comprise a higher number of seizures, e.g., more than 10. Seizure clustering may pose specific problems as many algorithms have prediction horizons extending over hours [10, 11], (see Chapter 18), minimal inter-seizure intervals have thus to be defined, which would typically be in the range of 30 minutes to 5 hours for most of the published seizure-prediction algorithms. In the case of more dense seizure clustering, specific solutions have to be defined. Taken together, these requirements on data content are met only by a small subgroup of patients undergoing intracranial EEG recordings.

20.3.1
Annotations to Raw Data

For an assessment of specificity and sensitivity of both detection and prediction algorithms, a valid and complete annotation of ictal events is critical. Presently, this requires the complete visual exploration of the data set by experienced electroencephalographers. Depending on the duration of recordings, this puts a considerable workload onto the evaluation team. In Freiburg, subclinical electrographic patterns are also annotated, as present-day performance limitations of prediction algorithms may be influenced by the occurrence of subclinical EEG patterns which may greatly prevail over manifest seizures in some patients (Figure 20.2). Depending on the specific purposes of evaluation, further analyses like sleep-staging, and statements regarding specific investigations, like electrical mapping, may be of interest.

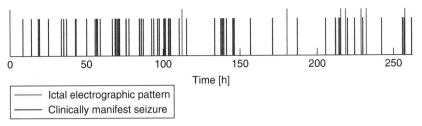

Fig. 20.2 Example of the temporal distribution of subclinical (gray shorter lines) and clinical (black bold lines) seizure patterns of a patient with mesiotemporal seizure origin and right hippocampal sclerosis over a period of 260 hours of continuous recordings. Note that more than 80% of electrographic ictal epileptic patterns were subclinical.

20.4
Metadata on Telemetry

Basic information on data acquisition includes.

- the type of recording, i.e., surface EEG/intracranial EEG using subdural or depth electrodes/simultaneous recordings of both surface and intracranial EEG data (see [5, 6, 12])
- the availability of additional physiological signals, like ECG
- the duration of EEG recordings
- the number of seizures recorded
- types of clinically manifest seizures
- data quality.

Data on recording type and the number of seizures during the recording period are crucial for the applicability of statistical analyses. Data quality needs to be indicated, e.g., by comments on the presence or absence of artifacts, gaps in the continuity of recordings, reduced data quality due to factors like subdural bleedings or other complications of intracranial recordings. Sampling frequencies are of critical importance if high-frequency signals are to be analyzed. In addition, the points in time when seizures occurred and their temporal distribution are important depending on the characteristics of the algorithm to be analyzed; e.g., it is problematic to apply algorithms working with long intervention times and occurrence periods (see [10]) if there is dense clustering of seizures.

There are other metadata which may be relevant but have not yet been analyzed sufficiently. These include the following.

- The localization of recording electrodes: the topographical position of intracranial electrodes may be of relevance for their possible use in seizure prediction both in relation to the brain structures recorded from, and in relation to, the epileptogenic area. Based on distinct anatomical connectivity, this may lead to differences in interdependencies of EEG channels, directionalities of interactions and mutual information exchange.
- The anatomical localization of seizure onset may have related implications.
- The morphology of electrographic seizure patterns may be related to mechanisms in seizure generation which are critical also for interictal–ictal transitions.
- The state of vigilance at seizure occurrence may also be related to the dynamics underlying seizure generation and has been shown to be of importance for the selection of appropriate reference periods for certain prediction algorithms [13].

- The type and level of anti-epileptic medication and the speed of their changes during the recording period may be a relevant factor influencing the stationarity of long-term EEG data and may thus influence the performance of algorithms which critically depend thereon. Integration of daily dosages may serve as a first approximation of active drug levels at binding sites.

20.5
Metadata on the Clinically Defined Epilepsy Syndrome

In addition to data characterizing the recording period and its circumstances, clinical data characterizing the type of epilepsy may be of relevance, e.g., for stratification of patient subgroups for certain types of analyses. These metadata encompass:

- the epilepsy syndrome (e.g., mesiotemporal epilepsy, frontal lobe epilepsy)
- the underlying etiology (e.g., hippocampal sclerosis, cortical dysplasia)
- seizure types and semiology
- type of surgical resection following telemetry
- outcome following surgery.

Etiology and epilepsy syndrome may have relevant implications for the process of epileptogenesis in general and for seizure generation in particular. Thus, algorithms capturing particular properties of EEG dynamics may have different performances depending on the individual aspects of interictal–ictal transitions related to the brain structure involved and on its pathological alteration. Seizure types reflect not only local aspects related to overt symptoms but also the propensity of the brain for various forms of propagation [14] and thus indicate the balance between inhibition and excitation in synaptically connected brain areas.

The surgical outcome serves as the gold standard for the correct identification of the epileptogenic area and thus contributes to the description of electrode positions in relation to seizure generators. This is of importance not only for univariate measures applied either to focal or extrafocal electrode contacts, but also for a multivariate analysis of interactions between the focus and extrafocal areas.

Although the selection of seizure-free patients offers specific advantages regarding the distinction between the epileptogenic zone and other brain areas, the inclusion of patients not achieving seizure freedom following epilepsy surgery and of patients with clear multifocal epilepsy may be of particular importance regarding the future clinical use of seizure-prediction systems, as patients without a surgical treatment option are particularly suitable as candidates for an application of prediction systems.

Further data of possible relevance may include gender, age at the time of recording, age at initial manifestation of epilepsy, the derived duration of

Fast access to clinical data from various sources

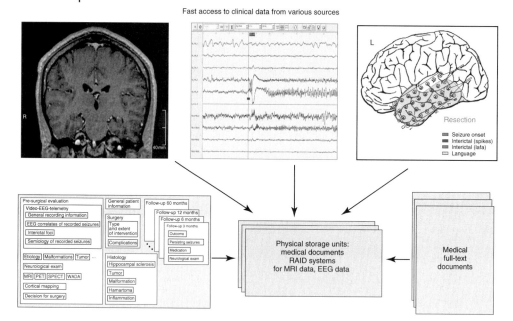

Fig. 20.3 Scheme of data types and relational connectivity in a multi-dimensional database containing raw data and metadata for epileptic seizure prediction. (Please find a color version of this figure on the color plates.)

epilepsy, and results from supplementary investigations such as functional imaging.

20.6
Database Structure

For the purpose of data management and retrieval, a relational database structure is suitable. This allows one to identify datasets based on certain features, e.g., seizure number, distribution, presence or absence of subclinical electrographic patterns. Additionally, it renders possible the stratification of patient subgroups according to the localization of seizure onset, the underlying pathology, the type and topography of electrodes or the postsurgical outcome.

More complex database structures may be useful if advanced types of analyses are to be performed. Thus, in a current European database project, data-driven analyses like semantic mining are planned, based on a data warehouse architecture.

Even major epilepsy centers are generally not able to collect enough data to support subgroup analyses in rigorously preselected patient groups, according to statistical demands like out-of-sample testing of algorithmic performance. Industrial companies thus use extended databases acquired at a number of recording sites. Presently, both in the US and in Europe, there are ongoing projects on extended databases using structures enabling multi-user access and contributions.

References

1 R. Aschenbrenner-Scheibe, T. Maiwald, M. Winterhalder, H. Voss, J. Timmer and A. Schulze-Bonhage, *Brain* **126**, 2616 (**2003**).

2 F. Mormann, T. Kreuz, C. Rieke, R. Andrzejak, A. Kraskov, P. David, C. Elger and K. Lehnertz, *Clin. Neurophysiol.* **116**, 569 (**2005**).

3 J. Martinerie, C. Adam, M. Le van Quyen, M. Baulac, S. Clemenceau, B. Renault and F. Varela, *Nature Medicine* **4**, 1173 (**1998**).

4 I. Osorio, M. Frei and S. Wilkinson, *Epilepsia* **39**, 615 (**1998**).

5 M. Le van Quyen, J. Martinerie, M. Baulac and F. Varela, *Neuroreport* **10**, 2149 (**1999**).

6 M. Le van Quyen, C. Adam, J. Martinerie, M. Baulac, S. Clemenceau and F. Varela, *Eur. J. Neurosci.* **12**, 2124 (**2000**).

7 M. Le van Quyen, J. Martinerie, V. Navarro, P. Boon, M. D'Have, C. Adam, B. Renault, F. Varela and M. Baulac, *Lancet* **357**, 183 (**2001**).

8 L. D. Iasemidis, P. Pardalos, J. C. Sackellares and D. Shiau, *J. Comb. Opt.* **5**, 9 (**2001**).

9 L. D. Iasemidis, D.-S. Shiau, W. Chaovalitwongse, J. C. Sackellares, P. M. Pardalos, J. C. Principe, P. R. Carney, A. Prasad, B. Veeramani and K. Tsakalis, *IEEE Trans. Biomed. Eng.* **50**, 616 (**2003**).

10 M. Winterhalder, T. Maiwald, H. U. Voss, R. Aschenbrenner-Scheibe, J. Timmer and A. Schulze-Bonhage, *Epilepsy Behav.* **4**, 318 (**2003**).

11 F. Mormann, R. G. Andrzejak, C. E. Elger and K. Lehnertz, *Brain* **130**, 314 (**2007**).

12 A. Schad, K. Schindler, B. Schelter, T. Maiwald, A. Brandt, J. Timmer and A. Schulze-Bonhage, *Clin. Neurophysiol.* **119**, 197 (**2008**).

13 B. Schelter, M. Winterhalder, T. Maiwald, A. Brandt, A. Schad, J. Timmer and A. Schulze-Bonhage, *Epilepsia* **47**, 2058 (**2006**).

14 K. Götz-Trabert, C. Hauck, K. Wagner, S. Fauser and A. Schulze-Bonhage, *Epilepsia* (**2008**) April 24 [Epub ahead of print].

21
Beyond Prediction – Focal Cooling and Optical Activation to Terminate Focal Seizures

Steven M. Rothman

21.1
Introduction

When the science of human seizure prediction and detection matures, there will be a plethora of practical questions about how to best implement these powerful techniques. While knowledge of the likelihood of an impending seizure would of course be very significant, indeed empowering, for patients, it would be even more desirable if some reliable method of seizure prevention could be rapidly activated. The known antiepileptic effect of cooling makes local cortical cooling one potential therapeutic strategy. Recent advances in the fabrication of small, electric cooling devices (thermoelectric or Peltier) has raised the possibility of incorporating an implantable cooling unit into a closed loop seizure detection system. Another possible strategy for focal seizure prevention or termination would be to combine a small, powerful ultraviolet light-emitting diode (UV LED) with local administration of a caged version of the inhibitory transmitter γ-aminobutyric acid (GABA). Seizure prediction could activate the LED, releasing GABA to suppress focal paroxysmal activity. Support for both of these strategies exists in a variety of experimental systems; adapting them for clinical use is the next challenge.

21.1.1
Scope of the Problem

The end of the twentieth century produced remarkable advances in the understanding and treatment of many of the epilepsies. Scientific speculation that some genetic epilepsies were caused by mutations in ion and voltage-gated channels was validated by a series of landmark genetic discoveries. New anticonvulsants, several with highly favorable therapeutic indices, were introduced. Sensitive magnetic resonance and positron-emitting based imaging tests that revealed the focal etiology of many complicated, symptomatic epilepsies emerged and evolved very rapidly. Finally, the once radical therapeutic option of surgical resection entered the epilepsy mainstream.

Seizure Prediction in Epilepsy. Edited by Björn Schelter, Jens Timmer and Andreas Schulze-Bonhage
Copyright © 2008 WILEY-VCH Verlag GmbH & Co. KGaA, Weinheim
ISBN: 978-3-527-40756-9

Despite this progress, many patients with epilepsy have been left behind. The focal cortical epilepsies have remained especially problematic. Accounting for up to half of patients with poorly controlled seizures; focal and multifocal seizures arising from the neocortex have proven extremely refractory to both conventional anticonvulsant therapy and the newer surgical approaches. Even with guidance from modern imaging techniques that allow functional anatomic correlation of seizure onset with surface rendering of the neocortex, the surgical treatment of cortical seizures is successful in less than 60% of cases [1]. There are at least three reasons why surgical management of neocortical epilepsy is so difficult. First, conventional neurophysiological techniques frequently have difficulty in localizing a patient's seizures. Second, identification of the specific neurological function(s) residing in the seizure onset zone, essential for an accurate prediction of adverse effects of resection, can be very difficult. Third the expectation that resection will leave a permanent neurological deficit precludes surgery altogether.

21.1.2
Alternatives to Permanent Resection for Neocortical Epilepsy

A variety of invasive but non-destructive strategies have recently been employed or proposed to reduce the frequency and severity of neocortical seizures. For the past 15 years, vagal nerve stimulation (VNS) has been used for a subset of children and adults with refractory epilepsy [2]. While clinical studies have validated its efficacy, the overall reduction in seizure frequency with VNS approximates 50%. This certainly represents an improvement in seizure control, but for many patients the improvement in quality of life is insubstantial. There is some experimental evidence that trigeminal nerve stimulation might provide better seizure control, but as yet, there have been no human studies of this modality. Transcranial magnetic stimulation for epilepsy has been described in several publications. The initial reports were optimistic, but subsequent controlled prospective trials have not shown significant reduction in seizure frequency [3]. There are also brain-slice studies indicating that the direct application of a constant or DC electrical field can diminish neuronal excitability and experimental seizure discharges [4].

There is limited experience with several variations of intermittent electrical stimulation for human epilepsy. Recent publications have identified subsets of patients who benefited from intermittent hippocampal or thalamic stimulation [5,6]. These results are still too premature to allow definite conclusions. The most sophisticated intervention under trial is a totally implantable, closed-loop feedback system for seizure detection and cortical stimulation to terminate focal seizures. Using the same electrodes for both recording and stimulation, in conjunction with a custom designed seizure-detection algorithm, this device has been able to abruptly terminate seizures by delivering a burst stimulation lasting about a tenth of a second [7]. At this time, the reliability of this device is being determined in prospective clinical trials at several North American epilepsy centers.

21.2
Cooling and the Brain

Another appealing, non-destructive strategy for terminating and possibly preventing focal seizures, is the application of focal cooling. There is an extensive literature documenting that cooling reduces synaptic transmission in the mammalian brain, and it should be possible to use new engineering technology to deliver very focal cooling. The first descriptions of the central neurological effects of focal cooling come from articles written by Stefani and Deganello at the end of the nineteenth century [8, 9]. A decade later, the German physiologist Trendelenburg began a systematic study of local hypothermia, investigating its effects on both brainstem autonomic reflexes and the neocortex [10]. While lacking any cellular insights, all three investigators concluded that cooling reduced neurological function at the system level. Throughout the rest of the 20th century, physiologists continued to use local cooling to investigate cortical and subcortical localization of specific brain functions.

Information about the neurobiological effects of cooling advanced over the last fifty years. Using intracellular recording at the frog neuromuscular junction, Katz and Miledi showed that cooling reduced end-plate potentials, likely by diminishing the probability of acetylcholine release from presynaptic terminals [11]. More recent experiments in mammalian tissue culture and brain slices have shown that cooling alters excitatory transmission by both pre- and post-synaptic mechanisms [12]. Cooling can augment the magnitude of neuronal action potentials and inhibits the sodium/potassium ATPase. Although the initial ATPase inhibition and concomitant cell depolarization may elicit transient hyperexcitability, cooling eventually reduces neuronal excitability. Our own recent observations indicate that hypothermia rapidly reduces transmitter release from presynaptic vesicles and suggest that this may be a dominant effect of rapid cooling on central neuronal excitability [13].

The potential clinical utility of cooling for neurological disease has been discussed for half a century. Fay began an extensive investigation of brain cooling in 1938 [14]. He suggested that either systemic hypothermia or local cooling of the brain with cooled fluid circulating through a sealed metal capsule might be an efficacious treatment for head trauma and intractable pain. He also thought reducing brain temperature might inhibit tumor growth and tried intracranial cooling for inoperable gliomas. More recent controlled studies have reported benefits of cooling patients after acute head trauma or asphyxia. Interestingly, the temperature reduction in these recent clinical reports is generally no more than $4\,^{\circ}C$, which would not be expected to have a very large effect on synaptic transmission. This is also a much smaller drop than the cooling required to terminate experimental seizures.

Physicians have been aware of a causal relationship between elevated temperature and seizures since Hippocrates. Moreover, numerous *in vitro* and *in vivo* experimental epilepsy studies have consistently demonstrated that cooling diminishes paroxysmal bursting and can reduce or stop seizure activity. There are encouraging clinical reports suggesting the utility of brain cooling for human seizures as well. A quarter of a century ago, two separate clinical investigations documented the efficacy of cooling in the therapy of a limited number of patients with acute, prolonged seizures or chronic, recurrent seizures. The initial paper showed that externally cooling (31–36 °C) six patients with refractory status epilepticus stopped seizures in five of them [15]. The second paper described 25 patients, ranging in age from 8 to 46, who had frequent major motor seizures poorly controlled on chronic anticonvulsant therapy [16]. While under general anesthesia, they were placed in an enclosed chamber and chilled to 29 °C with cold air. At that point, bilateral frontal burr holes were opened and the subarachnoid space or ventricles were irrigated with iced saline, which reduced cortical surface temperature below 24 °C in 21 of the patients. A total volume of 500–20 000 cc of saline was infused into the patients, who were slowly recovered from anesthesia. Of the 15 patients followed for one year, eleven showed a marked reduction in seizure frequency, including four who had been seizure free.

Two more recent descriptions of direct cooling of the human brain during operative neurosurgical mapping have confirmed that acutely lowering cortical temperature will terminate paroxysmal discharges [17, 18]. In these cases, focal spikes abruptly stopped when iced saline was applied to the neocortex in the operating room. Both sets of studies have encouraged the development of more convenient methods of brain cooling for epilepsy.

21.2.1
Methods for Cooling

The inconvenience of cooling with circulating cold water or conventional refrigeration devices has precluded the application of cooling for the chronic therapy of epilepsy. However, improvements in thermoelectric devices and the necessary supporting technology makes it essential to re-evaluate cooling as a therapy for some forms of epilepsy. Thermoelectric devices exploit Peltier's 1834 observation that a temperature gradient develops at the junction between two dissimilar conductors when an electric current is applied across them. The discovery in the 1920s that synthetic semiconductors were superior to metals as thermoelectric elements, hastened progress in this field.

The development of modern semiconductors, typically alloys of bismuth, tellurium, selenium, and antimony, has made possible the fabrication of small, light thermoelectric modules or Peltier devices, only a few mm in length and width and 1.5 mm thick (Figure 21.1a). In these modules, pairs of N- and P-type semiconductors are connected electrically in series and thermally in parallel, between two ceramic plates, to form a wafer. The newest thermoelectric devices are fabricated with thin-film technology developed for the microelectronics industry and are less than 200 μm thick. They have up to ten times the heat-pumping capability of conventional devices, making them ideal for medical devices (Figure 21.1b).

Thermoelectric devices are capable of generating temperature differentials of 70 °C, but for these differentials to result in cooling, heat must be efficiently

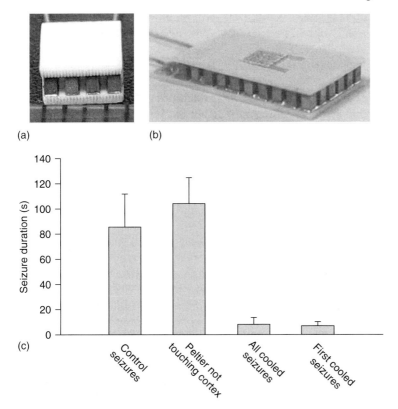

Fig. 21.1 Photographs of thermoelectric or Peltier devices. (a) Commercially available Peltier device showing the individual semiconductors connecting the two ceramic wafers. Scale below is in mm. (b) The figure shows a prototype thin-film device on top of a conventional TEC. (c) Cortical cooling significantly reduces the duration of seizures. Cooling shortened the duration of the entire group of cooled seizures as well as the first cooled seizure in each animal. Activation of the Peltier had no effect on seizure duration when the device did not directly contact the cortex. Cooled seizure durations were significantly different from both control groups ($p < 0.001$).

removed from the hot side. For most of our work, we have attached the hot side of the device to a copper rod, which both removes heat and acts as a convenient holder for a manipulator.

21.2.2
Results of Cooling Experimental Seizures

In 1999, we became aware of the attractive features of commercially available thermoelectric devices and began experiments to determine whether they could terminate acute seizures. Initially, we induced seizures in rat hippocampal brain slices by perfusing them with 4-aminopyridine, a blocker of voltage-gated potassium channels and a well-established convulsant. We positioned the slices on the surface of a thermoelectric device that had been machined into a plexiglass chamber. In this configuration, we were able to rapidly cool the slices at the onset of electrographic ictal-like behavior. We found that cooling to 20 °C quickly terminated this paroxysmal activity and also reversibly inhibited evoked field potentials [19].

These positive *in vitro* results encouraged us to attempt to suppress ictal activity *in vivo*. We found that we could reliably induce focal seizures in halothane-anesthetized rats with a 4-aminopyridine solution. The drug was inserted 0.5 mm below the pial surface of the motor cortex using a micropipette. The animals developed recurrent focal seizures within 20 minutes and continued to have electrographic seizures for approximately 2 hours. Untreated seizures lasted between 60 and 80 seconds. When we allowed a thermoelectric device to directly contact the neocortex immediately above the injection site and activated cooling to 20 °C at seizure onset, seizure duration was reduced to approximately 7 seconds (Figure 21.1c) [20].

This rapid effect is due to the local cooling, because the thermoelectric device did not influence seizure duration if it was not in direct contact with the cortical surface. In this preparation, there was a progressive decrease in seizure duration below 26 °C, but no difference between 22 and 20 °C [21]. We have not yet tried temperatures below 20 °C. We may even be overestimating the degree of cooling required for seizure termination. Another group has reported that paroxysmal discharges in a slice model of epilepsy were eliminated by rapid temperature reductions of only 1 °C [22].

We went on to develop a frequency-based seizure detection algorithm for the 4-aminopyridine seizures, which allowed us to use a closed-loop system to abort seizures. In order to minimize false positives, we set a relatively high threshold before cooling was activated. In these circumstances, our closed-loop system was as effective in terminating seizures as manual activation. Had we decided to accept more false positives, we could have initiated more rapid cooling and further reduced seizure duration [21].

Additional experiments have revealed other attractive aspects of cooling for focal seizures. Using a small thermocouple inserted into a 30 g needle, we mapped the cortex below our thermoelectric device and showed that cooling extends only about 4 mm below the surface. Thus, its effect should be localized to just a small region of neocortex below the pia. Recent work has explored the possibility of brain damage induced by local cooling. Thus far, we have not seen any evidence that cooling as low as 5 °C for 2 hours produces neuronal loss or activates apoptotic pathways [23]. While there is minimal gliosis close to the region of cortical contact with the Peltier, it is similar to the response provoked by any foreign body and cannot be attributed to cooling. Interestingly, when brain slices obtained from transgenic mice expressing the Green Fluorescent Protein were cooled to 5 °C, there was transient, but completely reversible blebbing of dendrite shafts and loss of spines. This is probably an effect of ion pump inhibition, because other investigators have made similar observations after treating brain slices with Na^+/K^+ ATPase inhibitors. We have also failed to observe any neuronal loss in cat neocortex, after 7–10 months of intermittent cooling during neurophysiological experiments.

These results make us optimistic that the effect of cooling will be exquisitely focal and well tolerated.

A series of more recent experiments demonstrated that cooling is capable of reducing the severity and length of other types of focal seizures. Both the grade and after discharge duration of kindled rat hippocampal seizures were reduced by cooling to between 23 and 26 °C with cold saline flowing through thin copper tubing positioned next to the dorsal hippocampus [24]. A limited set of observations on a monkey have verified that our thermoelectric device is capable of inhibiting function in the primate neocortex. When our thermoelectric device was placed in direct contact with the pia on the surface of the hand region of the precentral gyrus, there was a consistent, reversible impairment in skilled finger movements associated with a reduction in surface temperature. This indicates that the small thermoelectric devices have sufficient power to cool primate cortex. Of note, even with cooling to 10 °C, the impairment was not instantaneous and did not completely paralyze hand movements.

In order to move ahead with the design of a fully implantable cooling device for human focal epilepsy, we have begun to address the problem of heat dissipation from the hot side of thermoelectric devices. While a copper rod has worked well for experiments in anesthetized or immobilized animals, a more compact device will be required to tranfer heat from a clinical device. There is an extensive literature describing heat pipes, which are hollow, evacuated, wicked tubes that rapidly equilibrate temperature by allowing a liquid to alternately evaporate and recondense under reduced atmospheric pressure. Our engineering colleagues have designed a thin, bendable, laminar heat pipe composed of outer layers of copper foil and an inner layer of sintered copper columns sandwiched between two layers of sintered copper [25]. The charging fluid, water, flows between the copper columns. The mechanical properties of this heat pipe indicate that it should be capable of diffusing sufficient heat from a thermoelectric device in contact with the cortex to maintain a cold-side temperature of 20 °C without heating the adjacent brain above 38 °C. We envision positioning a similar heat pipe between the hot side of a thermoelectric device and dura, skull, or scalp, so that heat can be efficiently transferred to one of these highly vascular compartments.

21.2.3
Future Plans for Cooling

The most critical question at this time is the degree of cooling required to terminate or prevent human focal seizures; and rodents are unlikely to inform us about this question. There is no rodent model of chronic focal epilepsy that reliably reproduces the human condition, and the 4-AP model, which we have used in our studies, is much more severe than even the most refractory human epilepsy. The rats typically have 60–80 s seizures every 2–3 min, which would be unrealistic for almost all human epilepsies. Human seizures would be expected to be much less frequent. It seems likely, therefore, that rodent focal seizures will require a much larger temperature reduction than human seizures. This information is critical because the required degree of cooling will govern the current required to power the thermoelectric device. Moreover, the degree of cooling will determine the amount of heat that has to be dissipated through the heat pipe to the vasculature.

We believe that the most efficient way to sort this out will entail fabricating a cooling device that utilizes cold saline rather than a thermoelectric device. The device would combine a conventional silastic grid used for invasive monitoring and a bladder through which cold saline flows. After preliminary testing in a large animal to verify safety and cooling ability, the device could be temporarily implanted in focal seizure patients during monitoring, prior to surgical resection. It would permit determination of the degree of cooling required to stop seizures without having to cope with heat dissipation. Once the cooling parameters have been better defined, it will be possible to rationally design an implantable device using thermoelectric technology.

There are several clear obstacles and objections that need to be addressed before focal cooling can be used to treat human epilepsy. First, the convolutions of the human brain keep approximately two-thirds of the neocortex buried in sulci and inaccessible to surface cooling. However, it should still be possible to cool portions of the substantial area of exposed cortex that are responsible for seizure generation. In addition, cooling the lip of cortical sulci may block the spread of seizures arising within the sulci, a suspected mechanism of seizure generalization. It should even be feasible to connect thin, flexible heat pipes to thermoelectric devices to cool structures deeper in the brain.

Second, cooling may be as disruptive to normal brain function as some seizures. We are optimistic that there may be a temperature range that separates seizure control from disruption of normal cortical function. The results of Bakken and colleagues, that cooling below $10\,°C$ over eloquent cortex did not produce a complete anomia or non-fluent aphasia, support this hope [26].

Third, while we are concerned that the current requirements for present commercial thermoelectric devices exceed an ampere, the short-duty cycle may allow these devices to operate for long periods without recharge or replacement. Moreover, improvements in battery life and thin-film thermoelectric design, should soon make it possible to generate more efficient devices.

Fourth, while there is presently no fully validated algorithm for seizure detection or prediction, other sections of this monograph suggest that these remain attainable goals. Variations of these algorithms are currently undergoing trials in devices using either an external or implanted detection system linked to electrical stimulation. There is nothing about these systems that would preclude replacing stimulation with cooling as the efferent arm of the device.

Full implementation of cooling as a clinical therapy will still require further research. However, given the steady neurobiological progress over the last five years and the continuous progress in many phases of engineering, it would be surprising if some investigational application of cooling for epilepsy management was not available by the end of the decade.

21.3
Focal Uncaging for Epilepsy

We have also begun investigating the potential of small, optical devices to suppress the abnormal neuronal activity that is the concomitant of focal epilepsy. There is already a precedent for thinking along these lines. A California laboratory has modified the genes for potassium and glutamate-gated channels to confer light sensitivity to neurons [27]. However, in order for light to modify electrical activity, the altered gene has to be transfected into neurons. Nonetheless, investigators are already suggesting the possible clinical utility of this technology and it is being energetically investigated [28].

Rather than introduce new channels into neurons, we are exploring the possibility of using caged compounds to modulate neuronal activity [29]. This decision was based upon three recent developments: 1) the availability of at least one new caged GABA; 2) new information about tonic, $GABA_A$ receptors; and 3) new developments in UV LED technology. The new caged GABA is 4-[[(2H-1-benzopyran-2-one-7-amino-4-methoxy)carbonyl]amino] butanoic acid or BC204 [30]. This compound releases GABA when activated by relatively long wavelength ultraviolet radiation, 350 nm. Unlike other commercially distributed caged neurotransmitters, it is extremely stable in physiological preparations and lacks direct effects on $GABA_A$ receptors. When tested in hippocampal slices, it rapidly generated GABA when activated by millisecond flashes of a high intensity arc lamp.

The partial elucidation of tonic $GABA_A$ receptors has altered our understanding of chemical inhibition in various parts of the brain. These receptors are located away from the synaptic cleft and appear to be activated by GABA that spills over from synapses or is present in the brain extracellular space at very low concentration [31, 32]. They are unique in containing the $\alpha5$ or $\delta GABA_A$ receptor subunits, possessing high affinity for GABA, and lacking desensitization. This is relevant because the ambient concentration of GABA is in the low μM range, in contrast to synaptic GABA levels that approach mM concentration. Furthermore, the tonic GABA current is larger in animal epilepsy models and probably contributes more to inhibitory current flux than the phasic GABA current.

Advances in UV LED technology may allow us to uncage BC204 or analogs to activate these tonic $GABA_A$ receptors [33]. LEDs are semiconductor optoelectronic devices that emit incoherent narrow-spectrum light when a voltage is placed across them. Over the past two decades, semiconductor crystal-growing methods utilizing high-purity metallo-organic compounds have facilitated the fabrication of UV and visible LEDs. Nichia, the Japanese company with the most successful technique for LED crystal fabrication, has recently introduced highly efficient LEDs, capable of emitting light with wavelengths in the near-ultraviolet range. The quantum efficiency of the newer UV LEDs (the ratio of the number of photons of light emitted per second to the number of electrons injected per second into the LED), is around 30%, far higher than prior crystals. These new devices will significantly reduce heat production and should enable the development of implantable LEDs.

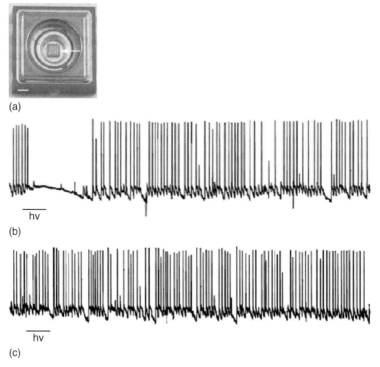

(a)

(b)

hv

(c)

hv

Fig. 21.2 Uncaging as a possible strategy for focal epilepsy. (a) Ultraviolet light emitting diode (UV LED) in 'packaging'. The actual diode is the central square crystal. Calibration line is 1 mm. (b) UV illumination for 4 s in the presence of BC204 (hv bar) stops rapid firing of cultured neurons made hyperexcitable by removal of extracellular magnesium. (c) Under otherwise identical circumstances, illumination had no effect when BC204 was not present in solution.

21.3.1
Early Results with Uncaging

Our initial experiments were designed to determine whether a Nichia UV LED (Figure 21.2a) would provide sufficient radiant power to uncage a detectable level of GABA from BC204 in dissociated hippocampal neurons. Neurons were voltage clamped at -60 mV in the presence of $30\,\mu M$ BC204 and illuminated with the power supply set to 200 mA for durations ranging between 1 and 15 seconds. This level of illumination, applied for as little as 1 s, led to a readily detectable current. When the exposure was lengthened, the amplitude of the current increased. A 4 s illumination generated a near-maximal current and was used in most of the subsequent experiments.

Having established that the UV LED could activate detectable currents in the presence of BC204, subsequent experiments examined GABA-mediated concentration response relations. We bath-applied GABA at 1, 4, 7 and $10\,\mu M$, to obtain a GABA response curve for *low* concentrations of GABA, the range of interest for these experiments. We then bath-applied BC204, first at 30 μM, and in a subsequent set of experiments at 10 μM, and measured the inward currents produced by 4 s illumination with decrementing UV LED power. The inward currents at each level of illumination and BC204 concentration were normalized to the current produced in the same cell by bath perfusion of GABA (10 μM). There was a clear relationship between the strength of illumination and the magnitude of the current elicited by uncaging BC204. When the BC204 concentration was 30 μM, LED currents ≥ 150 mA gave a response larger than the response produced by bath perfusion of 10 μM GABA, suggesting that a 4 s light application was capable of uncaging more than one-third of the BC204 in the vicinity of the recorded neuron. No current response was seen in any neuron in the absence of BC204. Combining the GABA and BC204 concentration responses showed that, in the presence of 30 μM BC204, a 25 mA LED current evoked neuronal currents corresponding to about 4 μM GABA.

Picrotoxin reduced the BC-204 induced inward currents by over 90% ($p = 0.013$ by paired t-test), establishing that GABA was the responsible ligand. Because caged compounds can act as antagonists at the targeted receptor [34, 35], responses of individual neurons ($n = 13$) to GABA (3 μM) in the presence and absence of BC204 (30 μM), were compared to determine if BC204 antagonized GABA in the absence of illumination. We did not observe any measurable antagonism of GABA by BC204, with the latter present in ten-fold excess ($p = 0.126$ by paired t-test).

21.3.2
Uncaging BC204 Suppresses 'Seizure-like' Activity

These preliminary experiments verified that our UV LED could induce robust GABAergic currents, even with only 10 μM BC204. There are a large number of reports of spontaneous paroxysmal, 'seizure-like' activity in cultures of central neurons, and this activity can be amplified if cultures are exposed to an extracellular solution lacking magnesium [36]. [37] Therefore, subsequent experiments were focused on examining the effects of uncaging BC204 on 'seizure-like' activity which appears when magnesium is removed from the perfusate.

When recording from cultured hippocampal neurons in the presence of extracellular calcium (2 mM) and magnesium (1 mM), there is often a low level of spontaneous activity. When the magnesium is removed, there is a dramatic increase in cell firing (Figure 21.2b,c). In some neurons, the firing is tonic and rapid enough to depolarize the neuron, while other cells develop phasic bursting, with brief periods when the cell returns to resting potential. Occasionally, firing rates are unaffected, but large increases in synaptic activity that destabilize the resting potential are observed. This complex behavior has been recognized previously by others and has been the focus of recent quantitative analyses [36, 38]. The spontaneous activity does not necessarily revert back to control levels when the magnesium concentration is normalized, because prolonged excessive firing can induce a permanent state of hyperexcitability in culture [39, 40].

When BC204 was added to the zero magnesium perfusate, there was no difference in the pattern of neuronal activity. However, when the UV LED was activated, there was a reduction or cessation of firing and the cell membrane potential stabilized (Figure 21.2b). The reduction in firing lasted for 10–20 s, likely reflecting the time required for the uncaged GABA to diffuse away from the recording site. Illumination in the absence of BC204 failed to affect the firing rate (Figure 21.2c). Similarly, if exogenous GABA (3–10 μM) was added while the culture was bathed in zero magnesium, there was a marked reduction in firing and synaptic activity (not shown).

Although the effects of activated BC204 or GABA on seizure-like activity were qualitatively obvious, we developed a method to quantify our observations. We allowed 1–2 minutes for solution changes, and, in most experiments, illuminated for 4 s, just at the onset of a 60 s recording period. We examined the 10 s epoch that included the last 2 s of BC204 illumination plus the following 8 s and measured the number of action potentials and the standard deviation of the baseline. We measured the same parameters in the final 10 s of each epoch, which allowed 40 s for any effect of uncaging to dissipate. When the same manipulation was repeated for a single cell, the measurements were averaged. When we examined 156 randomly selected epochs, we found that the two measures correlated ($r^2 = 0.66$).

When cultures lacking BC204 were illuminated with the UV LED, in standard or zero magnesium extracellular solution, there was no alteration in the firing pattern or baseline standard deviation. However, when we exposed the cultures to zero magnesium extracellular solution containing 30 μM BC204, illumination produced a marked decrease in the standard deviation of the baseline and the number of action potentials. This effect was evident with LED currents from 100–200 mA. When we reduced the BC204 concentration to 10 μM, but not lower, we found an almost identical result. In a subsequent set of experiments, we further reduced the current powering the UV LED to 25–100 mA and prolonged illumination to 8 s. We found that 100 and 50, but not 25 mA, significantly reduced both spike number and baseline standard deviation ($p < 0.05$ for spike number and baseline standard deviation with 50 and 100 mA v non illuminated).

When cultures were exposed to picrotoxin in the presence of BC204 (30 μM) and the absence of magnesium, 200 mA of UV LED current had no effect on spikes or baseline standard deviation ($n = 5$ cells, data not shown). Direct addition of GABA to zero magnesium perfusate suppressed paroxysmal activity. At 10 μM GABA, both baseline standard deviation and spike number were significantly reduced ($p < 0.02$ v zero magnesium for both parameters; $n = 9$ cells). At 3 μM, only the spike number reduction was significant ($p < 0.01$). There was no detectable effect of 1 μM GABA on either parameter.

21.3.3
Future Plans for *in vivo* Uncaging

In order to adapt the Nichia UV LEDs for *in vivo* use, they will be reconfigured and incorporated into *in situ* arrays tailored for the particular spatial requirements of the mammalian brain. Our ultimate goal is to combine a UV LED and BC204 to generate low levels of ambient GABA to activate tonic $GABA_A$ receptors in the cortex. It should be possible to implant a single LED or LED array close to the cortical surface above an epileptic focus. BC204 would then be locally applied into the subdural or subarachnoid space beneath the LED and allowed to passively penetrate into the cortex [41]. The LED could be powered at set intervals to continually suppress electrical activity in the underlying cortex. Alternately, it could be controlled by closed-loop feedback and powered only at time of seizure prediction, as described in other chapters in this volume.

We recognize that there are several practical issues that need to be addressed before this technology can be translated to clinical medicine. First, we need to optimize the configuration of the UV LED so that the maximum optical power is directed into the brain slice or cortex. There are ways to distribute several small UV LEDs over the intact cortical surface so that several cm^2 of tissue can be illuminated. Second, we need to establish that the UV LED has sufficient power to penetrate into the intact nervous system and uncage BC204. Based upon preliminary data, we anticipate that the light intensity will decrease by about 50% for every 200 μm of tissue thickness. We will need to determine the appropriate combination of light intensity and BC204 delivery that generates enough GABA to reduce paroxysmal activity.

21.4
Acknowledgments

The experimental work that formed the basis of this entry was supported by the Alafi Family Foundation and the NIH (R01 NS42936).

References

1 T. J. O'Brien, E. L. So, B. P. Mullan, G. D. Cascino, M. F. Hauser, B. H. Brinkmann, F. W. Sharbrough and F. B. Meyer, *Neurology* **55**, 1668 (**2000**).

2 K. O. Nakken, O. Henriksen, G. K. Røste and R. Lossius, *Seizure* **12**, 37 (**2003**).

3 W. H. Theodore, K. Hunter, R. Chen, F. Vega-Bermudez, B. Boroojerdi, P. Reeves-Tyer, K. Werhahn, K. R. Kelley and L. Cohen, *Neurology* **59**, 560 (**2002**).

4 B. J. Gluckman, E. J. Neel, T. I. Netoff, W. L. Ditto, M. L. Spano and S. J. Schiff, *J. Neurophysiol.* **76**, 4202 (**1996**).

5 J. F. Tellez-Zenteno, R. S. McLachlan, A. Parrent, C. S. Kubu and S. Wiebe, *Neurology* **66**, 1490 (**2006**).

6 R. P. Lesser and W. H. Theodore, *Neurology* **66**, 1468 (**2006**).

7 K. N. Fountas, J. R. Smith, A. M. Murro, J. Politsky, Y. D. Park and P. D. Jenkins, *Stereotact. Funct. Neurosurg.* **83**, 153 (**2005**).

8 A. Stefani, *Arch. Ital. Biol.* **24**, 424 (**1895**).

9 U. Deganello, *Arch. Ital. Biol.* **33**, 186 (**1900**).

10 V. Brooks, *Rev. Physiol. Biochem. Pharmacol.* **95**, 1109 (**1983**).

11 B. Katz and R. Miledi, *J. Physiol.* **181**, 656 (**1965**).

12 A. J. Trevelyan and J. Jack, *J. Physiol.* **539**, 623 (**2002**).

13 X. Yang, Y. Ouyang, B. Kennedy and S. Rothman, *J. Physiol.* **567**, 215 (**2005**).

14 T. Fay, *J. Neurosurg.* **16**, 239 (**1959**).

15 E. F. Vastola, R. Homan and A. Rosen, *Arch. Neurol.* **20**, 430 (**1969**).

16 K. Sourek and V. Travnicek, *J. Neurosurg.* **33**, 253 (**1970**).

17 C. J. Sartorius and M. S. Berger, *J. Neurosurg.* **88**, 349 (**1998**).

18 K. M. Karkar, P. A. Garcia, L. M. Bateman, M. D. Smyth, N. M. Barbaro and M. Berger, *Epilepsia* **43**, 932 (**2002**).

19 M. W. Hill, M. Wong, A. Amarakone and S. M. Rothman, *Epilepsia* **41**, 1241 (**2000**).

20 X. F. Yang and S. M. Rothman, *Ann. Neurol.* **49**, 721 (**2001**).

21 X.-F. Yang, D. W. Duffy, R. E. Morley and S. M. Rothman, *Epilepsia* **43**, 240 (**2002**).

22 G. K. Motamedi, P. Salazar, E. L. Smith, R. P. Lesser, W. R. S. Webber, P. I. Ortinski, S. Vicini and M. A. Rogawski, *Epilepsy Res.* **70**, 200 (**2006**).

23 X.-F. Yang, B. R. Kennedy, S. G. Lomber, R. E. Schmidt and S. M. Rothman, *Neurobiol. Dis.* **23**, 637 (**2006**).

24 J. M. Burton, G. A. Peebles, D. K. Binder, S. M. Rothman and M. D. Smyth, *Epilepsia* **46**, 1881 (**2005**).

25 J. K. Hilderbrand, G. P. Peterson and S. Rothman, *Heat Transfer Engineering*, **28**, 282 (**2007**).

26 H. E. Bakken, H. Kawasaki, H. Oya, J. D. W. Greenlee and M. A. Howard, *J. Neurosurg.* **99**, 604 (**2003**).

27 M. Volgraf, P. Gorostiza, R. Numano, R. H. Kramer, E. Y. Isacoff and D. Trauner, *Nat. Chem. Biol.* **2**, 47 (**2006**).

28 G. Miller, *Science* **314**, 1674 (**2006**).

29 H. A. Lester and J. M. Nerbonne, *Ann. Rev. Biophys. Bioeng.* **11**, 151 (**1982**).

30 B. Cürten, P. H. M. Kullmann, M. E. Bier, K. Kandler and B. F. Schmidt, *Photochem. Photobiol.* **81**, 641 (**2005**).

31 A. Scimemi, A. Semyanov, G. Sperk, D. M. Kullmann and M. C. Walker, *J. Neurosci.* **25**, 10016 (**2005**).

32 A. Semyanov, M. C. Walker, D. M. Kullmann and R. A. Silver, *Trends Neurosci.* **27**, 262 (**2004**).

33 E. F. Schubert and J. K. Kim, *Science* **308**, 1274 (**2005**).

34 W. Maier, J. E. T. Corrie, G. Papageorgiou, B. Laube and C. Grewer, *J. Neurosci. Methods* **142**, 1 (**2005**).

35 M. Canepari, L. Nelson, G. Papageorgiou, J. E. Corrie and D. Ogden, *J. Neurosci. Methods* **112**, 29 (**2001**).

36 D. A. Wagenaar, R. Madhavan, J. Pine and S. M. Potter, *J. Neurosci.* **25**, 680 (**2005**).

37 J. W. Gibbs, M. D. Shumate and D. A. Coulter, *J. Neurophysiol.* **77**, 1924 (**1997**).

38 D. A. Wagenaar, J. Pine and S. M. Potter, *BMC Neurosci.* **7**, 11 (**2006**).

39 H. P. Goodkin, J.-L. Yeh and J. Kapur, *J. Neurosci.* **25**, 5511 (**2005**).

40 J. W. Gibbs, S. Sombati, R. J. DeLorenzo and D. A. Coulter, *J. Neurophysiol.* **77**, 2139 (**1997**).

41 N. Ludvig, R. I. Kuzniecky, S. L. Baptiste, J. E. John, H. von Gizycki, W. K. Doyle and O. Devinsky, *Epilepsia* **47**, 1792 (**2006**).

22
Vagus Nerve and Hippocampal Stimulation for Refractory Epilepsy

Paul Boon, Veerle De Herdt, Annelies Van Dycke, Tine Wyckhuys, Liesbeth Waterschoot, Riem El Tahry, Dirk Van Roost, Robrecht Raedt, Wytse Wadman, Kristl Vonck

22.1
Introduction

Neurostimulation is an emerging treatment for neurological diseases. Electrical pulses are administered directly to or in the neighborhood of nervous tissue in order to manipulate a pathological substrate and to achieve a symptomatic or even curative therapeutic effect. Different types of neurostimulation exist mainly depending of the part of the nervous system that is being affected and the way in which this stimulation is being administered.

Electrical stimulation of the tenth cranial nerve or *vagus nerve stimulation (VNS)* is an extracranial form of stimulation that was developed in the eighties and is currently routinely available in epilepsy centers around the world. Through an implanted device and electrode, electrical pulses are administered to the afferent fibers of the left vagus nerve in the neck. It is indicated in patients with refractory epilepsy who are unsuitable candidates for epilepsy surgery or who have had insufficient benefit from such a treatment [1]. Another form of extracranial neurostimulation consists of *transcranial magnetic stimulation (TMS)*. A coil that transmits magnetic fields is held over the scalp and allows a non-invasive evaluation of separate excitatory and inhibitory functions of the cerebral cortex. In addition, repetitive TMS (rTMS) can modulate the excitability of cortical networks [2]. This therapeutic form of TMS is currently being investigated as a treatment option for refractory epilepsy but it has not been widely used, unlike VNS.

Intracerebral neurostimulation requires accessing the intracranial nervous system as stimulation electrodes are inserted into intracerebral targets for *deep brain stimulation (DBS)* or placed over the cortical convexity for *cortical stimulation (CS)*. These modalities of neurostimulation are not novel for neurological indications. Some have been extensively used, e.g., for movement disorders and pain [3, 4]. Moreover, several new indications such as obsessive compulsive behavior and cluster headache are being investigated with promising results [5, 6]. In the past DBS and CS of different brain structures such as the cerebellum, the locus

Seizure Prediction in Epilepsy. Edited by Björn Schelter, Jens Timmer and Andreas Schulze-Bonhage
Copyright © 2008 WILEY-VCH Verlag GmbH & Co. KGaA, Weinheim
ISBN: 978-3-527-40756-9

coeruleus and the thalamus have already been performed. This was done mostly in patients with spasticity or psychiatric disorders who also had epilepsy but the technique was not fully explored or developed into an efficacious treatment option [7–10]. The vast progress in biotechnology along with the experience in other neurological diseases in the past ten years has led to a renewed interest in intracerebral stimulation for epilepsy. A few epilepsy centers around the world have recently re-initiated trials with deep-brain stimulation in different intracerebral structures such as the thalamus, the subthalamic nucleus and the caudate nucleus [11–15].

This manuscript will focus on vagus nerve and hippocampal stimulation as a treatment for epilepsy. For both treatment modalities unresolved clinical and basic research questions require further attention.

The precise mechanism of action by which VNS exerts its antiepileptic effect is unknown. Optimal candidates for VNS have not been identified. About one-third of patients undergoing VNS will eventually not benefit from the intervention. Increased insight into the mechanism of action may help to identify responders and increase clinical efficacy. The reverse reasoning holds true as well. Identification of predictive factors for a positive clinical outcome may further elucidate the mechanism of action of VNS.

DBS is a more recently explored field in epilepsy. Compared to VNS it is a more invasive treatment option. As for VNS, the precise mechanism of action and the ideal candidates for this treatment option are currently unidentified. Moreover, it is unknown which intracerebral structures should be targeted to achieve optimal clinical efficacy. Two major strategies can be distinguished. One approach is to target crucial central nervous system structures that are considered to have a 'pacemaker', 'triggering' or 'gating' role in the epileptogenic networks that have been identified, such as the thalamus or the subthalamic nucleus [16]. Another approach is to interfere with the ictal onset zone itself. This implies the identification of the ictal onset zone, a process that sometimes requires implantation with intracranial electrodes [17]. At Ghent University Hospital this latter approach was chosen.

For both VNS and DBS there is an essential and specific issue related to the fact that these treatments require the use of electronic devices. Electrical pulses are defined by different characteristics and the way they are applied to human tissue makes use of certain stimulation parameters (output current, frequency, pulse width, duty cycle) that have to be decided upon. These parameters are mainly dependent on what is known to be safe and, secondly, on what is believed to be potentially efficacious. For VNS so-called 'optimal'/'standard' stimulation parameters have been identified, mainly on the basis of animal experiments [18]. The fairly large number of non-responders is a possible reflection of the fact that different stimulation paradigms using different combinations of stimulation parameters may be superior.

For DBS very little information on efficacious stimulation parameters is available from human or animal epilepsy experiments. From the large number of studies in patients with movement disorders, ranges of safe stimulation parameters have been identified [19]. Potentially efficacious stimulation parameters were deducted from previous DBS studies in other intracerebral targets, and applied in the pilot trial in patients [7–15].

22.2
Vagus Nerve Stimulation

22.2.1
Clinical Efficacy and Safety

22.2.1.1   Randomised Controlled Trials

The first descriptions of the implantable VNS Therapy™ system for electrical stimulation of the vagus nerve in humans appeared in the literature in the early nineties [20, 21].

At the same time, initial results from two single-blinded pilot clinical trials (phase-1 trials EO1 and EO2) in a small group of patients with refractory complex partial seizures who had been implanted since November 1988 in three epilepsy centers in the USA, were reported [21–25]. In 9/14 patients, treated for 3–22 months, a reduction in seizure frequency of at least 50 % was observed [21]. One of the patients was seizure-free for more than seven months. Complex partial seizures, simple partial seizures as well as secondary generalized seizures, were affected [23]. It was noticed that a reduction in frequency, duration and intensity of seizures lagged 4–8 weeks after the initiation of treatment [22].

In 1993, Uthman et al. reported on the long-term results from the EO1 and EO2 study [26]. Fourteen patients had now been treated for 14–35 months. There was a mean reduction in seizure frequency of 46 %. Five patients had a seizure reduction of at least 50 %, of whom two experienced long-term seizure freedom. In none of the patients did VNS induce seizure exacerbation. In the meantime, two prospective multicenter ($n = 17$) double-blind randomized studies (EO3 and EO5) were started, including patients from centers in the USA ($n = 12$), Canada ($n = 1$) as well as in Europe ($n = 4$) [27–31]. In these two studies, patients over the age of 12 with partial seizures were randomized to a HIGH or LOW stimulation paradigm. The parameters in the HIGH stimulation group (output: gradual increase up to 3.5 mA, 30 Hz, 500 µs, 30 s on, 5 min off) were those believed to be efficacious, based on animal data and the initial human pilot studies. Because patients can sense stimulation, the LOW stimulation parameters (output: single increase to point of patient perception, no further increase, 1 Hz, 130 µs, 30 s on, 3 hours off) were chosen to provide some sensation to the patient in order to protect the blinding of the study. LOW stimulation parameters were believed to be less efficacious and the patients in this group represented an active control group. The results of EO3 in 113 patients were promising with a decrease in seizures of 24 % in the HIGH stimulation group versus 6 % in the LOW stimulation group after three months of treatment [28–30]. The number of patients was insufficient to achieve Food and Drug Administration (FDA) approval leading to the EO5 study in the USA including 198 patients. 94 patients in the HIGH stimulation group had a 28 % decrease in seizure frequency versus 15 % in patients in the LOW stimulation group [31].

22.2.1.2 Clinical Trials with Long-term Follow-up

The controlled EO3 and EO5 studies had their primary efficacy end-point after 12 weeks of VNS. Patients who ended the controlled trials were offered enrolment in a long-term (1–3 years of FU) prospective efficacy and safety study. Patients belonging to the LOW stimulation groups were crossed-over to HIGH stimulation parameters. In all published reports on these long-term results, increased efficacy with longer treatment was found [32–36]. In these open extension trials the mean reduction in seizure frequency increased up to 35 % at one year and up to 44 % at two years of FU. After that, improved seizure control reached a plateau [35].

In the following years, other large prospective clinical trials were conducted in different epilepsy centers worldwide. In Sweden, long-term follow-up in the largest patient series ($n = 67$) in one center not belonging to the sponsored clinical trials at that time, reported similar efficacy rates with a mean decrease in seizure frequency of 44 % in patients treated up to five years [37]. A joint study of two epilepsy centers in Belgium and the USA included 118 patients with a minimum follow-up duration of six months. They found a mean reduction in monthly seizure frequency of 55 % [38]. In China a mean seizure reduction of 40 % was found in 13 patients after 18 months of VNS [39].

An open-label retrospective study evaluated long-term outcome in seven different epilepsy centers in Belgium. 138 patients with a follow-up of at least 12 months had an overall reduction in mean monthly seizure frequency of 51 % and a responder rate of 59 % [40].

At Ghent University Hospital successful treatment of status epilepticus (SE) with VNS was reported [41]. A seven-year-old girl presented with a refractory non-convulsive SE. A vagus nerve stimulator was placed after 11 days of thiopental-induced coma. Electroencephalography showed normalization one week following the start of VNS and she experienced a sustained seizure-free outcome after a follow-up of >13 months. A few case-reports describing the use of VNS for refractory SE in pediatric and adult patients are available in the literature. Malik et al. reported on three children with pharmacoresistant SE who were successfully treated with VNS [42]. It was not specified whether the status was convulsive or non-convulsive in these patients. Winston et al. reported a case of a 13-year old boy in whom VNS interrupted a convulsive SE immediately after stimulation was started [43]. Patwardhan et al. described a case of a 30-year old man with medically intractable seizures due to severe allergic reactions to multiple AEDs with subsequent evolvement into refractory SE. He underwent VNS treatment after nearly nine days of barbiturate-induced coma. Stimulation was initiated in the operating room. In the following days EEG revealed resolution of previously observed periodic lateralized epileptiform discharges with the stimulator programmed at 1 mA and a duty cycle of 30 s on and 3 min off. The patient became seizure-free [44]. Zimmerman et al. reported on three adult patients in whom refractory non-convulsive SE due to AED withdrawal was treated with VNS. After implantation of the device, stimulation output was rapidly increased to 3 mA in the three patients. The time to termination of the SE was 3–5 days [45].

22.2.2
Safety, Side-effects and Tolerability

Safety concerns with regard to VNS treatment can be approached from different angles. As the device needs to be implanted, a surgical intervention is required. The effects of delivering current to nervous tissue need to be considered as this might provoke changes in innervated organs and result in acute or chronic side-effects. Patients with refractory epilepsy are often young people. The implanted device and wires have to be examined for MRI compatibility.

The classical surgical technique has been described in detail by several authors [46–48]. Cosmetic side-effects have already been improved since the production of the smaller Model 101 and will be greatly improved once the Model 103 Generator Demipulse and Model 104 Generator Demipulse Duo become widely available.

22.2.2.1 Ramping up and Long-term Stimulation

For therapeutic purposes, VNS aims at stimulating vagal afferents. There are widespread connections from the vagus nerve to the central nervous system. Through these connections efficacious stimulation parameters may also induce other central nervous system side-effects. Moreover, selectively stimulating efferents is difficult and approximately 20 % of the fibers in the cervical part of the vagus nerve are efferent fibers. These fibers innervate thoracoabdominal organs, which explains the potential serious side-effects when these fibers are stimulated [29]. Certain side-effects related to undesired stimulation of nerve fibers might be immediately perceptible by the patient. The main efferent innervation of the vagus nerve serves to monitor and modulate visceral activity. These autonomic processes are usually not perceived by the patient. There may also be side-effects specifically related to chronic stimulation that will cause symptoms and become clinically apparent only after long-term treatment.

The most prominent and consistent sensation in patients when the vagus nerve is stimulated for the first time, even at low output current levels, is a tingling sensation in the throat and hoarseness of the voice. The tingling sensation may be due to secondary stimulation of the superior laryngeal nerve that branches off from the vagus nerve superior to the location of the implanted electrode, but travels along the vagus nerve in the carotid sheath [49].

The superior laryngeal nerve carries sensory fibers to the laryngeal mucosa. Stimulation of the recurrent laryngeal nerve that branches off distally from the location of the electrode and carries motor (Aα) fibers to the laryngeal muscles causes the stimulation-related hoarseness [50, 51]. Fiberoptic laryngoscopy and videostroboscopic examination have shown left vocal cord adduction (tetanic contraction) during stimulation at 30 Hz or higher [51–55]. These stimulation-related side-effects are dose dependent, which means that higher amplitudes, higher frequencies and wider pulse widths are associated with more intense sensations and voice changes [26].

With regard to side-effects related to stimulation of vagal efferents, the effect on heart rate has been a major concern. The stimulation electrode is always implanted on the left vagus nerve, which is believed to contain fewer sinoatrial fibers than the right.

In the long-term extension trials, the most frequent side-effects were hoarseness in 19 % of patients and coughing in 5 % of patients at two years' follow-up; shortness of breath in 3 % of patients at three years [35]. There was a clear trend towards diminishing side-effects over the three-year stimulation period. 98 % of the symptoms were rated mild or moderate by the patients and the investigators [56]. Side-effects can usually be resolved by decreasing stimulation parameters. Central nervous system side-effects, typically seen with AEDs, were not reported. After three years of treatment, 72 % of the patients were still on the treatment [35]. The most frequent reason for discontinuation was lack of efficacy. Holter monitoring in a sample of patients of the EO4 study showed no clinically symptomatic changes. Pulmonary function testing was performed in 124 patients with no change between baseline and long-term treatment [34].

Despite the fact that the initial studies showed no clinical effect on heart rate, occurrence of bradycardia and ventricular asystole during intra-operative testing of the device (stimulation parameters: 1 mA, 20 Hz, 500 μs, 17 s) have been reported in a few patients. None of the reported patients had a history of cardiac dysfunction, nor did they show abnormal cardiac testing after surgery. Tatum et al. reported on four patients who experienced ventricular asystole intraoperatively during device testing [57]. In three patients, the implantation procedure was aborted. In one patient a rechallenge of stimulation with incremental increases from 0.25 to 1 mA did not reveal a reappearance of bradycardia. Implantation was completed and no further cardiac events were noticed after the start of VNS. Asconape et al. reported on a single patient who developed asystole during intra-operative device testing. After removal of the device, the patient recovered completely [58]. Ali et al. described three similar cases. Cardiac rhythm strips were available and showed a regular 'p'-wave (atrial rhythm) with no ventricular activity, indicating a complete AV nodal block. In two of these patients the device was subsequently removed. In one patient the device was left in place without experiencing any other adverse events after the start of VNS [59]. Andriola et al. reported on three patients who experienced an asystole during intraoperative lead testing and who were subsequently chronically stimulated [60]. Ardesch et al. reported on three patients with intraoperative bradycardia and subsequent uneventful stimulation [61]. Possible hypotheses with regard to the underlying cause are inadvertent placement of the electrode on one of the cervical branches of the vagus nerve or indirect stimulation of these branches, reversal of the polarities of the electrode which would lead to primary stimulation of efferents instead of afferents, indirect stimulation of cardiac branches, activation of afferent pathways affecting higher autonomic systems or of the parasympathetic pathway with an exaggerated effect on the atrioventricular node, technical malfunctioning of the device, or idiosyncratic reactions. The contributing role of specific AEDs should be further investigated. Suggested steps to be taken in the operating room in the case of bradycardia or asystole during generator and lead-impedance testing have been formulated by Asconape et al. [58]. Adverse cardiac complications at the start or during ramping-up of the stimulation intensity have not been observed [29]. Very recently, one case report described a late-onset bradyarrhythmia after two years of vagus nerve stimulation [62].

22.2.2.2 MRI

Most patients with refractory epilepsy who are treated with VNS have previously undergone MRI during the presurgical evaluation. It is not uncommon for such patients to require MRI after VNS implantation in order to further monitor underlying neurological diseases, e.g., in case of unexplained seizure frequency increase, follow-up of intracranial lesions, or for MRI indications as in the general population. Based on laboratory testing using a phantom to simulate a human body, the VNS Therapy™ system device is labelled MRI compatible when used with a send and receive head coil [63]. In addition to the safety issues, there was no significant image distortion [64]. A retrospective analysis of 27 MRI scans, performed in 25 patients in 12 different centers, confirmed the findings from the experimental set-up in a clinical series. All patients were scanned on a 1.5 Tesla machine. On one occasion a body coil was used. Three scans were performed with the stimulator in the on-mode. One patient reported a mild voice change for several minutes; one child reported chest pain and claustrophobia. Twenty-three patients reported no discomfort around the lead or the generator. It was concluded that MRI is safe as long as guidelines stated in the physician's manual of the implanted device are followed.

In these guidelines it is suggested to program the pulse generator output current and the magnet output current to 0 mA. On one occasion this has led to the occurrence of a generalized status epilepticus in a patient who was well controlled with an output current of 2 mA [65]. The authors of the report recommend that intravenous access should be obtained and a benzodiazepine should be either available or preadministered in patients with a well-defined response who undergo elective MRI and in whom the generator is acutely programmed to 0 mA.

Functional MRI (fMRI) is a recently developed technique that allows non-invasive evaluation of cerebral functions such as finger movements and language [66]. It has been widely used for research but is currently increasingly applied to evaluate functional tissue in the neighborhood of lesions before resective surgery and also for assessing language dominance in the presurgical evaluation of epilepsy patients [66, 67]. When fMRI in patients with VNS is used for research purposes to evaluate VNS-induced changes in cerebral blood flow, scanning should be performed in the on-mode. To prevent the device from being turned off during scanning, an adjustment in the surgical positioning of the device is necessary. The device should be positioned so that the electrode pins that are plugged into the generator are parallel instead of perpendicular to the long axis of the body [68]. There have been several studies in patients treated with VNS for epilepsy as well as for depression showing that fMRI is safe and feasible [69–73]. These studies were performed to elucidate the mechanism of action of VNS and will be discussed later in this study.

The use of body coils may be indicated in patients requiring spinal MRI. Placing a cold pack (water and ammonium nitrate) over the left side of the patient's neck protected the vagus nerve, in three children, from heat that can theoretically be generated when using the body coil [74].

When removal of the electrode is indicated, e.g., due to insufficient efficacy, complete removal is recommended over cutting the distal edges and leaving the electrode in place [75]. Full removal allows potential future MRI with body coils. Heating of the electrode is related to the lead length. If full removal of the electrode is difficult the leads should be cut to less than 10 cm.

In several of our patients uneventful MRI was performed according to the pre-scribed precautions. In one patient with frequent simple partial seizures successful and uneventful fMRI was performed with the stimulator in the off-mode.

22.2.3
Mechanism of Action

Since the first human implant of the VNS Therapy™ device in 1989, over 50 000 patients have been treated with VNS worldwide. As for many antiepileptic treatments, clinical application of VNS preceded the elucidation of its mechanism of action (MOA). Following a limited number of animal experiments in dogs and monkeys, investigating safety and efficacy, the first human trial was performed [22]. The basic hypothesis on the MOA was based on the knowledge that the tenth cranial nerve afferents have numerous projections within the central nervous system and that, in this way, action potentials generated in vagal afferents have the potential to affect the entire organism [76]. To date, the precise mechanism of action of VNS and how it suppresses seizures remains to be elucidated.

Research directed towards investigating the antiseizure, antiepileptic and po-tential antiepileptogenic properties of VNS, as well as towards the identification of involved fibers, intracranial structures and neurotransmitter systems, has been performed. Animal experiments and research in humans treated with VNS have comprised electrophysiological studies (EEG, EMG, EP), functional anatomic brain imaging studies (PET, SPECT, fMRI, c-fos, densitometry), neuropsychological and behavioral studies. Also, from the extensive clinical experience with VNS, interesting clues concerning the MOA of VNS have arisen. More recently the role of the vagus nerve in the immune system has been investigated.

From the extensive body of research on the MOA, it has become conceivable that effective stimulation in humans is primarily mediated by afferent vagal A- and B-fibers [77,78]. Unilateral stimulation influences both cerebral hemispheres, as shown in several functional imaging studies [79, 80]. Crucial brainstem and intracranial structures have been identified and include the locus coeruleus, the nucleus of the solitary tract, the thalamus and limbic structures [81–84]. Neurotransmitters playing a role may involve the major inhibitory neurotransmitter GABA, but also serotoninergic and adrenergic systems [85,86]. More recently, Neese et al. found that VNS, following experimental brain injury in rats, protects cortical GABAergic cells from death [87]. A SPECT study in humans before and after one year of VNS, showed a normalisation of $GABA_A$ receptor density in the individuals with a clear therapeutic response to VNS [88]. Follesa et al. showed an increase of norepinephrine concentration in the prefrontal cortex of the rat brain after acute VNS [1]. An increased norepinephrine concentration after VNS has also been measured in the hippocampus [89] and the amygdala [90]. Currently, VNS is being explored as a neuroimmunomodulatory treatment. As the vagus nerve plays a critical role in the signalization and modulation of inflammatory processes, the so-called anti-inflammatory pathway, this could represent a new modality in the MOA of VNS for epilepsy [91, 92].

Early animal experiments in acute seizure models suggest an anti-seizure effect of VNS. In our own group, VNS significantly increased the seizure threshold for focal motor seizures in the cortical stimulation model [93]. Also, in the human literature, evidence exists that VNS may exert an acute abortive effect. The magnet feature allows a patient to terminate an upcoming seizure [94]. Also, a few case reports describe the use of VNS for refractory SE in pediatric and adult patients [41, 42]. A recent study investigated the effects of acute VNS on cortical excitability by using transcranial magnetic stimulation (TMS) [95].

The fact that seizures reoccur after the end of battery life has been reached is a strong argument against VNS having an antiepileptogenic effect. However, as progress in the development of more relevant animal models for epilepsy has been made, the antiepileptogenic potential of neurostimulation in general is being fully explored and some promising results have been reported, eg., in the kindling model [96, 97]. The basis for the combined acute and more chronic effects of VNS most likely involves recruitment of different neuronal pathways and networks. The more chronic effects are thought to be a reflection of modulatory changes in subcortical site-specific synapses with the potential to influence larger cortical areas. In the complex human brain these neuromodulatory processes require time to build up. Once installed, certain antiepileptic neural networks may be more easily recruited, e.g., by changing stimulation parameters that may then be titrated to the individual need of the patient. This raises hope for potential anti-epileptogenic properties of VNS using long-term optimized stimulation parameters that may affect and potentially reverse pathological processes that have been installed over a long period of time.

22.3
Hippocampal Stimulation

22.3.1
Clinical Efficacy and Safety

Electrical seizure onset in the amygdala and hippocampus is the key feature of the medial temporal lobe epilepsy syndrome [98]. Acute DBS in medial temporal lobe structures for control of seizures has been described [99]. In a small number of patients with complex partial seizures requiring invasive video-EEG monitoring for localizing purposes, unilateral DBS decreased interictal and ictal epileptic activity during a two-week period using temporary depth electrodes. The recording electrodes that were used for invasive video-EEG monitoring are unsuitable for long-term DBS and had to be removed. Subsequently, all patients underwent a temporal lobectomy. Animal studies have shown abortive effects on epileptic activity when electrical fields were applied to hippocampal slices [100]. *In vivo* studies in rats showed that electrical stimuli applied following a kindling stimulus ('quenching') can delay the development of the kindling process [101]. Bragin et al. and Velisek et al. found that repeated stimulation of the hippocampal perforant path in the kainate rat model significantly reduced seizures [102, 103]. Performing chronic DBS implies removal of recording electrodes and replacement by chronic DBS electrodes. Because the purpose is to stimulate the ictal onset zone, replacement of electrodes should be anatomically as accurate as possible. Even with currently available neuronavigation technology, positioning of a second electrode in exactly the same position as the initial one is difficult. We have therefore studied the feasibility of recording intracranial EEG activity for localizing purposes and subsequent long-term DBS of the identified ictal onset zone using the same electrodes with the aim to evaluate the long-term efficacy and safety of chronic DBS in medial temporal lobe structures, and to investigate the feasibility of using chronic DBS electrodes for the localization of the ictal onset zone prior to DBS to avoid an additional invasive procedure. An initial pilot study was performed to demonstrated proof-of-concept [104]. The number of patients in the pilot study was increased and long-term follow-up was reported [105]. The study prospectively evaluated the efficacy of long-term deep-brain stimulation in medial temporal lobe structures in patients with MTL epilepsy. Twelve consecutive patients with refractory MTL epilepsy were included. The protocol included invasive video-EEG monitoring for ictal onset localization and evaluation for subsequent stimulation of the ictal onset zone. Side-effects and changes in seizure frequency were carefully monitored: 10/12 patients underwent chronic MTL DBS; 2/12 patients underwent temporal lobectomy. After a mean follow-up of 31 months (range: 12–52 months) 1/10 stimulated patients is seizure free (>1 year), 1/10 patients has a >90% reduction in seizure frequency; 5/10 patients have a seizure frequency reduction of >50%; 2/10 patients have a seizure frequency reduction of 30–40%; 1/10 patients is a non-responder. None of the patients reported side-effects. In one patient MRI showed asymptomatic intracranial hemorrhages along the trajectory of the DBS electrodes. None of the patients showed significant changes in clinical neurological and neuropsychological testing. Patients who underwent temporal lobectomy are seizure free (>1 year), AEDs are unchanged and no side-effects have occurred.

This open pilot study demonstrates the potential efficacy of long-term DBS in MTL structures that should now be further confirmed by multicenter randomized controlled trials. CoRaStiR is an ongoing multicenter study in Belgium and Germany including patients with unilateral hippocampal sclerosis who will be randomized to unilateral hippocampal DBS or amygdalohippocampal resection.

22.3.2
Mechanism of Action

The mechanism of action (MOA) of DBS in reducing seizures remains unclear. Some support the hypothesis that actual stimulation is not necessary to achieve efficacy and claim that efficacy is based on the lesion provoked by the insertion of the electrode ('microthalamotomy' effect) [15]. Furthermore, prolonged seizure control in patients who underwent invasive recording with conventional electrodes has been described [106]. Blinded randomization of patient to 'on' and 'off' stimulation paradigms following implantation during follow-up >6 months, may clarify this issue and may also simultaneously clarify the potential effect of sham stimulation due to an implanted device. DBS may also act through local inhibition induced by current applied to nuclei that are involved in propagating, sustaining or triggering of epileptic activity in a specific CNS structure ('reversible functional lesion'). Apart from this 'local' inhibition, the MOA of DBS may be based on the effect on projections leaving from the area of stimulation to other central nervous structures. This may be the most likely hypothesis when crucial structures in epileptogenic networks are involved. However, considering that the medial temporal lobe structures are also potentially involved in these networks it may be that targeting the ictal focus may also affect the epileptogenic network.

22.4
Conclusion

Patients with refractory epilepsy present a particular challenge to new therapies. VNS has proved to be an efficacious and safe treatment. The efficacy of VNS in less severely affected populations remains to be evaluated. The current consensus on efficacy is that 1/3 of patients have a considerable improvement in seizure control with a reduction in seizure frequency of at least 50 %, 1/3 of patients experience a worthwhile reduction of seizure frequency between 30 and 50 %. In the remaining 1/3 of the patients there is little or no effect. VNS seems equally efficient for children. The degree of improvement in seizure control from VNS remains comparable to new antiepileptic drugs. Patients appear willing to undergo surgery for improvements in this range in order to avoid the usual undesirable effects of antiepileptic medication. Contrary to treatment with AEDs, efficacy has a tendency to improve with longer duration of treatment up to 18 months postoperatively. Analysis of larger patient groups and insight into the mode of action may help to identify patients with epileptic seizures or syndromes that respond better to VNS, and guide the search for optimal stimulation parameters. Further improvement of clinical efficacy may result from this.

Deep-brain stimulation for epilepsy is beyond the stage of proof-of-concept but still needs thorough evaluation in confirmatory pilot studies before it can be offered to a larger patient population. The most adequate targets and stimulation parameters need to be identified. For patients who are less suitable candidates for epilepsy surgery, DBS may become a valuable alternative. Randomized and controlled studies in larger patient series are ongoing and will need to identify the potential treatment population and optimal stimulation paradigms.

References

1 E. Ben-Menachem, *Lancet Neurol.* **1**, 477 (**2002**).

2 C. A. Tassinari, M. Cincotta, G. Zaccara and R. Michelucci, *Clin. Neurophysiol.* **114**, 777 (**2003**).

3 P. Pollak, V. Fraix, P. Krack, E. Moro, A. Mendes, S. Chabardes, A. Koudsie and A.-L. Benabid, *Mov. Disord.* **17** Suppl. 3, 75 (**2002**).

4 J. P. Nguyen, J. P. Lefaucher, C. L. Guerinel, J. F. Eizenbaum, N. Nakano, A. Carpentier, P. Brugières, B. Pollin, S. Rostaing and Y. Keravel, *Arch. Med. Res.* **31**, 263 (**2000**).

5 B. Nuttin, P. Cosyns, H. Demeulemeester, J. Gybels and B. Meyerson, *Lancet* **354**, 1526 (**1999**).

6 M. Leone, A. Franzini, G. Broggi and G. Bussone, *Neurol. Sci.* **24** Suppl. 2, 143 (**2003**).

7 I. S. Cooper, *Cerebellar Stimulation in Man.* (Raven Press Books Ltd. New York **1978**).

8 G. D. Wright, D. L. McLellan and J. G. Brice, *J. Neurol. Neurosurg. Psychiatry* **47**, 769 (**1984**).

9 A. R. Upton, I. S. Cooper, M. Springman and I. Amin, *Int. J. Neurol.* **19–20**, 223 (**1985**).

10 B. Feinstein, C. A. Gleason and B. Libet, *Stereotact. Funct. Neurosurg.* **52**, 26 (**1989**).

11 R. S. Fisher, S. Uematsu, G. L. Krauss, B. J. Cysyk, R. McPherson, R. P. Lesser, B. Gordon, P. Schwerdt and M. Rise, *Epilepsia* **33**, 841 (**1992**).

12 F. Velasco, M. Velasco, A. L. Velasco, F. Jiménez, I. Marquez and M. Rise, *Epilepsia* **36**, 63 (**1995**).

13 S. A. Chkhenkeli and I. S. Chkhenkeli, *Stereotact. Funct. Neurosurg.* **69**, 221 (**1997**).

14 S. Chabardès, P. Kahane, L. Minotti, A. Koudsie, E. Hirsch and A.-L. Benabid, *Epileptic Disord.* **4** Suppl. 3, 83 (**2002**).

15 M. Hodaie, R. A. Wennberg, J. O. Dostrovsky and A. M. Lozano, *Epilepsia* **43**, 603 (**2002**).

16 M. J. Iadarola and K. Gale, *Science* **218**, 1237 (**1982**).

17 D. King and S. Spencer, *J. Clin. Neurophysiol.* **12**, 32 (**1995**).

18 J. W. Woodbury and E. W. Kinghorn, *Epilepsia.* **39** Suppl. 6, 194 (**1998**).

19 A. L. Benabid, A. Koudsie, A. Benazzouz, L. Vercueil, V. Fraix, S. Chabardes, J. F. Lebas and P. Pollak, *J. Neurol.* **248** Suppl. 3, 37 (**2001**).

20 R. Terry, W. B. Tarver and J. Zabara, *Epilepsia* **31** Suppl. 2, 33 (**1990**).

21 R. S. Terry, W. B. Tarver and J. Zabara, *Pacing Clin. Electrophysiol.* **14**, 86 (**1991**).

22 B. M. Uthman, B. J. Wilder,
    E. J. Hammond and S. A. Reid,
    Epilepsia **31** Suppl. 2, 44 (**1990**).

23 J. K. Penry and J. C. Dean,
    Epilepsia **31** Suppl. 2, 40 (**1990**).

24 E. J. Hammond, B. M. Uthman, S. A.
    Reid, B. J. Wilder and R. E. Ramsay,
    Epilepsia **31** Suppl. 2, 51 (**1990**).

25 B. J. Wilder, B. M. Uthman and
    E. J. Hammond, *Pacing Clin.
    Electrophysiol.* **14**, 108 (**1991**).

26 B. M. Uthman, B. J. Wilder, J. K.
    Penry, C. Dean, R. E. Ramsay,
    S. A. Reid, E. J. Hammond,
    W. B. Tarver and J. F. Wernicke,
    Neurology **43**, 1338 (**1993**).

27 L. K. Holder, J. F. Wernicke
    and W. B. Tarver, *Pacing Clin.
    Electrophysiol.* **15**, 1557 (**1992**).

28 E. Ben-Menachem,
    R. Mañon-Espaillat, R. Ristanovic,
    B. J. Wilder, H. Stefan, W. Mirza,
    W. B. Tarver and J. F. Wernicke,
    Epilepsia **35**, 616 (**1994**).

29 R. E. Ramsay, B. M. Uthman,
    L. E. Augustinsson, A. R. Upton,
    D. Naritoku, J. Willis, T. Treig,
    G. Barolat and J. F. Wernicke, *Epilep-
    sia* **35**, 627 (**1994**).

30 The Vagus Nerve Stimulation Study
    Group, *Neurology* **45**, 224 (**1995**).

31 A. Handforth, C. M. DeGiorgio, S. C.
    Schachter, B. M. Uthman, D. K.
    Naritoku, E. S. Tecoma, T. R. Henry,
    S. D. Collins, B. V. Vaughn, R. C.
    Gilmartin et al., *Neurology* **51**, 48
    (**1998**).

32 L. K. Holder, J. F. Wernicke and
    W. B. Tarver, *Journal of Epilepsy* **6**,
    206 (**1993**).

33 R. George, M. Salinsky, R. Kuzniecky,
    W. Rosenfeld, D. Bergen, W. B.
    Tarver and J. F. Wernicke, *Epilepsia*
    35, 637 (**1994**).

34 M. C. Salinsky, B. M. Uthman, R. K.
    Ristanovic, J. F. Wernicke and W. B.
    Tarver, *Arch. Neurol.* **53**, 1176 (**1996**).

35 G. L. Morris and W. M. Mueller,
    Neurology **53**, 1731 (**1999**).

36 C. M. DeGiorgio, S. C. Schachter,
    A. Handforth, M. Salinsky,
    J. Thompson, B. Uthman, R. Reed,
    S. Collins, E. Tecoma, G. L. Morris
    et al., *Epilepsia* **41**, 1195 (**2000**).

37 E. Ben-Menachem, K. Hellström,
    C. Waldton and L. E. Augustinsson,
    Neurology **52**, 1265 (**1999**).

38 K. Vonck, V. Thadani, K. Gilbert,
    S. Dedeurwaerdere, L. De Groote,
    V. De Herdt, L. Goossens,
    F. Gossiaux, E. Achten, E. Thiery
    et al., *J. Clin. Neurophysiol.* **21**, 283
    (**2004**).

39 A. Hui Che Fai, J. Lam Man Kuen,
    W. Ka Shing, R. Kay and
    P. Wai Sing, *Chin. Med. J. (Engl)*
    117, 58 (**2004**).

40 V. De Herdt, P. Boon, B. Ceulemans,
    H. Hauman, L. Lagae, B. Legros,
    B. Sadzot, P. Van Bogaert,
    K. Van Rijckevorsel, H. Verhelst
    et al., *Eur. J. Paediatr. Neurol.* **11**, 261
    (**2007**).

41 P. Boon, V. De Herdt,
    L. Waterschoot, H. Verhelst,
    A. De Jaeger, R. Van Coster,
    D. Van Roost and K. Vonck, *Vagus
    nerve stimulation for refractory
    status epilepticus.*, Abstract pre-
    sented at the 27th International
    Epilepsy congress, Singapore (**2007**).

42 S. Malik and A. Hernandez,
    Epilepsia **45** Suppl. 7, 155 (**2004**).

43 K. R. Winston, P. Levisohn, B. R.
    Miller and J. Freeman, *Pedi-
    atr. Neurosurg.* **34**, 190 (**2001**).

44 R. V. Patwardhan, J. Dellabadia,
    M. Rashidi, L. Grier and A. Nanda,
    Surg. Neurol. **64**, 170 (**2005**).

45 R. Zimmerman, J. Sirven,
    J. Drazkowski, J. Bortz and
    D. Shulman, *Epilepsia* **43** Suppl. 7,
    286 (**2002**).

46 A. A. Kemeny, Surgical technique
    in vagus nerve stimulation in
    Vagus Nerve Stimulation (Martin
    Dunitz, 2003), pp. 33–48, 2nd edn.

47 S. A. Reid, *Epilepsia* **31** Suppl. 2, 38
    (**1990**).

48 H. J. Landy, R. E. Ramsay, J. Slater,
    R. R. Casiano and R. Morgan,
    J. Neurosurg. **78**, 26 (**1993**).

49 J. Claes and P. Jaco, *Acta Otorhi-
    nolaryngol. Belg.* **40**, 215 (**1986**).

50 R. B. Banzett, A. Guz, D. Paydarfar,
    S. A. Shea, S. C. Schachter and R. W.
    Lansing, *Epilepsy Res.* **35**, 1 (**1999**).

51 S. J. Charous, G. Kempster,
E. Manders and R. Ristanovic,
Laryngoscope **111**, 2028 (**2001**).

52 R. K. Ristanovic, D. Bergen,
P. Szidon and M. J. Bacon, *Epilepsia* **33** Suppl. 3, 101 (**1992**).

53 D. S. Lundy, R. R. Casiano, H. J.
Landy, J. Gallo, B. Gallo and R. E.
Ramsey, *J. Voice* **7**, 359 (**1993**).

54 D. Zumsteg, D. Jenny and H. G.
Wieser, *Neurology* **54**, 1388 (**2000**).

55 W. Kersing, P. H. Dejonckere,
H. E. van der Aa and H. P. J.
Buschman, *J. Voice* **16**, 251 (**2002**).

56 E. Ben-Menachem, *J. Clin. Neurophysiol.* **18**, 415 (**2001**).

57 W. O. Tatum, D. B. Moore,
M. M. Stecker, G. H. Baltuch,
J. A. French, J. A. Ferreira, P. M.
Carney, D. R. Labar and F. L.
Vale, *Neurology* **52**, 1267 (**1999**).

58 J. J. Asconapé, D. D. Moore, D. P.
Zipes, L. M. Hartman and W. H.
Duffell, *Epilepsia* **40**, 1452 (**1999**).

59 I. I. Ali, N. A. Pirzada, Y. Kanjwal,
B. Wannamaker, A. Medhkour,
M. T. Koltz and B. V. Vaughn,
Epilepsy Behav. **5**, 768 (**2004**).

60 M. R. Andriola, T. Rosenweig,
S. Vlay and S. Brook, *Epilepsia* **41** Suppl. 7, 223 (**2000**).

61 J. J. Ardesch, H. P. J. Buschman,
P. H. van der Burgh, L. J. J. C.
Wagener-Schimmel, H. E. van der Aa
and G. Hageman, *Clin. Neurol. Neurosurg.* **109**, 849 (**2007**).

62 P. Amark, T. Stodberg and
L. Wallstadt, *Epilepsia* **48**, 1023 (**2007**).

63 J. A. Nyenhuis, J. D. Bourland,
K. S. Foster, G. P. Graber, R. S.
Terry and R. A. Adkins, *Epilepsia* **38** Suppl. 8, 140 (**1997**).

64 S. R. Benbadis, J. Nyhenhuis, W. O.
Tatum, F. R. Murtagh, M. Gieron
and F. L. Vale, *Seizure* **10**, 512 (**2001**).

65 F. Beitinjaneh, M. Guido and M. R.
Andriola, *Epilepsia* **43**, 337 (**2002**).

66 E. Achten, G. D. Jackson, J. A.
Cameron, D. F. Abbott, D. L. Stella
and G. C. Fabinyi, *Radiology* **210**, 529 (**1999**).

67 K. Deblaere, W. H. Backes,
P. Hofman, P. Vandemaele, P. A.
Boon, K. Vonck, P. Boon, J. Troost,

J. Vermeulen, J. Wilmink et al.,
Neuroradiology **44**, 667 (**2002**).

68 A. Maniker, W. C. Liu, D. Marks,
K. Moser and A. Kalnin, *Surg. Neurol.* **53**, 178 (**2000**).

69 R. Sucholeiki, T. Alsaadi, G. L.
Morris and B. Biswal, *Epilepsia* **40** Suppl. 7, 181 (**1999**).

70 D. E. Bohning, M. P. Lomarev,
S. Denslow, Z. Nahas, A. Shastri and
M. S. George, *Invest. Radiol.* **36**, 470 (**2001**).

71 R. Sucholeiki, T. M. Alsaadi, G. L.
Morris, J. L. Ulmer, B. Biswal and
W. M. Mueller, *Seizure* **11**, 157 (**2002**).

72 J. T. Narayanan, R. Watts, N. Haddad,
D. R. Labar, P. M. Li and C. G.
Filippi, *Epilepsia* **43**, 1509 (**2002**).

73 M. Lomarev, S. Denslow, Z. Nahas,
J.-H. Chae, M. S. George and
D. E. Bohning, *Journal of Psychiatric Research* **36**, 219 (**2002**).

74 A. A. Wilfong, *Epilepsia* **43** Suppl. 7,
347 (**2002**).

75 J. Espinosa, M. T. Aiello and D. K.
Naritoku, *Surg. Neurol.* **51**, 659 (**1999**).

76 H. R. Berthoud and W. L. Neuhuber,
Auton. Neurosci. **85**, 1 (**2000**).

77 A. Zagon and A. A. Kemeny, *Epilepsia* **41**, 1382 (**2000**).

78 M. S. Evans, S. Verma-Ahuja, D. K.
Naritoku and J. A. Espinosa, *Acta Neurol. Scand.* **110**, 232 (**2004**).

79 T. R. Henry, R. A. Bakay, J. R.
Votaw, P. B. Pennell, C. M. Epstein,
T. L. Faber, S. T. Grafton and J. M.
Hoffman, *Epilepsia* **39**, 983 (**1998**).

80 K. Van Laere, K. Vonck, P. Boon,
J. Versijpt and R. Dierckx, *J. Nucl. Med.* **43**, 733 (**2002**).

81 D. K. Naritoku, W. J. Terry and R. H.
Helfert, *Epilepsy Res.* **22**, 53 (**1995**).

82 S. E. Krahl, K. B. Clark, D. C. Smith
and R. A. Browning, *Epilepsia* **39**, 709 (**1998**).

83 V. Osharina, V. Bagaev, F. Wallois
and N. Larnicol, *Auton. Neurosci.* **72** 126–27 (**2006**).

84 J. T. Cunningham, S. W. Mifflin,
G. G. Gould and A. Frazer, *Neuropsychopharmacology* (**2007**), Epub ahead of print.

85 E. J. Hammond, B. M. Uthman,
B. J. Wilder, E. Ben-Menachem,
A. Hamberger, T. Hedner and
R. Ekman, *Brain Res.* **583**, 300 (**1992**).

86 E. Ben-Menachem, A. Hamberger,
T. Hedner, E. J. Hammond,
B. M. Uthman, J. Slater, T. Treig,
H. Stefan, R. E. Ramsay and J. F.
Wernicke, *Epilepsy Res.* **20**, 221
(**1995**).

87 S. L. Neese, L. K. Sherill, A. A. Tan,
R. W. Roosevelt, R. A. Browning,
D. C. Smith, A. Duke and R. W.
Clough, *Brain Res.* **1128**, 157 (**2007**).

88 F. Marrosu, A. Serra, A. Maleci,
M. Puligheddu, G. Biggio and
M. Piga, *Epilepsy Res.* **55**, 59 (**2003**).

89 R. W. Roosevelt, D. C. Smith, R. W.
Clough, R. A. Jensen and R. A.
Browning, *Brain Res.* **1119**, 124
(**2006**).

90 D. L. Hassert, T. Miyashita and C. L.
Williams, *Behav. Neurosci.* **118**, 79
(**2004**).

91 L. V. Borovikova, S. Ivanova,
M. Zhang, H. Yang, G. I. Botchkina,
L. R. Watkins, H. Wang,
N. Abumrad, J. W. Eaton and K. J.
Tracey, *Nature* **405**, 458 (**2000**).

92 T. Hosoi, Y. Okuma and Y. Nomura,
*Am. J. Physiol. Regul. Integr.
Comp. Physiol.* **279**, 141 (**2000**).

93 V. De Herdt, J. Dewaele, J. Van Aken,
K. Vonck, J. Delbeke, R. Raedt,
W. Wadman and P. Boon, *Epilep-
sia* **47** Suppl. 4, 300 (**2006**).

94 P. Boon, K. Vonck, P. V. Walleghem,
M. D'Havé, L. Goossens,
T. Vandekerckhove, J. Caemaert and
J. D. Reuck, *J. Clin. Neurophysiol.* **18**,
402 (**2001**).

95 V. Di Lazzaro, A. Oliviero, F. Pilato,
E. Saturno, M. Dileone, M. Meglio,
G. Colicchio, C. Barba, F. Papacci
and P. A. Tonali, *Neurology* **62**, 2310
(**2004**).

96 D. K. Naritoku and J. A. Mikels,
Epilepsia **37** Suppl. 5, 75 (**1996**).

97 A. Fernández-Guardiola,
A. Martínez, A. Valdés-Cruz, V. M.
Magdaleno-Madrigal, D. Martínez
and R. Fernández-Mas, *Epilepsia* **40**,
822 (**1999**).

98 S. S. Spencer, P. Guimaraes, A. Katz,
J. Kim and D. Spencer, *Epilepsia* **33**,
537 (**1992**).

99 M. Velasco, F. Velasco, A. L. Velasco,
B. Boleaga, F. Jiménez, F. Brito and
I. Marquez, *Epilepsia* **41**, 158 (**2000**).

100 J. Lian, M. Bikson, C. Sciortino,
W. C. Stacey and D. M. Durand,
J. Physiol. **547**, 427 (**2003**).

101 S. R. Weiss, X. L. Li, J. B. Rosen,
H. Li, T. Heynen and R. M.
Post, *Neuroreport* **6**, 2171 (**1995**).

102 A. Bragin, C. L. Wilson and J. Engel,
Exp. Brain Res. **144**, 30 (**2002**).

103 L. Velísek, J. Velísková and P. K.
Stanton, *Neurosci. Lett.* **326**, 61 (**2002**).

104 K. Vonck, P. Boon, E. Achten,
J. D. Reuck and J. Caemaert,
Ann. Neurol. **52**, 556 (**2002**).

105 P. Boon, K. Vonck, V. De Herdt,
A. Van Dycke, M. Goethals,
L. Goossens, M. V. Zandijcke,
T. De Smedt, I. Dewaele, R. Achten
et al., *Epilepsia* **48**, 1551 (**2007**).

106 N. M. Katariwala, R. A. Bakay, P. B.
Pennell, L. D. Olson, T. R. Henry
and C. M. Epstein, *Neurology* **57**, 1505
(**2001**).

23
Responsive Neurostimulation for the Treatment of Epileptic Seizures

Gregory K. Bergey

23.1
Introduction

Despite the introduction of multiple new antiepileptic drugs (AEDs) over recent years, many patients remain refractory to medications. Patients with partial seizures (with or without secondary generalization) make up over half of patients with seizure disorders [1] and only about half of these patients will have their seizures controlled with available AEDs. If patients with partial seizures fail to be controlled with good trials of three different AEDs, the chance of seizure control with additional medication trials may be less than 5 % [2]. Seizure surgery remains a highly effective and underutilized treatment modality for a subset of these refractory patients, but not all patients are good surgical candidates. Patients with multifocal partial seizures or seizures originating from eloquent brain areas are examples of patients who are not optimal surgical candidates. Some patients with non-lesional neocortical epilepsy may also not have the same chance for seizure freedom with surgery as do patients with mesial temporal sclerosis or lesional epilepsy. Although new AEDs continue to be developed, alternative means of treatment of epilepsy are needed. Neurostimulation offers the potential benefits of mechanisms of action distinct from AEDs and stimulation also avoids the potential toxic, cognitive, and idiosyncratic side-effects of medications.

23.2
Characteristics of Partial Seizures

Partial seizures originate from focal epileptogenic regions. Partial seizures can be either simple partial (with no alteration of consciousness) or complex partial (with alteration of consciousness). Simple partial seizures (e.g., an aura) can evolve to complex partial seizures, and complex partial seizures can secondarily generalize. Partial seizures originating from a single focus are typically clinically similar in a given patient. Partial seizures are transient events, typically lasting less than 120 seconds plus any postictal period [3]. Seizures spontaneously terminate, often

Seizure Prediction in Epilepsy. Edited by Björn Schelter, Jens Timmer and Andreas Schulze-Bonhage
Copyright © 2008 WILEY-VCH Verlag GmbH & Co. KGaA, Weinheim
ISBN: 978-3-527-40756-9

synchronously in all brain regions. While AEDs may reduce seizure number and secondary generalization, there is no evidence that AEDs shorten partial seizure duration or alter intrinsic seizure dynamics [4].

Application of sophisticated time-frequency analyses of intracranial recordings of seizures reveals that the intrinsic dynamics of partial seizures are remarkably similar for all seizures from a given focus in a given patient, particularly at seizure onset [5, 6]. This similarity in seizure onset dynamics greatly facilitates early and accurate seizure detection.

23.3
Types of Neurostimulation

Neurostimulation can be either chronic, programmed stimulation or responsive stimulation. Chronic stimulation involves regular periodic stimulation that is independent of seizure occurrence. Programmed stimulation can be applied to brain regions (e.g., thalamus, hippocampus) or to extracranial sites (e.g., vagus nerve). Vagus nerve stimulation (VNS) is the one approved therapy for the treatment of partial seizures [7]. About 40–50 % of patients with partial seizures will have a 50 % reduction in their seizures with VNS therapy, but few (<5 %) become seizure free. Vagus nerve stimulation has the option for patients to activate stimulation with a magnet. While there are anecdotal reports of VNS activation at seizure onset benefiting certain patients, controlled trials of the effects of patient activation of the VNS have not been published. Because the VNS does not benefit all patients and because few patients have their seizures controlled, other programmed stimulation paradigms are being investigated, involving intracranial stimulation, in the hope that this will provide improved efficacy over VNS. Other chapters in this volume will discuss VNS, chronic anterior thalamic stimulation, and programmed hippocampal stimulation and these will therefore not be discussed further here.

Responsive neurostimulation (RNS) is designed to stimulate the brain shortly after seizure onset. In contrast to programmed stimulation (e.g., VNS, thalamic stimulation) that are open-loop stimulation systems, responsive stimulation employs a closed-loop system. Although the potential exists in the future to apply RNS using seizure-prediction algorithms, the present RNS applications use seizure-detection algorithms. These can utilize various methods or combinations of detection methods such as line length, area under the curve, or half wave. Seizure detection then triggers the closed-loop responsive stimulation directed at or near the seizure focus. For RNS to provide meaningful benefit to the patient, seizures must be detected with high sensitivity, and the applied therapy should terminate the seizure before it evolves to a disabling seizure. While AEDs and VNS are designed to reduce seizure frequency, RNS is applied to alter seizure dynamics after the seizure has begun. As mentioned above, complex partial seizures are of relatively short duration. If RNS only reduces a 75 second complex partial seizure to 60 seconds, then this may provide no meaningful benefit to the patient if alteration of consciousness still occurs. If, however, RNS termi-nates the seizure after 2–5 seconds, when the patient has either had no clinical manifestations or perhaps only a brief aura, then such therapy would be very beneficial.

The consideration of application of electrical stimulation directly to brain regions raises several potential concerns. The first is that such excitatory stimuli may be ineffective or in fact even cause seizures. There is concern that such stimulation might be painful, but the brain parenchyma is pain insensitive and such stimulation is not only painless, but typically the patient is not aware of the therapy (in contrast to VNS where there is some awareness of the stimulation). Another question is whether repetitive stimulation might kindle the human brain, but this has not been observed.

The concept of application of electrical stimulation to the brain to control seizures might at first appear counterintuitive since epileptic seizures are periods of temporary increased synchronous neuronal network activity. In the hippocampus 80–90 % of neurons are excitatory; any applied stimulus would stimulate these neurons as well as any inhibitory interneurons. In neuronal network models, seizure termination can be produced by excitatory stimuli [8, 9]. In these same network models, seizure termination does not require inhibition (i.e., inhibitory neurons can be effectively removed, with synaptic weights reduced to zero). These studies suggest that stimulation with excitatory stimuli can be reasonably expected to alter seizure dynamics.

The application of RNS is distinct from cardiac defibrillation. Treatment of ventricular cardiac arrhythmias is done with a high-intensity stimulus designed to repolarize the entire heart, after which the normal pacemaker activity returns. The hypothesis underlying RNS therapy is to apply small currents (e.g., ≤ 12 mA) to focal regions of the brain and disrupt the abnormal synchronous activity, producing earlier termination than would otherwise occur. Cardiac arrhythmias left untreated typically do not terminate spontaneously whereas epileptic seizures are intrinsically transient events.

Requirements for responsive stimulation are several. It appears to be ben-eficial to know the region of seizure onset, since the purpose of RNS is to disrupt partial seizures early, before regional propagation. However, since epilep-tic seizures are network phenomena, it is possible that stimulation at more remote sites will provide benefit, but this is not known. The seizures must be able to be detected with high sensitivity. The fact that complex partial seizures from a specific focus in a given patient have very similar dynamics, makes it relatively straightforward to tune a detection algorithm to be very sensi-tive, particularly when using intracranial electrodes which avoid the artifacts present with scalp recordings. The device must be well tolerated and long-term stimulation must be safe. Ultimately the desire is to be able to modify (e.g., terminate) further seizure activity and evolution, specifically to prevent subclinical (i.e., electrical) or simple partial seizures from evolving to seizures that produce alteration or loss of consciousness (i.e., complex partial or secondarily generalized seizures).

23.4
Current Status of Investigations of Responsive Neurostimulation

Prior to beginning investigations with the implantable RNS, several preliminary studies were performed. The first proof of principle study was to determine whether stimulation at modest currents (≤12 mA) could terminate seizure activity. These trials were performed in patients with subdural grids undergoing functional mapping as part of their presurgical evaluations. During this mapping occasionally the stimulation will produce afterdischarges. Trials were done to confirm that subsequent stimulation could terminate the afterdischarges [10]. Independently another group also determined that stimulation could terminate these afterdischarges [11].

The next phase of RNS evaluation was the external RNS (eRNS). In these investigations, patients undergoing intracranial recordings for presurgical evaluations were connected to the eRNS, a device that duplicated the implantable technology. Connection was done so as not to compromise the presurgical evaluations. These studies confirmed that detection algorithms could be accurately tuned to accurately detect seizures from intracranial electrodes. In some patients, if there was a brief period (typically 12–48 hrs) following completion of the presurgical evaluation but before actual scheduled surgery (removal of intracranial electrodes with, at times, resective surgery), the device was switched to a closed-loop responsive system. These studies, while not designed to determine efficacy, suggested that responsive stimulation could terminate seizures in some patients [12] (cf. Figure 23.1) and that the stimulations were well tolerated. Only a rare patient had any perception of the stimulation. These closed-loop eRNS studies were only done if there was a window prior to surgery; surgery was not delayed in any patient to do closed-loop studies. The favorable results from these preliminary studies led to trials of the implantable RNS.

The Neuropace® RNS system is the only device for responsive neurostimulation currently undergoing controlled trials in humans. The implantable RNS contains a battery powered microprocessor controlled device that can deliver brief electrical stimuli through implanted intracranial leads. The present design allows for two intracranial electrodes to be connected to the device (more can be implanted), either depth arrays or subdural strips. Placement of these arrays is based on previous ictal recordings during intensive monitoring; not all patients require previous intracranial monitoring (Figure 23.2). The device is extradural and is placed in a tray recessed in and anchored to the skull and covered by muscle and skin. The RNS device can be programmed painlessly through the skin with a wand connected to a portable computer. Similarly, real time EEG and checks of electrode impedance can be made with the wand. Stored data, including EEG, can be downloaded by the patient or investigator and sent via internet to the Neuropace server where they can be reviewed by investigators. Although a record of all detections and delivered therapies is recorded, actual recorded EEG epochs are limited by the storage capabilities of the small device. Options exist to prevent overwriting; more frequent patient downloading can increase the number of detailed epochs available for analysis.

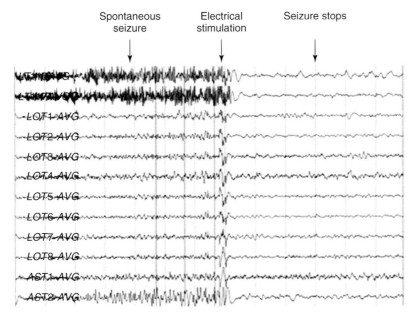

Fig. 23.1 Example of seizure termination produced by closed-loop external responsive neurostimulation (eRNS) in a patient with intractable epilepsy undergoing intracranial monitoring with subdural grid arrays for presurgical evaluation ([12], Neurospace, used with permission).

Stimulation parameters for the RNS device allow up to 12 mA total current to be delivered; this may be divided among multiple contacts. Different contacts or the device itself can serve as anode or cathode. Typical pulse durations are in the range of 160 μs with stimulus durations of 100–200 ms. Stimulus frequencies are variable, but 100–200 Hz are most commonly employed. Up to five therapies (responsive stimulation) can be delivered for each detected event.

The eRNS trial, which was not designed to demonstrate efficacy, showed that the detection algorithms could be tuned to provide accurate detection of seizure onset within two seconds. Some therapy was delivered during the closed-loop phase and some seizures appeared to be terminated with therapy.

Following the eRNS study, a safety and feasibility trial was conducted with the implantable RNS device. There were efficacy evaluation periods, but the study was not blinded unless a given center had more than four patients entered, in which case additional patients were randomized to stimulation on or off for a 28 day period after which all patients received therapy. The open extension of these patients continues. This study of 65 patients at 12 centers was completed in 2005. There were no serious surgical complications. The RNS implantation is well tolerated; patients typically spend at most one overnight stay in the hospital following implantation, and are fully active the first post-operative day. There were no unanticipated serious device related adverse events. While this study was

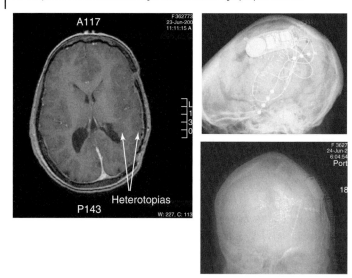

Fig. 23.2 A patient with left dominant temporal cortical dysplasia and unilateral periventricular heterotopias as shown on the MRI (T1 weighted image) on the left prior to RNS implantation. The skull films on the right show the RNS device and the implanted subdural strips and a depth electrode array targeting the periventricular nodules. The seizures in this patient originated from the periventricular nodular heterotopias. She has had a >90 % seizure reduction with the RNS; implantation was three years ago.

designed for assessment of safety and tolerability of RNS implantation and not efficacy, some assessments of the effects on seizure frequency were done. Complex partial, generalized tonic-clonic, and total disabling seizures were significantly reduced by 40 %, 55 %, and 41 %, respectively, (not all patients had GTCS). Some patients who did not benefit from RNS were subsequently determined upon review to have multifocal seizures or to have electrode placements that were not close to the seizure focus [13, 14]. Other unblended preliminary reports of small numbers of patients have been published [15, 16].

Following the safety and feasibility RNS trial, a pivotal blinded trial, has begun and is ongoing. Following a baseline period (patients need to have at least three complex partial seizures per month) the RNS is implanted, patients are randomized to therapy on or off for four months and then all patients can enter the open label period when therapy is enabled. Thirty centers are participating and it is expected that recruitment will be completed in late 2008. It is hoped that, with appropriate patient selection and improved therapy paradigms, better efficacy will be seen than in the safety and feasibility trial.

Although certain parameters (see above) are employed for setting the RNS, there are no standard stimulation parameters that are employed in all patients and the stimulation parameters can be adjusted during the blinded period by the treatment physician (who is not blinded). Equal time is spent with patients with therapy enabled and those not being stimulated so that patients cannot determine whether therapy is enabled or off during the blinded phase. Since there are no cognitive or systemic side-effects from RNS, in contrast to AED trials, and since patients are not aware of RNS (in contrast to VNS stimulation), the blind can in fact be preserved, perhaps even more easily than in AED or VNS trials where side-effects or stimulation may be appreciated by the patients.

23.5
Conclusion

Neurostimulation for the treatment of epilepsy remains a promising therapeutic modality with the benefits of mechanisms distinct from AEDs and the advantage of a lack of drug-related side-effects. Chronic stimulation paradigms have produced seizure reduction in patients, but rarely have resulted in seizure control, the goal of treatment. The concept of responsive neurostimulation is an attractive one, in that the stimulus is delivered in 'response' to seizure activity at or near the seizure focus, with the goal to disrupt or terminate electrical or simple partial seizures before they propagate to become disabling seizures with alteration or loss of consciousness.

The technology for the RNS is actually well developed. The device can be tuned to detect seizure activity in each patient with high sensitivity and detection can be accurately done within seconds of electrical seizure onset. The implanted device is well tolerated and there have been no safety concerns. There is no evidence that multiple stimulations kindle the human brain, one of the early theoretical concerns. Programming can be easily done transcutaneously and seizures can be downloaded for analysis. Improved storage of events and batteries with longer life (or rechargeable) will enhance the RNS capabilities.

Interestingly, tuning the RNS for early detection results in responses not to just clinical seizures or long electrical events, but to interictal electrical activity and indeed any activity that fulfills the detection criteria. Patients may have many detections and delivered therapies per day, many more than would be expected based on clinical seizure numbers. Total therapy per day is still only seconds because of the brief duration of the stimuli. These are not false positive detections, since they are detections based on the programmed criteria and the desire for early detection. Clearly, however, many of the detections and many of the delivered therapies are directed to activity that would not evolve to become disabling seizures. Whether therapy directed to this interictal activity is beneficial is not determined; it may be. If, however, the results of the pivotal trial demonstrate efficacy, as is hoped, it may be difficult to determine whether these effects are solely due to effects on seizure onset or whether benefits might result from the many stimulations triggered by other activity that would not have evolved into clinical seizures even without therapy.

There are other unanswered questions. As with all types of neurostimulation, the optimal stimulation parameters are not yet clearly established with RNS. It appears that proximity of the stimulating electrodes near the seizure focus is desirable. It is possible that regional network seizure activity could be disrupted by more remote stimulation, although if early termination is the goal then one wants to stimulate before there is extensive seizure propagation. Whether certain patient subgroups (e.g., mesial temporal, neocortical, etc.) will benefit from RNS remains to be determined. The conclusion of the pivotal RNS trial will provide insights into the efficacy of the RNS, but many of these questions may remain unanswered until there is more extensive experience.

References

1 W. A. Hauser, J. F. Annegers and W. A. Rocca, *Mayo Clinic Proc.* **71**, 576–86 (**1996**).

2 P. Kwan and M. J. Brodie, *N. Engl. J. Med.* **342**, 314–9 (**2000**).

3 P. Afra, C. C. Jouny, G. K. Bergey, *Epilepsia* **49**, 677–84 (**2008**).

4 B. Adamolekun, C. C. Jouny, P. J. Franszczuk and G. K. Bergey, *Epilepsia* **45** (Suppl. 7), 58 (**2004**).

5 C. C. Jouny, P. J. Franszczuk and G. K. Bergey, *Clin. Neurophysiol.* **114**, 426–37 (**2003**).

6 C. C. Jouny, B. Adamolekun, P. J. Franaszczuk and G. K. Bergey, *Epilepsia* **48**, 297–304 (**2007**).

7 G. L. Morris, 3rd and W. M. Mueller, *Neurology* **53**, 1731–5 (**1999**).

8 P. J. Franaszczuk, P. Kudela and G. K. Bergey, *Epilepsy Res.* **53**, 65–80 (**2003**).

9 W. S. Anderson, P. Kudela, J. Cho, G. K. Bergey and P. J. Franascczuk, *Biol. Cybern.* **97**, 173–94 (**2007**).

10 S. C. Karceski, M. J. Morrell, R. Emerson and T. Thompson, *Epilepsia* **41** (Suppl. 7), 202 (**2000**).

11 R. P. Lesser, S. H. Kim, L. Beyderman, D. L. Miglioretti, W. R. Webber, M. Bare, B. Cysyk, G. Krauss and B. Gordon, *Neurology* **53**, 2073–81 (**1999**).

12 G. K. Bergey, J. W. Britton, G. D. Cascino, H.-M. Choi, S. C. Karceski, E. H. Kossoff, K. J. Meador, E. K. Ritzl, D. Spencer, S. Spencer, P. Ray, D. Greene and J. Greenwood, *Epilepsia* **43** (Suppl. 7), 191 (**2002**).

13 G. K. Bergey, G. Worrell, D. Chabolla, R. Zimmerman, D. Labar, R. Duckrow, A. Murro, M. Smith, D. Vossler, G. Barkley and M. Morrell, *Neurology* **66** (Suppl. 2), A387 (**2006**).

14 G. L. Barkley, B. Smith, G. K. Bergey, G. Worrell, D. Chabolla, J. Drazkowski, D. Labar, R. Duckrow, A. Murro, M. Smith, R. Gwinn, B. Fisch, L. Hirsch and M. Morrell, *Annual Meeting American Epilepsy Society* (**2006**).

15 E. H. Kossoff, E. K. Ritzl, J. M. Politsky, A. M. Murro, J. R. Smith, R. B. Duckrow and G. K. Bergey, *Epilepsia* **45**, 1560–7 (**2004**).

16 K. N. Fountas, J. R. Smith, A. M. Murro, J. Politsky, Y. D. Park and P. D. Jenkins, *Stereotact. Funct. Neurosurg.* **83**, 153–8 (**2005**).

24
Chronic Anterior Thalamic Deep-brain Stimulation as a Treatment for Intractable Epilepsy

Richard Wennberg

24.1
Introduction

Neurostimulation as an alternate form of treatment for intractable epilepsy was first considered based on observations dating from the 1940s that electrical stimulation of various subcortical structures could modify the cortical EEG, with high-frequency stimulation 'desynchronizing' the EEG and low-frequency stimulation 'synchronizing' the EEG [1–4]. Increased cortical synchrony mediated by low-frequency stimulation was demonstrated to be 'proepileptic' while cortical desynchronization mediated by high-frequency stimulation was shown to be 'antiepileptic' [5,6]. Experimental or clinical antiepileptic properties have since been reported with chronic electrical stimulation of a number of different central and peripheral nervous system sites, including the cerebellum, hypothalamus (mamillary nuclei), vagus nerve, trigeminal nerve, caudate nucleus, substantia nigra, centromedian thalamus, anterior thalamus, subthalamic nucleus, and direct epileptic focus stimulation in the amygdalohippocampal region [7–39]. Peripheral vagus nerve stimulation is the only form of neurostimulation currently licensed for treatment of patients with refractory epilepsy: controlled trials and subsequent widespread clinical usage have demonstrated significant, albeit modest, reduction in seizure frequencies [40,41].

In the hope that direct stimulation of central nervous system structures might provide additional, more robust, benefit in terms of seizure control, there has been a renewed interest in performing clinical trials of deep-brain stimulation (DBS) for epilepsy, especially given the modern successes of DBS in the treatment of various movement disorders. The first clinical trials of DBS in epilepsy were performed in the 1970s and 1980s, mainly using cerebellar and, to a lesser extent, anterior thalamic stimulation or caudate stimulation. Most of these studies were uncontrolled and, although a majority of patients were described to benefit from DBS treatment, the details were not always clear in the reports and the few controlled studies performed showed little benefit [7–11,16].

Seizure Prediction in Epilepsy. Edited by Björn Schelter, Jens Timmer and Andreas Schulze-Bonhage
Copyright © 2008 WILEY-VCH Verlag GmbH & Co. KGaA, Weinheim
ISBN: 978-3-527-40756-9

DBS of the centromedian thalamic nucleus (CM) for intractable epilepsy has been reported in most detail in the literature, with beneficial results described in terms of seizure control for a majority of patients studied [17–20]. However, one controlled study of CM stimulation for epilepsy did not find the observed reduction in seizures to achieve statistical significance [24].

The mechanisms by which DBS may control seizures are largely hypothetical and unproven. CM stimulation, acting via the widely projecting 'non-specific' thalamic system, is hypothesized to act through induction of cortical desynchronization, preventing seizure propagation and generalization [17, 24]. Based on experimental data describing a 'nigral control of epilepsy' system in rodents, controlled in large part by activity in the substantia nigra pars reticulata (SNpr) [42], subthalamic nucleus stimulation has been proposed to act through disfacilitation of SNpr neurons [28], although there is no direct evidence for such a control system in primates. The anterior thalamus has been demonstrated to be involved in seizure propagation, both experimentally and clinically, and stimulation or lesioning of the anterior nucleus (AN) or its afferent pathways has been shown experimentally to have antiepileptic properties [8, 12–16]. The dorsomedial nucleus of the thalamus (DM), situated posterior and inferior to AN, has also been shown to be involved in the maintenance and propagation of seizures, specifically those involving limbic brain structures [43, 44]. Anterior thalamus stimulation, aimed especially at AN, is thus hypothesized to act through blockade of corticothalamic synchrony, similar to CM. All of these proposed mechanisms are strictly hypothetical and, in fact, even the local effects of DBS are poorly understood. In a broad sense, most of the clinical effects of DBS can be considered to result from local 'inhibition' of function, in that the effects are typically mimicked by lesions or application of inhibitory neurochemicals. Nevertheless, the mechanisms of local inhibition are unresolved and it is possible that some effects of DBS could result from local neuronal or axonal excitation.

The exact parameters necessary to optimally alter the relevant corticothalamic networks with electrical stimulation are unknown, apart from the need for high-frequency stimulation (for example, greater than or equal to 100 Hz). In the experimental models cited above, it is only high-frequency stimulation that shows antiepileptic properties, usually attributed to a cortical desynchronizing effect. In contrast, low-frequency stimulation tends to be proepileptic in experimental models, an observation conceptually linked to the increased synchronization in cortex that can be demonstrated through induction of the so-called recruiting rhythm with low-frequency thalamic stimulation.

24.2
Anterior Thalamus DBS for Epilepsy

As described above, one of the regions suggested for therapeutic DBS in epilepsy is the anterior thalamus. The more specific target typically discussed is AN, which is an area relatively accessible to neurosurgical DBS electrode implantation and a nuclear complex with widespread corticothalamic connectivity, situated in an optimal position to interrupt and influence the development and propagation of epileptic discharges arising elsewhere in the brain. The AN is the crucial thalamic outflow point in the circuit of Papez, connecting the hippocampal formation, via the diencephalon, with the mesial frontal lobe structures and cingulum, areas which in turn send inputs back to the mesial temporal limbic structures. Lesions in AN or its proximate input pathways have been shown to have antiepileptic effects in experimental models. Likewise, local application of inhibitory chemical neurotransmitters and the application of high-frequency electrical stimulation has mimicked the antiepileptic effects of lesioning in the same models [12–15].

AN DBS could beneficially affect epilepsy in either of two ways. The local 'lesion-mimicking' effects could act to block propagation of cortically-generated seizures, helping to prevent diffuse spread of seizure activity and, in so doing, minimize the severity of seizures. In addition, it is possible that the neuromodulatory effects of chronic AN DBS might beneficially alter the level of cortical 'excitability' in a fashion resulting in a decreased propensity for seizure occurrence, thus decreasing seizure frequency.

Results of pilot trials of chronic AN DBS for epilepsy have shown a decrease in seizure frequency in a majority of patients [8, 16, 30, 32, 33, 37, 38], although this has not been clearly linked to the stimulation itself, as opposed to some other effect of the procedure, such as microthalamotomy, surgical placebo, etc., [30, 33]. Decreases in seizure severity have been reported in some patients but this variable has been harder to quantify with certainty to date. A multi-center clinical trial of AN DBS for epilepsy ('SANTE' – Stimulation of the Anterior Nucleus of the Thalamus for Epilepsy [45]) involving more than one hundred patients in the United States is nearing completion at the time of writing. The results from this trial, which includes an initial three-month period of blinded, randomized assignment of patients to receive either true stimulation or sham stimulation, will provide important information to supplement and hopefully clarify much that has been described in the different pilot studies.

The stereotactic techniques for implantation of the thalamic DBS electrodes are described in the original papers, as are the details of the individual pilot study protocols and stimulation parameters used. Readers interested in these details are referred to the original reports. It should be emphasized that none of these trials used any type of 'responsive' electrical stimulation – i.e., stimulation was not delivered in response to seizure detection, but rather stimulation was delivered chronically, usually intermittently in a fashion analagous to vagus nerve stimulation, though trials of continuous stimulation were also carried out in some patients.

A summary of the clinical results of the different pilot studies published in recent years is presented in Table 24.1. As can be seen from the table, a majority of patients overall showed a decrease in seizure frequency with AN DBS that was

Table 24.1 Summary of results from pilot studies of chronic AN DBS for intractable epilepsy.

Study	Patients	MTLE[2]	Baseline period	AED[3] changes	Mean/median stimulation frequency (range)	Blinded stimulation OFF/ON testing	Seizure frequency			
							1 year[5]	2 years	Stimulation OFF	Last visit (years)
Hodaie 2002 [30]; Andrade 2006 [33]	6	0	3 months	None for ≥2 years post-op	100 Hz (100–185)	Yes				
AN1							-29%	-29%	-38%	-90% (5)
AN2							-22%	-1%	-7%	-62% (7)
AN3							-52%	NA[4]	-51%	-63% (4)
AN4							-90%	-87%	-90%	-78% (5)
AN5							-73%	-78%	-62%	-91% (5)
AN6							-3%	-24%	-24%	+3% (4)
Kerrigan 2004 [32]	5	1?	2 months (1 patient) others?	None for ≥3 months post op	100 Hz	No			Unblinded reports	

			Yes?								Post-op 'sham' period	
	K1								-30%	NA	(Worse)	NA
	K2								-2%	NA	(Worse)	NA
	K3								-2%	NA	(Worse)	NA
	K4								+41%	NA	(Worse)	NA
	K5								+1%	NA	NA	NA
Lim 2007 [37]		4	0	≥9 months	None for ≥6 weeks post-op	100 Hz (90–180)	No (but OFF 2–4 weeks post-op)					
	L1								-65%	-25%	-69%	-30% (4)
	L2								-35%	-55%	-94%	-15% (4)
	L3								-60%[6]	-100%	-61%	-100% (3.5)
	L4								+40%[7]	-45%	-44%	-55% (3)
Osorio 2007 [38]		4	4	6 months	None	157 Hz (145-170)	No				Un-blinded reports	

continued overleaf

Table 24.1 continued.

Study	Patients	MTLE[2]	Baseline period	AED[3] changes	Mean /median stimulation frequency (range)	Blinded stimulation OFF/ON testing	Seizure frequency			
							1 year	2 years	Stimu-lation OFF	Last visit (years)
A		Yes					−93%	−94%[8]	(Worse)	−89% (3)
B		Yes					−56%	−64%	NA	−77% (3)
C		Yes					−78%	−93%	NA	−91% (3)
D		Yes					+18%[9]	−100%	NA	−70% (3)
Total	19	5					−35% (n = 19)	−61% (n = 13)	−48% (n = 10)	−65% (n = 14)

2 MTLE = Mesial temporal lobe epilepsy
3 AED = Antiepileptic drug
4 NA = Not available
5 Seizure frequency values at 1 year, 2 years and last visit for Lim 2007 study [37] estimated from 'Postimplantation Seizure Frequency vs Baseline' graph (Figure 1 in [37])
6 Right DBS electrode removed 5 months post-op; Left DBS electrode removed 10 months post-op
7 Stimulator accidentally OFF 7–12 months
8 Left DBS electrode disconnected temporarily
9 Question of possible atypical seizures, first year

sustained during long-term follow-up. Although the study of Kerrigan et al. [32] did not find a significant decrease in seizure frequency, these authors did report a significant improvement with respect to seizure severity in their patients. Seizure reduction with AN DBS was greater in the series of Osorio et al. [38] than in the other studies, which the authors speculate may be due to the higher mean stimulation frequency used in their study and/or the fact that all of their patients suffered from documented mesial temporal lobe epilepsy (MTLE). Theoretically, one might expect AN DBS to be most effective in MTLE, given the intimate neuroanatomical connections between the hippocampal formation and anterior thalamic nuclei in the aforementioned circuit of Papez.

The series of Kerrigan et al. [32] and Osorio et al. [38] did not include blinded analyses of sham stimulation. However, both of these studies reported multiple instances of worsened seizure control in patients whose stimulators were incidentally discovered to have been inactivated. The series described in Hodaie et al. [30], Andrade et al. [33] and Lim et al. [37] included formal analyses of seizure control post-implantation of the DBS electrodes, prior to active stimulation, and the first two also included periods of blinded sham stimulation carried out at various points during follow-up. In contrast to the unblinded reports, these series could not demonstrate a difference between stimulation ON and stimulation OFF.

It is to be hoped that the forthcoming results from the SANTE trial will illuminate: (a) whether beneficial effects of AN DBS may be most pronounced in the subset of patients with MTLE; and (b) whether active stimulation provides unequivocal added benefit beyond that seen with simple insertion of the DBS electrodes.

24.3
EEG Recordings

Combined scalp-thalamic EEG recording of interictal spikes is a complicated issue. In brief, typical scalp EEG spikes and sharp waves are recorded synchronously and with opposite polarity from the thalamic DBS electrode contacts, with a small but reproducible amplitude decrement present at each electrode contact more distant from the cortical source. A detailed analysis of these intracranial waveforms indicates that the DBS electrode-recorded potentials, seen in synchronous association with EEG scalp spikes, represent volume conduction from epileptiform discharges generated in the neocortex rather than locally generated activity resulting from cortical-subcortical neural propagation. These findings have been discussed and presented in detail elsewhere [46, 47].

Combined scalp-thalamic EEG recording of ictal activity (seizures) is also complicated but not necessarily limited by the intracranial volume-conduction issues seen with the high-amplitude interictal spike recordings. Seizures have been recorded from the DBS electrodes during postoperative recordings carried out before electrode internalization, and early ictal recruitment of the anterior thalamic structures has been seen even in a patient with extratemporal, suprasylvian neocortical partial epilepsy, despite the remoteness of seizure onset from the circuit of Papez structures [48].

Postoperative scalp EEG recording to document so-called recruiting rhythm generation in response to low-frequency AN stimulation has been carried out as a means of assessing functional thalamocortical connectivity between the implanted anterior thalamic regions and the superficial cortex. This issue has been studied in detail recently and interested readers are referred to the paper of Zumsteg et al. [49]. For practical purposes, the usefulness of rhythmic EEG synchronization induced by low-frequency thalamic DBS as a predictor of clinical efficacy appears questionable at this time.

EEG source localization studies and combined DBS/depth electrode studies performed in rare patients have proven useful in demonstrating the cortical activation patterns seen in response to anterior thalamic DBS. Specifically, AN stimulation affects primarily the ipsilateral cingulate gyrus, insular cortex and mesial and lateral temporal structures, whereas DM stimulation (performed through the deepest contacts of the thalamic DBS electrodes, which come to rest inferior and posterior to AN, in DM) affects primarily the ipsilateral orbitofrontal, mesial and lateral frontal areas, but also the mesial temporal structures [50–52].

24.4
Conclusions

It is too early to know whether chronic AN DBS will prove to be an effective therapy for intractable epilepsy in humans. Although conceptually appealing and supported by some experimental data, the procedure to date cannot really be considered to have been proven effective enough to warrant its invasiveness. More restricted patient selection, for example limiting the procedure to patients with temporolimbic (and perhaps mesial frontal) epilepsies, may define a subset of patients most likely to benefit, but this remains to be seen. Hopefully, the forthcoming results from the multi-center SANTE trial will provide guidance in the near future.

The lack of demonstrable added benefit of stimulation in the studies that included a blinded sham stimulation component is concerning. The results from the SANTE trial will be informative on this count too. It is possible that the optimal stimulation parameters have yet to be found, but it would be comforting at least to be able to identify a difference between stimulation OFF and stimulation ON effects as a first step. Furthermore, as with vagus nerve stimulation, a true surgical sham control would be necessary to ascertain how much of any benefit may be related to placebo.

With reference to this conference's overarching theme of seizure prediction, or at least seizure detection, it is possible that thalamic DBS time-locked to seizure onset might prove more clearly beneficial in the long run [53]. Work on this front is ongoing and will hopefully prove fruitful. Notwithstanding, it is also conceivable that the actual site of stimulation and the specific stimulation parameters used may turn out to be relatively unimportant. Epilepsy, with its dynamic, episodic nature, is a condition very different from the 'hard-wired', non-episodic movement disorders and other conditions commonly treated with DBS. In the absence of definitive evidence to date for a specific, linear, 'neuroanatomical' mechanism underlying the antiepileptic effects of DBS, it is plausible to consider that minimal stimulation perturbations of the nervous system in any of a variety of sites could all have similar, if limited, beneficial effects. Indeed, stopping a chronic stimulation paradigm might be as effective as starting a stimulation paradigm, with both changes potentially providing a perturbation sufficient to modulate the likelihood of spontaneous transitions to epileptic seizures.

References

1 E. W. Dempsey and R. S. Morison, *Am. J. Physiol.* **135**, 293 (**1942**).
2 R. S. Morison and E. W. Dempsey, *Am. J. Physiol.* **135**, 281 (**1942**).
3 R. S. Morison and E. W. Dempsey, *Am. J. Physiol.* **138**, 297 (**1943**).
4 G. Moruzzi and H. W. Magoun, *Electroenceph. Clin. Neurophysiol.* **1**, 455 (**1949**).
5 H. H. Jasper and J. Droogleever-Fortuyn, *Res. Publ. Assoc. Res. Nerv. Ment. Dis.* **26**, 272 (**1947**).
6 J. Hunter and H. H. Jasper, *Electroenceph. Clin. Neurophysiol.* **1**, 305 (**1949**).
7 I. S. Cooper, I. Amin, M. Riklan, J. M. Waltz and T. P. Poon, *Arch. Neurol.* **33**, 559 (**1976**).
8 I. S. Cooper, A. R. Upton and I. Amin, *Appl. Neurophysiol.* **43**, 244 (**1980**).
9 M. Sramka, G. Fritz, M. Galanda and P. Nádvornik, *Acta. Neurochir.*(Wien) Suppl. 23, 257 (**1976**).
10 G. D. Wright, D. L. McLellan and J. G. Brice, *J. Neurol. Neurosurg. Psychiatry* **47**, 769 (**1984**).
11 I. S. Cooper and A. R. Upton, *Biol. Psychiatry* **20**, 811 (**1985**).
12 M. A. Mirski and J. A. Ferrendelli, *Brain Res.* **399**, 212 (**1986**).
13 M. A. Mirski and J. A. Ferrendelli, *J. Neurosci.* **7**, 662 (**1987**).
14 M. A. Mirski and R. S. Fisher, *Epilepsia* **35**, 1309 (**1994**).
15 M. A. Mirski, L. A. Rossell, J. B. Terry and R. S. Fisher, *Epilepsy Res.* **28**, 89 (**1997**).
16 A. R. Upton, I. Amin, S. Garnett, M. Springman, C. Nahmias and I. S. Cooper, *Pacing Clin. Electrophysiol.* **10**, 217 (**1987**).
17 F. Velasco, M. Velasco, C. Ogarrio and G. Fanghanel, *Epilepsia* **28**, 421 (**1987**).
18 M. Velasco, F. Velasco, A. L. Velasco, M. Luján and J. V. del Mercado, *Epilepsia* **30**, 295 (**1989**).
19 F. Velasco, M. Velasco, A. L. Velasco and F. Jiménez, *Epilepsia* **34**, 1052 (**1993**).
20 F. Velasco, M. Velasco, A. L. Velasco, F. Jiménez, I. Marquez and M. Rise, *Epilepsia* **36**, 63 (**1995**).
21 M. Velasco, F. Velasco, A. L. Velasco, B. Boleaga, F. Jiménez, F. Brito and I. Marquez, *Epilepsia* **41**, 158 (**2000**).
22 F. Velasco, J. D. Carrillo-Ruiz, F. Brito, M. Velasco, A. L. Velasco, I. Marquez and R. Davis, *Epilepsia* **46**, 1071 (**2005**).
23 A. L. Velasco, F. Velasco, M. Velasco, D. Trejo, G. Castro and J. D. Carrillo-Ruiz, *Epilepsia* **48**, 1895 (**2007**).
24 R. S. Fisher, S. Uematsu, G. L. Krauss, B. J. Cysyk, R. McPherson, R. P. Lesser, B. Gordon, P. Schwerdt and M. Rise, *Epilepsia* **33**, 841 (**1992**).
25 S. A. Chkhenkeli and I. S. Chkhenkeli, *Stereotact. Funct. Neurosurg.* **69**, 221 (**1997**).
26 L. Vercueil, A. Benazzouz, C. Deransart, K. Bressand, C. Marescaux, A. Depaulis and A. L. Benabid, *Epilepsy Res.* **31**, 39 (**1998**).
27 E. E. Fanselow, A. P. Reid and M. A. Nicolelis, *J. Neurosci.* **20**, 8160 (**2000**).
28 T. Loddenkemper, A. Pan, S. Neme, K. B. Baker, A. R. Rezai, D. S. Dinner, E. B. Montgomery and H. O. Lüders, *J. Clin. Neurophysiol.* **18**, 514 (**2001**).

29 S. Chabardès, P. Kahane, L. Minotti, A. Koudsie, E. Hirsch and A.-L. Benabid, *Epileptic Disord.* 4 Suppl. 3, 83 (2002).

30 M. Hodaie, R. A. Wennberg, J. O. Dostrovsky and A. M. Lozano, *Epilepsia* 43, 603 (2002).

31 K. Vonck, P. Boon, E. Achten, J. D. Reuck and J. Caemaert, *Ann. Neurol.* 52, 556 (2002).

32 J. F. Kerrigan, B. Litt, R. S. Fisher, S. Cranstoun, J. A. French, D. E. Blum, M. Dichter, A. Shetter, G. Baltuch, J. Jaggi *et al.*, *Epilepsia* 45, 346 (2004).

33 D. M. Andrade, D. Zumsteg, C. Hamani, M. Hodaie, S. Sarkissian, A. M. Lozano and R. A. Wennberg, *Neurology* 66, 1571 (2006).

34 C. M. DeGiorgio, A. Shewmon, D. Murray and T. Whitehurst, *Epilepsia* 47, 1213 (2006).

35 J. F. Tellez-Zenteno, R. S. McLachlan, A. Parrent, C. S. Kubu and S. Wiebe, *Neurology* 66, 1490 (2006).

36 P. Boon, K. Vonck, V. D. Herdt, A. V. Dycke, M. Goethals, L. Goossens, M. V. Zandijcke, T. D. Smedt, I. Dewaele, R. Achten *et al.*, *Epilepsia* 48, 1551 (2007).

37 S.-N. Lim, S.-T. Lee, Y.-T. Tsai, I.-A. Chen, P.-H. Tu, J.-L. Chen, H.-W. Chang, Y.-C. Su and T. Wu, *Epilepsia* 48, 342 (2007).

38 I. Osorio, J. Overman, J. Giftakis and S. B. Wilkinson, *Epilepsia* 48, 1561 (2007).

39 J. Vesper, B. Steinhoff, S. Rona, C. Wille, S. Bilic, G. Nikkhah and C. Ostertag, *Epilepsia* 48, 1984 (2007).

40 The Vagus Nerve Stimulation Study Group., *Neurology* 45, 224 (1995).

41 A. Handforth, C. M. DeGiorgio, S. C. Schachter, B. M. Uthman, D. K. Naritoku, E. S. Tecoma, T. R. Henry, S. D. Collins, B. V. Vaughn, R. C. Gilmartin *et al.*, *Neurology* 51, 48 (1998).

42 M. J. Iadarola and K. Gale, *Science* 218, 1237 (1982).

43 E. H. Bertram, D. X. Zhang, P. Mangan, N. Fountain and D. Rempe, *Epilepsy Res.* 32, 194 (1998).

44 E. H. Bertram, P. S. Mangan, D. Zhang, C. A. Scott and J. M. Williamson, *Epilepsia* 42, 967 (2001).

45 N. M. Graves, A. M. Lozano, R. A. Wennberg, I. Osorio, S. Wilkinson, G. Baltuch, J. A. French, J. F. Kerrigan, A. Shetter and R. S. Fisher, *Epilepsia* 48 Suppl. 7, 148 (2004).

46 R. Wennberg, B. Pohlmann-Eden, R. Chen and A. Lozano, *Clin. Neurophysiol.* 113, 1867 (2002).

47 R. A. Wennberg and A. M. Lozano, *Clin. Neurophysiol.* 114, 1403 (2003).

48 R. A. Wennberg, B. Pohlmann-Eden and A. M. Lozano, *Epilepsia* 43 Suppl 7, 55 (2002).

49 D. Zumsteg, A. M. Lozano and R. A. Wennberg, *Clin. Neurophysiol.* 117, 2272 (2006).

50 D. Zumsteg, A. M. Lozano, H. G. Wieser and R. A. Wennberg, *Clin. Neurophysiol.* 117, 192 (2006).

51 D. Zumsteg, A. M. Lozano and R. A. Wennberg, *Clin. Neurophysiol.* 117, 1602 (2006).

52 D. Zumsteg, A. M. Lozano and R. A. Wennberg, *Epilepsia* 47, 1958 (2006).

53 I. Osorio, M. G. Frei, S. Sunderam, J. Giftakis, N. C. Bhavaraju, S. F. Schaffner and S. B. Wilkinson, *Ann. Neurol.* 57, 258 (2005).

25
Thoughts about Seizure Prediction from the Perspective of a Clinical Neurophysiologist

Demetrios N. Velis

25.1
Introduction

Present-day good clinical practice in the diagnosis and treatment of epilepsy relies on the correct classification of epilepsy syndromes. In turn, the present ILAE classification emphasizes the correct use of seizure classification which is indispensable without having access to seizure documentation, including description of electroclinical correlations. Such correlations are commonly obtained by means of long-term combined video/EEG seizure monitoring, of which several variants exist, all of which make heavy use of at least scalp EEG and sometimes invasive EEG recordings. In all cases clinical neurophysiology bears the brunt of seizure documentation, often in a clinical setting of the so-called Epilepsy Monitoring Unit (EMU).

It stands to reason that the EMU setting has traditionally been the testing ground of various algorithms developed for the purpose of quantifying the EEG record for detection purposes, be those for interictal epileptiform paroxysms or for the identification of epileptic seizure episodes. Long-term video-EEG recording in the course of presurgical evaluation is eminently suited for the purpose of validating spike and seizure-detection algorithms. It follows that the clinical neurophysiologist is keenly interested in an adequate anticipation of seizure events. Long-term video-EEG seizure monitoring is a labor-intensive, and consequently costly, diagnostic technique, which greatly stands to benefit from signal-analysis techniques that may indicate imminent seizure onset. Analysis of the running EEG signal and quantitative studies of interictal activity incidence and distribution, although highly valid and largely validated for the purpose of helping localize the irritative zone, have proven singularly unsuccessful in seizure anticipation. In fact, subjective information supplied by the patient, such as the occurrence of prodromi, is often quite reliable in this respect rather than the quantification of the interictal event itself.

Detection of the ictal event, on the other hand, whether on the basis of the classical visual interpretation of the running or recorded EEG signal, or on the various signal analytical techniques applied for that purpose, is largely successful; in

Seizure Prediction in Epilepsy. Edited by Björn Schelter, Jens Timmer and Andreas Schulze-Bonhage
Copyright © 2008 WILEY-VCH Verlag GmbH & Co. KGaA, Weinheim
ISBN: 978-3-527-40756-9

the case of intracranial EEG recordings obtained from the zone of ictal onset, approaching near-perfect specificity and sensitivity.

With the notable exception of reflex epilepsies, catamenial epilepsy and the occasional anecdotal reference of a case report of chronobiologically consistent seizure clustering, epileptic seizure occurrence has been, as far as the clinical neurophysiologist is concerned, largely relegated to the realm of forecasting uncertainty. Earlier attempts at reliably predicting or even anticipating seizure occurrence have been met with skepticism. These results have often not been replicated [1] owing to initial flaws in methodology as has already been pointed out [2, 3] and fuelled by the discussion on whether the ictal event may ever be predictable [4–6]. The reader is referred to Chapter 7 by Lopes da Silva et al. in this volume, based on the keynote lecture of this workshop, and the published literature [7, 8].

Thus the clinical neurophysiologist is becoming aware of the intractability of the problem on seizure prediction, not only in cases of intractable epilepsy, and may become just a bit skeptical on whether this problem may indeed prove as ill-posed as the inverse problem of electro- and magneto-encephalography. Inference on studies derived from active stimulation paradigms in reflex epilepsy characterized by visual sensitivity [9, 10] suggests that changes occurring in the brain en route to a provoked (e.g., in visual sensitivity) or spontaneously occurring seizure (e.g., in mesial temporal lobe epilepsy) may in fact be gleaned on the basis of changes in phase-clustering of the obtained response after an appropriate perturbing stimulus has been administered. Nevertheless the phase-clustering index remains a statistical tool yielding probability values that an ictal event is more or less likely to occur rather than when it is to occur or not to occur, as the case may be.

In what theoretically may be seen as a mutually beneficial contest between signal analysis physicists or engineers and clinical neurophysiologists the latter have for upwards of ten years now motivated patients and staff at the EMU alike poring over days and weeks of continuous multi-channel scalp or intracranial EEG records in an effort to obtain as artifact-free a record as possible for adequate signal analysis. They have braved institutional review boards and medical ethics committee grilling, contributed text to endless grant applications, dealt with countless mathematical formulas and equations, dabbled in uni- and multivariate analysis techniques of linear and nonlinear measures, tagged along to many a special interest group meeting, and have yet to hear an encouraging word as to whether it has all been worthwhile. In order to establish a modicum of opinion-forming for the purposes of this symposium, the author undertook a short questionnaire-based survey among his fellow full-time clinical neurophysiologists employed as such in The Netherlands, a country in Western Europe boasting an unusually high percentage of such professionals among its neurologists and hosting two large dedicated institutes for the diagnosis and treatment of epilepsy, one of which was established 125 years ago. A total of 65 questionnaires were mailed to an address on file in early 2007; two were returned to sender as undeliverable. There were 37 responders, including one incomplete response. A total of 36 responses were scored. The results of that survey were presented during the Freiburg Workshop and are cited in the Appendix. As witnessed by this survey, the equivocal results so far gleaned from the published literature have apparently led to a considerable degree of doubt among the clinical neurophysiological community in The Netherlands. Indeed, the discussions which took place during the Freiburg Workshop corroborated the impression that the level of discourse had hitherto been hindered by poorly conceived and relatively poorly executed, mostly underpowered studies primarily as a result of lack of consensus as to how to approach the occurrence of the interictal and the ictal event, not only in terms of statistical inference [11, 12] but also in terms of their significance with respect to the basic mechanisms of epileptogenicity and epileptogenesis [13, 14].

One of our first priorities in the community of theoreticians, physicists, engineers, epileptologists and clinical neurophysiologists should be to come up with a commonly accepted definition as to what constitutes anticipation versus prediction of an ictal event. In doing so we shall have to deal with the significance of the epileptiform paroxysm. For that purpose we as EEGers can hardly expect that the rest of the signal analysis community may accept the existing IFSECN classification of what constitutes an epileptiform paroxysm which states that a paroxysm is a phenomenon with abrupt onset, rapid attainment of a maximum and sudden termination, distinguished from background activity. This term is commonly used to refer to epileptiform patterns and to seizure patterns. In distinguishing between an interictal epileptiform pattern and a seizure pattern, the IFSECN guidelines indicate that the former applies to distinctive waves or complexes, distinguished from background activity and resembling those recorded in a proportion of human subjects suffering from epileptic disorders. Epileptiform patterns include spikes and sharp waves, alone or accompanied by slow waves, occurring singly or in bursts lasting at most a few seconds. The IFSECN recognizes that seizure patterns in the EEG may not be accompanied by clinical manifestations, in which case these are called subclinical seizure patterns. To distinguish between interictal and ictal EEG patterns, the IFSECN recommends that the term epileptiform be used in conjunction with the interictal event [15].

A challenge ahead of us is to deal with the so-called electrographic seizure event, with which most EEGers will denote a paroxysmally occurring event of several seconds in duration during which no clinically manifest behavioral changes may be noted in a patient who, when engaged in an interactive task, will not only be oblivious to its occurrence but also not be hindered by it. The EEG of the electrographic seizure event, for all intents and purposes indistinguishable from the EEG at the onset of a clinically manifest seizure event, is easily detected by most of the commercially available seizure detection algorithms. In both cases the location and wave morphology may be identical. The question that the EEGer would like to have answered is whether this event may be of significance in the cascade leading to a clinically manifest epileptic seizure or whether it may be relegated to the realm of the relatively non-contributory event of the interictal paroxysm. Should it be significant in term of the system generating the signal we measure as EEG then we shall have to see whether trying to interfere with the system rather with the event itself may lead to diminution of the chance of seizure occurrence [3, 16].

Another equally important challenge the EEGer faces is to document the time slot which is most appropriate for the administration of whatever countermeasures we may devise to help prevent the occurrence of clinically manifest epileptic seizures in a patient on the basis of the running EEG, irrespective of whether this should be in terms of aborting a seizure or diminishing its chance of occurring altogether. Aborting seizures by means of electrical counterstimulation is an ancient art, known since Roman times, in this case relying on giving a timely jolt delivered to the patient by an electric eel [17]. In our times this has been reported in both epileptic network models [18, 19] and in patients [20, 21]. The distinction of what constitutes open versus closed-loop electrical stimulation has been made on the basis of discussions carried out among the participants of the Second International Workshop on Seizure Prediction, held in 2006 [22]. Closed-loop electrical stimulation based on a perturbation pattern, and an open-loop system monitoring subtle changes en route to a seizure, has been advocated in patients implanted with invasive electrodes during video/EEG seizure monitoring. Such dynamical system changes may be inferred from the measured fluctuations of the EEG phase-clustering index, indicating a high likelihood of seizure occurrence [23], but have yet to be validated in clinical practice.

While appropriate stimulation may bring about a seizure in certain forms of reflex epilepsy, the first report of therapeutic sensory-motor counter stimulation has been attributed to John Hughlings Jackson who ostensibly described how a patient of his could arrest a Jacksonian fit by forcibly manipulating the thumb in which the clonic movement occurred before these spread to the rest of the hand and lower arm [24]. Local application of toxins such as that of the cone snail seems to be effective at least in experimental animal models [25]. All such interventions have one thing in common, which is that they require timely application of a counter stimulus. Other techniques in seizure prevention such as lowering of the temperature at the zone of ictal onset in the hippocampus [26] have been reported to be effective and may be expected to be less demanding than the application of a counter stimulus.

Undoubtedly, if not for want of a better intellectual challenge, the sceptical EEGer may still be won over by pondering the practical consequences of effective seizure anticipation in a clinical setting: benefits may be derived in the Epilepsy Monitoring Unit (EMU) from adequate use of techniques offering a reliable estimate for impending seizure occurrence. In the most practical of situations the level of monitoring may be tailored on the seizure likelihood, freeing resources and cutting costs [27]. Imaging modalities such as reliably performing an ictal SPECT study [28] may become feasible for a considerably larger percentage of the EMU population. The effect of any type of intervention meant to reduce the chance of seizure occurrence may be readily measured, particularly for techniques with a relatively difficult-to-prove effectiveness such as vagal nerve stimulation [29], transcranial magnetic stimulation [30, 31], and deep-brain stimulation [32] in epilepsy. The fact that the perfect study, rejecting the null hypothesis that epileptic seizures in humans may never be reliably anticipated if not predicted, has yet to be performed (B. Litt, this workshop). This implies that world-wide collaboration in the community of basic scientists, mathematicians, engineers, and clinical neurophysiologists who are addressing the issue of seizure anticipation will continue for the foreseeable future.

25.2
Appendix: Does the EEGer Need Seizure Prediction?

The Dutch Full-time Clinical Neurophysiologists' Survey

1. Is seizure prediction possible now?
   Yes   0
   No   22
   Do not know   14

2. Will seizure prediction ever be possible?
   Yes   5
   No   20
   Do not know   11

3. Does the EEGer need seizure prediction now?
   Yes   5
   No   31
   Do not know   0

4. Will the EEGer ever need seizure prediction?
   Yes   9
   No   20
   Do not know   7

5. Would you recruit patients for seizure-prediction trials on a compassionate-use basis of seizure-prevention methods?
   Yes   7
   No   28
   Do not know   1

6. Would you recruit patients for seizure-prediction trials on an intention-to-treat basis in an effort to develop new therapy strategies?
   Yes   0
   No   29
   Do not know   17

7. Would you consider it a major improvement in your EEG practice if seizure prediction were to be reliably carried out?
   Yes   15
   No   21
   Do not know   0

8. Would patients consider it a major improvement in their treatment if seizure prediction were to be reliably carried out?

Yes	23
No	5
Do not know	8

9. Would you recommend to patients that they make use of seizure-prediction systems if seizure prediction were to be reliably carried out?

Yes	30
No	1
Do not know	5

10. Who should foot the bill for the use of seizure-prediction systems if seizure prediction were to be reliably carried out?

Patient	2
3rd Party payer	11
Both 30-70	5
Both 50-50	9
Both 70-30	4
Do not know	5

11. Should the EEGer need to think about seizure prediction?

Yes	20
No	1
Do not know	15

References

1 R. Aschenbrenner-Scheibe, T. Maiwald, M. Winterhalder, H. U. Voss, J. Timmer and A. Schulze-Bonhage, *Brain* **126**, 2616–26 (**2003**).

2 R. G. Andrzejak, F. Mormann, T. Kreuz, et al., *Phys. Rev. E* **67**, 010901 (**2003**).

3 K. Lehnertz, F. Mormann, H. Osterhage, et al., *J. Clin. Neurophysiol.* **24**, 147–53 (**2007**).

4 F. Mormann, R. G. Andrzejak, C. E. Elger and K. Lehnertz, *Epilepsia* **46**, 1335–6 (**2005**).

5 F. Mormann, R. G. Andrzejak, C. E. Elger and K. Lehnertz, *Brain* **130**, 314–33 (**2007**).

6 F. Mormann, T. Kreuz, C. Rieke, et al., *Clin. Neurophysiol.* **116**, 569–87 (**2005**).

7 F. H. Lopes da Silva, W. Blanes, S. N. Kalitzin, J. Parra, P. Suffczynski and D. N. Velis, *Epilepsia* **44** Suppl. 12, 72–83 (**2003**).

8 F. H. Lopes da Silva, W. Blanes, S. N. Kalitzin, J. Parra, P. Suffczynski and D. N. Velis, *IEEE Trans. Biomed. Eng.* **50**, 540–8 (**2003**).

9 S. N. Kalitzin, J. Parra, D. N. Velis and F. H. Lopes da Silva, *IEEE Trans. Biomed. Eng.* **49**, 1279–86 (**2002**).

10 J. Parra, S. N. Kalitzin and F. H. Lopes da Silva, *Curr. Opin. Neurol.* **18**, 155–9 (**2005**).

11 T. Kreuz, R. G. Andrzejak, F. Mormann, et al., *Phys. Rev. E Stat. Nonlin. Soft Matter Phys.* **69**, 061915 (**2004**).

12 B. Schelter, M. Winterhalder, T. Maiwald, et al., *Epilepsia* **47**, 2058–70 (**2006**).

13 J. Engel, Jr., C. Wilson and
A. Bragin, *Epilepsia* **44** Suppl. 12,
60–71 (**2003**).

14 J. Engel, Jr., *Suppl. Clin. Neurophysiol.* **57**, 392–9 (**2004**).

15 Anonymous, *Electroencephalogr.
Clin. Neurophysiol.* Suppl. **52**, 1–304
(**1999**).

16 E. L. Ohayon, S. N. Kalitzin,
P. Suffczynski, et al., *J. Physiol. Paris.*
98, 507–29 (**2004**).

17 W. H. Theodore and R. S. Fisher,
Lancet Neurol. **3**, 111–8 (**2004**).

18 P. Suffczynski, S. N. Kalitzin
and F. H. Lopes Da Silva, *Neuroscience* **126**, 467–84 (**2004**).

19 P. Suffczynski, F. H. Lopes da Silva,
J. Parra and D. N. Velis, *J. Clin.
Neurophysiol.* **22**, 288–99 (**2005**).

20 I. Osorio, M. G. Frei, S. Sunderam,
et al., *Ann. Neurol.* **57**, 258–68 (**2005**).

21 R. P. Lesser, S. H. Kim,
L. Beyderman, et al., *Neurology* **53**,
2073–81 (**1999**).

22 I. Osorio and M. G. Frei, *Epilepsia* **48**,
406 (**2007**).

23 S. N. Kalitzin, D. N. Velis,
P. Suffczynski, J. Parra and

F. H. Lopez da Silva, *Clin. Neurophysiol.* **116**, 718–28 (**2005**).

24 G. Holmes, *Lancet*, **i**, 957–73 (**1927**).

25 J. A. Haack, J. Rivier, T. N. Parks,
E. E. Mena, L. J. Cruz, B. M.
Olivera, T. Conantokin, *J. Biol.
Chem.* **265**, 6025–9 (**1990**).

26 G. K. Motamedi, P. Salazar, E. L.
Smith, et al., *Epilepsy. Res.* **70**,
200–10 (**2006**).

27 D. N. Velis, P. Plouin, J. Gotman and
F. H. Lopez da Silva, *Epilepsia* **48**,
379–84 (**2007**).

28 W. Van Paesschen, *Epilepsia*
45 Suppl. 4, 35–40 (**2004**).

29 A. Barnes, R. Duncan, J. A.
Chisholm, K. Lindsay, J. Patterson
and D. Wyper, *Eur. J. Nucl. Med.
Mol. Imaging* **30**, 301–5 (**2003**).

30 R. Cantello, S. Rossi, C. Varrasi,
et al., *Epilepsia* **48**, 366–74 (**2007**).

31 K. J. Werhahn, J. Lieber, J. Classen
and S. Noachtar, *Epilepsy Res.* **41**,
179–89 (**2000**).

32 W. H. Theodore and R. Fisher, *Acta
Neurochir.* Suppl. **97**, 261–72 (**2007**).

26

State of Seizure Prediction: A Report on Informal Discussions with Participants of the Third International Workshop on Seizure Prediction

Hitten P. Zaveri, Mark G. Frei, Ivan Osorio

26.1
Introduction

Seizure prediction using EEG signals has been attempted for over two decades now. Past and current work in this field can be broadly categorized into three phases. A conceptual advance was achieved in the first phase through the definition of the problem and initial attempts to predict the onset of seizures from the EEG. Several studies have been reported since these first efforts. The description of chaos in deterministic systems heralded a second phase of work (from approximately the mid-1980s), one which remains active to the present time. The efforts that comprise this second phase of work consider seizures to be an expression of chaos and seek to detect markers that represent transition to this phase. More recently, there has been a broadening of the field which may be considered to be a third phase of the effort. In this phase of work a broader set of time series analysis methods, including those drawn from classical time series analysis, are being employed to detect pre-seizure cursors. While the motivations for the use of these measures tend to be diverse, a few common threads can be discerned. Central among these are a desire to measure changes in cortical excitation and synchronization that would allow detection of the transition from the background state to a preictal state, without which prediction may not be possible.

Several other characteristics are also coming to typify emergent work in this field, irrespective of the time series measures being employed. The increased capability of instrumentation systems to collect longer-term data and the increased capacity of digital data storage systems and computer processing power is allowing greater ability to collect and analyze longer data sets. This is evident in recent publications. Clearer statistical definitions have been proposed that allow determination of whether or not an approach results in the prediction of seizures and quantifies the improvement in prediction performance beyond simply chance. An adoption of some of these standards is also evident in more recent reports. There has also been an evolution in the model for seizure predictability from a naïve deterministic model for state transition to one that is probabilistic in nature. There are increasingly

Seizure Prediction in Epilepsy. Edited by Björn Schelter, Jens Timmer and Andreas Schulze-Bonhage
Copyright © 2008 WILEY-VCH Verlag GmbH & Co. KGaA, Weinheim
ISBN: 978-3-527-40756-9

sophisticated studies, as well, with animal and slice models of epilepsy suggesting an influence of research initiated in humans, to experiments by basic scientists in order to better understand pre-seizure changes and seizure generation.

Central to work in this field are the efforts of an interdisciplinary group of scientists. Members of this group include engineers, physicists, mathematicians, epileptologists, neurosurgeons, and basic scientists. These scientists have come together to understand the issues involved in seizure prediction, and beginning with the second workshop on seizure prediction in 2006, the scope of this effort has been expanded to include seizure generation and seizure control. This interdisciplinary team and the support of its work in multiple countries by various national and international organizations and also the support of its meetings and the attempt to create a common language of goals and metrics, as well as the involvement of academia and industry, all help to provide confidence that progress, both iterative and revolutionary, is possible in this field.

This section summarizes an informal attempt to gauge the opinion of some of these scientists regarding the future of this effort. Those involved in the discussion reported here were all participants of the Third International Workshop on Seizure Prediction. Scientists were asked two questions during informal discussions, during lunch, dinner or coffee breaks, in either individual or group conversations. The questions asked were the following: (1) What single factor represents the greatest obstacle to the advance of seizure prediction? (2) What single factor, not necessarily related to the first factor identified, provides the greatest hope for an advance in the field? The following participants discussed the two questions:

1. Gregory Bergey, John Hopkins University, USA

2. Anatol Bragin, UCLA, USA

3. Paul Carney, University of Florida, USA

4. Piotr Franaszczuk, John Hopkins University, USA

5. Mark Frei, Flint Hills Scientific, USA

6. Jean Gotman, Montreal Neurological Institute, Canada

7. Matia Gotman, SWR Orchestra, Freiburg, Germany

8. Nina Graves, Medtronic, USA

9. Stiliyan Kalitzin, SEIN, Netherlands

10. Anna Korzeniewska, John Hopkins University, USA

11. Fernando Lopes da Silva, SEIN, Netherlands

12. Ivan Osorio, University of Kansas, USA

13. Joelle Pineau, McGill University, Canada

14. Justin Sanchez, University of Florida, USA

15. Andreas Schulze-Bonhage, University of Freiburg, Germany

16. Steven Schiff, Pennsylvania State University, USA

17. Yitzhak Schiller, Rambam Medical Center, Israel

18. Randall Stewart, NINDS, USA

19. Piotr Suffczynski, Warsaw University, Poland

20. Demetrios Velis, SEIN, Netherlands

21. Fabrice Wendling, Universite De Rennes, France

22. Hitten Zaveri, Yale University, USA.

The discussions were often wide-ranging. Some of the topics raised are beyond the scope of this report and have not been included here. A few common threads could be drawn from the discussions. It was considered that these thoughts may have value for those interested in seizure-prediction research, and for the planning and organization of future workshops on seizure prediction. These are presented below.

26.2
Modality

While it is widely recognized that the intracranial EEG (icEEG) contains considerable information relevant for attempts at seizure prediction, concerns were expressed about this modality. There is a perceived limitation of the icEEG in capturing markers of a pre-seizure state, due to the fact that it provides information at a single scale, while many believe that seizure generation and seizures are multi-scale phenomena. Others expressed concern that the icEEG was but one modality, and we would need to study several modalities, such as neurochemistry, ionic currents, pH, and temperature either in conjunction or separately to capture and understand pre-seizure changes and seizure generation. Another concern was the lack of a physiological model for the icEEG. Additionally, there were concerns regarding numerous shortcomings of icEEGs obtained during intracranial monitoring for epilepsy surgery. These include, but are not limited to, trauma caused by electrode placement, effects of anesthesia and of AED taper.

26.3
Seizure Generation and Models

A general concern, which appeared to encompass several related concerns, centered on seizure generation. One of these concerns was that the epilepsy patient population was heterogeneous. Given that epilepsy is a collection of disorders, and is the end result of a number of factors which impact the brain, it is not clear that there will be commonality in the pre-seizure state among different etiologies. This raises questions about the ability of one or even of a few seizure-prediction approaches to capture the heterogeneity found in the epilepsy patient population. The point being expressed is that there may be several routes, and not only one route to seizure generation, and that the inability of any particular approach to successfully predict seizures across a large group of subjects may simply reflect the diversity of the epilepsy patient population.

A fair amount of the discussion also centered on models. There was concern on the lack of actionable information on seizure generation which could be incorporated within useful biophysical models; that is, models which could capture the essence of seizure generation, and which could provide greater insight into the process by which seizures are generated. A general suggestion was that the field would benefit from the use of models along the lines of those which exist for our planetary system, which can be used to calculate the position of planets to a degree of accuracy. It was argued that, in the absence of accurate models of brain function, even simple models which allowed a first-level approximation, for example, such as that provided by the Bohr model of the atom, could be useful. It was also argued that models must not only be useful, they must be mechanistic, and based on observations at a mechanistic level (e.g., cell-to-cell communication and ion shifts).

26.4
Academia and Industry

An increased commercial interest in the field was evident from several aspects of the meeting and its organization. This increased interest was considered to present both a challenge and a source of hope. Commercial interest, it was felt, indicated a greater overall interest in the field, the maturing of knowledge within the field, and the hope that this knowledge could be translated into better treatment for patients. Concerns were raised, however, that commercial entities operate on very different, and shorter, timelines, and that, if there was not clear success within a short time, it could spell a possible setback for the field. A second concern was that the group had not adequately addressed issues involving intellectual property (IP) generation and protection. It was indicated that issues concerning IP could set up difficulty in partnerships between academic groups, and between academic and industry groups. It was strongly suggested that the group raise awareness on IP issues, and discuss openly the differences in approaches and goals of academia and industry-based groups.

26.5
The Question of Seizure Prediction and its Prioritization

Still other views were expressed regarding the question of seizure prediction and the prioritization of efforts towards seizure prediction. The first concern, stated bluntly, is that it is not necessarily possible to predict future states of a dynamical system on the basis of a limited amount of measurements. Thermal fluctuations, external input and dynamic instability may be limiting factors. More research is needed to understand the influence of those and possibly other factors on our ability to predict epileptic seizures. This reservation was stated in the context of the arguments, presented above, for improved measurement of the brain with multiple scales of measurement and the use of icEEG in conjunction with other modalities. This reservation also conveys the skeptic's view of seizure prediction. A second viewpoint is that the field has tried to proceed too far too quickly. The concern is that the definition of seizure prediction as a metric of progress for this field is overly ambitious and unrealistic. What we can achieve with the EEG and the current definition of seizures, however, is seizure detection. It was suggested that we should further improve detection, with efforts being focused on earlier detection of seizures. It was argued that this may allow better implementation of means for automated seizure control and possibly lead to an advance in seizure prediction.

26.6
Summary

These notes reflect informal discussions during the Third International Workshop on Seizure Prediction centered on the two questions presented above. The questions and resultant discussions were not intended as a formal poll and they were not considered, at the time, to constitute material for a report. This was also not an attempt to discuss and report the views of only a particular subset of the participants of the workshop, nor are the views reported here held by all those who participated in the discussions. The role of the questions was to help organize informed discussion between scientists from diverse backgrounds and interests. In this they worked well, as they led to considerable discussion during breaks. In retrospect, however, the questions and discussions can be considered to be incomplete. For example, due to a lack of time the discussions often did not reach the point where the second question could be addressed. Also, the questions do not touch upon seizure control, which is an integral part of the work of many scientists within the group. Further, the participants were asked to identify only a single challenge and a single factor which provided hope. Such limitations possibly restricted the discussion. Still, if the information presented here is considered within the context from which it has been derived, it may have value in that it helps gauge the views of this diverse group of scientists, and may help guide part of the planning for future meetings.

There was remarkably little discussion of specific algorithms or time series measures which have fueled work in the previous phases of effort on seizure prediction. The concerns, rather, focus on different aspects of the phenomenon being studied, how best to measure it and how best to model it. Some of the issues raised during discussions are open questions. For example, the lack of a physiological model for the genesis of the EEG has existed for the tenure of our knowledge of this signal. Other issues are the subject of intense direct and overlapping research by multiple groups. For example, there are multiple efforts to better measure brain activity at different scales and incorporate this knowledge within our understanding of seizure generation. There are also attempts to improve the recording of the icEEG, and to accumulate enough long-term data to help address the challenge of the heterogeneity of the epilepsy patient population.

Scientists expressed an opinion that progress, though slow, was being made; that insight was being generated through the use of time series analysis of the EEG, and that this insight could help guide experiments being conducted by basic scientists. Some were pleased that attempts were being made to record seizures at multiple scales and incorporate this information within our understanding of seizures and attempts to predict the onset of seizures. Others were pleased at efforts to collect and test on longer data sets. Hope was expressed that, with traction, more powerful theory and methodology could be brought to bear on the problem. Some scientists expressed a growing flexibility to work with partial knowledge to understand seizures and intervene to control seizures. Finally some of the discussion touched upon one of the most essential characteristics of this effort, its remarkable interdisciplinary nature. It was emphasized that the key to progress may lie in better integration, across various investigational approaches, of extant information on seizure generation. That is, the requisite information may already exist, but sufficient effort may not have been expended to synthesize it. Others found hope in the very existence of an international interdisciplinary effort, spanning academia and industry, believing this effort would be able to make progress towards understanding seizure generation, and towards predicting and controlling seizures.

26.7
Acknowledgement

We thank Piotr Franaszczuk, Stiliyan Kalitzin, Justin Sanchez and Steven Schiff for helpful discussions during the preparation of this report.

Index

Seizure Prediction in Epilepsy. Edited by Björn Schelter, Jens Timmer and Andreas Schulze-Bonhage
Copyright © 2008 WILEY-VCH Verlag GmbH & Co. KGaA, Weinheim
ISBN: 978-3-527-40756-9